Proceedings of the First US/Japan Conference on the Frontiers of Statistical Modeling: An Informational Approach

Volume 3

Engineering and Scientific Applications

Proceedings of the First US/Japan Conference on the Frontiers of Statistical Modeling: An Informational Approach

Volume 3

Engineering and Scientific Applications

Edited by

H. BOZDOGAN

Department of Statistics,
The University of Tennessee,
Knoxville, TN, U.S.A.

Co-Edited by

S. L. SCLOVE

Information & Decision Sciences,
University of Illinois at Chicago,
Chicago, IL, U.S.A.

A. K. GUPTA

Department of Mathematics & Statistics,
Bowling Green State University,
Bowling Green, OH, U.S.A.

D. HAUGHTON

Department of Mathematical Sciences,
Bentley College,
Waltham, MA, U.S.A.

G. KITAGAWA

The Institute of Statistical Mathematics,
4–6–7 Minami-Azabu, Minato-Ku,
Tokyo, Japan

T. OZAKI

The Institute of Statistical Mathematics,
4–6–7 Minami-Azabu, Minato-Ku,
Tokyo, Japan

K. TANABE

The Institute of Statistical Mathematics,
4–6–7 Minami-Azabu, Minato-Ku,
Tokyo, Japan

SPRINGER SCIENCE+BUSINESS MEDIA, B.V.

A C.I.P. Catalogue record for this book is available from the Library of Congress.

ISBN 978-94-010-4368-7 ISBN 978-94-011-0854-6 (eBook)
DOI 10.1007/978-94-011-0854-6

Printed on acid-free paper

CONTENTS OF VOLUME 3

LIST OF CONTRIBUTORS TO VOLUME 3

Hirotugu Akaike
Director General
The Institute of Statistical Mathematics
4-6-7 Minami-Azabu, Minato-Ku
Tokyo, JAPAN 106

Emanuel Parzen
Department of Statistics
Texas A & M University
College Station, TX 77843 U. S. A.

Hideo Nakamura
Bailey Japan Co., Ltd.
51 Baraki, Nirayama-cho
Tagata-gun, Shizuoka-ken
JAPAN 410-21

Kohei Ohtsu
Tokyo University Merchantile Marine
2-1-6 Etchujima
Koto-ku, Tokyo, JAPAN 135

Genshiro Kitagawa
The Institute of Statistical Mathematics
4-6-7 Minami-Azabu, Minato-Ku
Tokyo, JAPAN 106

David Vere-Jones
Institute of Statistics & Operations Research
Victoria University of Wellington
P. O. Box 600, Wellington
NEW ZEALAND

Yosihiko Ogata
The Institute of Statistical Mathematics
4-6-7 Minami-Azabu, Minato-Ku
Tokyo, JAPAN 106

Will Gersch and David Stone
Information & Computer SciencesUniversity of Hawaii
Honolulu, HI 96822, U. S. A.

LIST OF CONTRIBUTORS TO VOLUME 3

Tomoyuki Higuchi
The Institute of Statistical Mathematics
4-6-7 Minami-Azabu, Minato-Ku
Tokyo, JAPAN 106

Tohru Ozaki
The Institute of Statistical Mathematics
4-6-7 Minami-Azabu, Minato-Ku
Tokyo, JAPAN 106

Masami Hasegawa
The Institute of Statistical Mathematics
4-6-7 Minami-Azabu, Minato-Ku
Tokyo, JAPAN 106

Tatjana Lange
Department of Electrical Engineering
College of Technology and Economics
Berlin Lichtenberg, GERMANY

D. Pados, P. Papantoni-Kazakos, D. Kazakos, and A. Koyiantis
Department of Electrical Engineering
University of Virginia
Charlottesville, VA 22903, U. S. A.

Masaharu Tanemura
The Institute of Statistical Mathematics
4-6-7 Minami-Azabu, Minato-Ku
Tokyo, JAPAN 106

Alexander M. Shurygin
Statistical Laboratory
Department of Mechanics & Mathematics
Moscow State University
Moscow 119899, RUSSIA

"Professor Akaike is quietly assembling his own theory of statistical estimation based on entropy, information and likelihood, centered around the Akaike's Information Criterion (AIC). This theory is more likely to survive than most, being based on data and common sense rather than dogma."

Paraphrased from
L.A. Baxter
Journal of the Royal Statistical Society, A, 1991, 154, Part 2, pp. 356-366.

EDITOR'S GENERAL PREFACE

On May 24 to 29, 1992, The US/Japan Conference on the Frontiers of Statistical Modeling: An Informational Approach, was held at the Department of Statistics, University of Tennessee, Knoxville, to commemorate the sixty-fifth birthday of Professor Hirotugu Akaike and to honor him for his revolutionary contributions to modern statistical theory and practice. This conference was originally conceived while I was a Visiting Research Fellow and Visiting Associate Professor at the Institute of Statistical Mathematics, Tokyo, Japan during January-August 1988. A steering committee was formed with several colleagues from the Institute of Statistical Mathematics, and the initial plans of the conference were laid out at that time.

The proceedings of this conference are being published in three volumes under the names of the three scientific core areas:

I. Theory and Methodology of Time Series Analysis,
II. Multivariate Statistical Modeling, and
III. Engineering and Scientific Applications.

The conference theme was the implementation of statistical modeling through an informational approach to complex, real-world problems.

Over the past twenty years, modern statistical activity has been moving away from traditional formal methodologies based on test theory. The traditional methodologies based on narrowly specified hypotheses have become difficult in solving complex, large scale real-world problems, and are woefully inadequate to meet the challenges of our societal needs. Current efforts among the most sophisticated are to identify instantaneously the best fitting model among a class of competing models for a given complex data structure with computational ease, interpretability, and elegance using a new and novel informational approach. This richer approach links statistics as an experimental science with high speed computation and supercomputer technology, and bridges the conventional frequentist and the Bayesian schools of thought in statistics. With this new approach traditional table look up is eliminated. Total flexibility and versatility are provided to practitioners, researchers, and students in solving their statistical problems.

This new approach, was originally pioneered by Professor Hirotugu Akaike starting as early as 1971. Akaike (1981) writes:

"In 1970, I received an invitation to the Second International Symposium on Information Theory, which was to be held in Tsahkadsor, Armenia, USSR. At that time, I was

interested in extending FPE to the determination of the number of factors in a factor analysis model, a statistical model originally developed in Psychology. However, it was not at all clear what the prediction error of this model was. The pressure of the impending deadline for the submission of the conference paper was increasing and this caused several weeks of sleepless nights.

On the morning of March 16, 1971, while I was taking a seat in a commuter train, I suddenly realized that the parameters of the factor analysis model were estimated by maximizing the likelihood and that the mean value of the logarithmus of the likelihood was connected with the Kullback-Leibler information number. This was the quantity that was to replace the mean squared error of prediction. A new measure of the badness of a statistical model with the parameters determined by the method of maximum likelihood was then defined by the formula

AIC=(-2)log(maximum likelihood) + 2 (number of parameters)."

The development and the introduction of Akaike's (1973) Information Criterion, "AIC", marked the beginning of the era of systematic approach to present-day statistics, namely model evaluation and selection. AIC is clearly one of the most interesting and important developments in the field of statistics in recent years. See, for example, Kotz and Johnson (Eds.) (1992, p. 599): *Breakthroughs in Statistics Volume I: Foundations and Basic Theory*, Springer-Verlag, New York.

This development represents an important advance over the conventional framework of statistics as developed by R. A. Fisher and many others following in his footsteps.

By combining ideas related to what is now called *"predictive efficiency"* with the notion of Kullback-Leibler information, Akaike arrived at AIC for evaluating alternative statistical models which cleverly combines a measure of goodness-of-fit of the model with a term relating to the number of parameters used to achieve that fit.

Akaike's AIC provides an answer to the question of how much improvement in fit should an additional parameter make before it is included in the model and on what scale should that improvement be measured?

Of course, important fundamental work like this answers some questions and raises many others. A by-product of Akaike's work is that he has directed the attention of other researchers to important problems. As a result, many important statistical modeling techniques have been developed in various fields of statistics, biomedicine, control theory, ecology, econometrics, engineering, psychometrics, sociology, and in many other fields. Voluminous research papers, books, research monographs, and Ph.D. theses have been written using AIC, including that of this editor. Further new model selection criteria have been developed based on Akaike's fundamental work using the fascinating informational, or entropic approach as the underlying sound analytical building block.

Professor Akaike has also been inspirational as far as encouraging us to have the courage to participate in consulting even when sometimes at first we may not feel up to the challenge. Presently his approach and philosophy to statistics have become a new and modern way of statistical activity and research with very successful industrial and scientific applications.

Therefore, the Profession of Statistics is greatly in debt to Akaike for his fundamental contribution in this new trend of informational approach to statistics, and for repeatedly calling our attention to the very important model evaluation and selection problem.

We proudly dedicate these three volumes of the Proceedings of the US/Japan Conference to Professor Akaike as our token gesture of deep appreciation of his outstanding contribution to the field of "Informational Modeling" on the occasion of celebration of his sixty-fifth birthday.

More than 100 participants from the US, Japan, Canada, Denmark, Germany, Israel, New Zealand, Switzerland, United Kingdom, and the Netherlands attended this important conference. About 40 very distinguished statisticians who made their contributions to these proceedings gave the invited lectures during a five day very intensive conference.

The conference was sponsored by the Department of Statistics, which is a part of the College of Business Administration, by the Office of the Dean of the College of Business Administration (CBA), and the Office of Research Administration, all at the University of Tennessee, Knoxville; and the Institute of Statistical Mathematics of Tokyo, Japan. The American Statistical Association (ASA) was co-sponsor along with Japan Statistical Society (JSS), Japan Society of Applied Statistics (JASAS), Classification Society of North America (CSNA), Psychometric Society, Engineering Physics and Mathematics Division of Oak Ridge National Laboratory, and well known professional organizations and research centers.

The result was a very pleasant and stimulating conference.

It is with great pleasure and gratitude that I acknowledge the help of the co-editors both from the US and Japan, the invited contributors, the referees from around the world, the chairs of the sessions, all the financial sponsors and the co-sponsors of the conference. I am deeply grateful to Dean C. Warren Neel of the CBA and Professor David Sylwester, the Head of the Statistics Department, for providing me the opportunities and the financial support to execute this prestigious conference. I am grateful to Professor Kunio Tanabe, my co-chair from Japan, and the Director of Prediction and Control at the Institute of Statistical Mathematics, Tokyo, Japan for helping me to successfully realize the conference from the beginning to the end, and also sponsoring all the Japanese researchers.

I wish to thank Professor Manny Parzen of the Department of Statistics at Texas A & M University who accepted to deliver the entertaining banquet speech on the evening of May 28, 1992 with his "change jokes", and "hammers and nails".

As always, there are key people behind the scenes in putting every successful conference. Without the dedication and help of these key individuals almost surely one can expect disasters in holding conferences. In my case, I had the personal privilege of working with Judy Snow, our Departmental Head Secretary, as my Conference Coordinator. The success of the conference was largely due to Judy's exemplary attitudes, congeniality, high professional standards and ability to relate to the diverse needs of the participants. Without my looking over her shoulder, Judy managed the conference single handedly on a daily basis and made all the arrangements. She made life easy so that I could spend most of my time in putting the scientific program together. We worked as a team. I specially thank Judy from the bottom of my heart for making the conference a success. I wish also to thank Alberta Randles, our Departmental Secretary, and eight Volunteer Graduate Students: Jim Minesky, Yuehui Fan, Paul Bowen, Gary Schneider, Martha Nunn, Rob Collignon, Frank Van Manen, and Chen Xi for their wonderful help in driving the vans, setting up the coffee-breaks, the bulletin boards, escorting the participants, and so forth, throughout the conference. I am grateful to all of them.

My colleagues in the Department of Statistics provided a congenial atmosphere and acted as my local organizers. I wish to thank every single one of them for their help and encouragement all the way through. I can still hear the voice of Dr. Richard Sanders shouting from the corridor : "Ham go home!" as I was working in the late hours. Dr. Mary Sue Younger chaired the local Organizing Committee, and helped to design several field trips and excursions with a variety of interests in mind to have a break and change the pace in the five-day intensive meeting. These included taking a dinner Cruise on the Star of Knoxville on the Tennessee River, hiking in the Great Smoky Mountains

National Park, and dining in the picturesque "Old City" of Knoxville. Her help is also gratefully acknowledged. Pandora Riggs, Editor of Publications Center, and Designer Hugh Bailey at the University of Tennessee worked very hard with me in capturing my conceptualized ideas in preparing the design of the conference posters and the brochures. They are both acknowledged also.

I wish to thank Dr. Joseph (Joe) Johnson, the President, Professor William Snyder, Chancellor, Dr. Sheadrick Tillman, Assistant Vice Chancellor of Research, Dean C. Warren Neel, the Dean of CBA, and Dean Michael Stahl, Associate Dean of CBA, and Professor David Sylwester, Head of the Statistics Department of the University of Tennessee in participating in the opening welcome and greetings of the conference.

Dr. David Larner, Publisher of Science and Technology Division of Kluwer Academic Publishers in the Netherlands visited the University of Tennessee prior to the US/Japan Conference and during the five-day conference to discuss in detail the publishing plans and the format of the proceedings. These three volumes are the final product of his keen interest in the subject matter as a former Physics Professor himself. I am grateful to his continued support, interest, and persistence throughout the project. I am also grateful to Mr. John Martindale, Editor of the Science and Technology Division of Kluwer in the North American Continent, for his understanding, patience, and support in the final production of the three-set volume. Myself, and the co-editors thank both Dr. Larner and Mr. Martindale for bringing out these volumes.

Last but not least, I would like to thank my wife Nancy and two sons Bedros and Kirkor for their love and patience during the planning stages of the conference, and during the editing of these three volumes when I was not around to participate in family activities. I am grateful for their understanding and sacrifice.

It is hoped that the reader will find many interesting ideas and challenges in these resulting proceedings volumes, and that the contributions will stimulate further research in many complex, real-life applications of statistics in meeting our societal needs using the "Informational Modeling" approach.

Knoxville, July 1993

Hamparsum Bozdogan
The University of Tennessee

References

Akaike, H. (1973). Information Theory and an Extension of the Maximum Likelihood Principle, in *Second International Symposium on Information Theory*, B. N. Petrov and F. Csaki (Eds.), Budapest: Academiai Kiado, 267-281.

Akaike, H. (1981). A New Look at the Statistical Model Identification, in *This Week's Citation Classic*, Institute for Scientific Information, CC/Number 51, December 21.

Kotz, S. and Johnson, N. L. (Eds.) (1992). *Breakthroughs in Statistics Volume I: Foundations and Basic Theory*, Springer-Verlag, New York.

PREFACE TO VOLUME 3

This is the third volume of the proceedings of The US/Japan Conference on the Frontiers of Statistical Modeling: An Informational Approach. Papers included in this volume deal with engineering and scientific applications.

As a result of Akaike's (1973) AIC, many important statistical modeling techniques have been developed in engineering and scientific applications. Since 1987 in Japan, many supercritical thermal power plants are governed by dynamic statistical controllers. On line multivariate autoregressive time series models are used to maximize the capacity of the production of electricity and to minimize the cost of production. Other "Informational Modeling" techniques have been used to control ship's motions and marine engines for energy conservation, in studying prediction and occurrences of earthquakes, and in data analysis taken aboard a spinning spacecraft such as the Pioneer Venus for signal extraction.

The first paper is another keynote lectures given by Professor Akaike and discusses the implications of the informational viewpoint for statistical science. The rest of the papers included in this volume deal with the applications of model selection and evaluation in optimal control of thermal power plants, seismological applications, multichannel studies, nonlinear filter designs, inference of evolutionary trees from DNA and protein sequence data, group method of data analysis, neural binary classification, studying directional interactions, and stable estimation techniques.

It is with great pleasure and gratitude that the editor and co-editors acknowledge the authors and the referees who contributed to this volume, and the chairs of this scientific core of the conference.

Knoxville, July 1993

Hamparsum Bozdogan
The University of Tennessee

IMPLICATIONS OF INFORMATIONAL POINT OF VIEW ON THE DEVELOPMENT OF STATISTICAL SCIENCE

H. AKAIKE
The Institute of Statistical Mathematics
4-6-7 Minami-Azabu, Minato-ku
Tokyo, Japan 106

SUMMARY

In this paper the informational approach to statistics based on an informational interpretation of log likelihood of a statistical model is contrasted with the conventional Fisherian framework. It is pointed out that for the recognition of practical importance of statistical science more emphasis should be placed on the development of new models than the refinement of mathematical analysis of existing procedures, although the latter activity should never be ignored. This is based on the observation that it is the contribution of statistics to the development of new hypotheses for the interpretation of data that creates the importance of the subject. It is argued that the informational approach prepares a basis for the rapid expansion of the area of application of statistics by providing a theoretical foundation for the development and evaluation of new statistical models in diverse fields.

The paper is intended specifically for the discussion of following subjects.

1. Review of Fisherian statistics

It is quite notable that R.A. Fisher introduced his original concept of log likelihood in his first paper. Apparently the idea was motivated by the necessity of fitting a parametric frequency function to observed data and a criterion of fit which was free from the problem of choice of the interval lengths in the computaion of theoretical frequencies was conceived.

Fisher later developed detailed theoretical analysis of the likelihood function and discussed the foundation of mathematical statistics. He classified the problems of statistics into three types; the problems of specification, estimation and distribution. This produced a conceptual framework for later development of statistics. However, when we recognize that the power of statistics derives from its effectiveness in developing new hypotheses, the separation of the problem of specification of thedistribution simply as a matter of practical statisticians seems to have produced too much restrictive influence on later development of statistics. In particular, it placed theproblem of comparison of distributions of different functional forms outside the main interest of theoretical statisticians. The contribution of a statistician is evaluated through his ability of developing a useful hypothesis for the solution of a problem in the real world. Here proper choice of the basic distribution plays the central role in the effort for a successful solution. This fact must be taken very seriously for proper development of research and education of statistics.

2. Informational view of statistics

When the author started the research of statistics he had a very discouraging image of

H. Bozdogan (ed.), Proceedings of the First US/Japan Conference on the Frontiers of Statistical Modeling: An Informational Approach, 1–25.

statistics from the stereotyped and non-constructive description of statistical procedures in ordinary text books. Experience taught that almost each practical problem demanded the development of a new model that properly represented the characteristics of the object under consideration. When the use of a flexible model was considered, such as the use of AR model in time series analysis, the necessity of a systematic procedure for the evaluation and comparison of models of different type became quite clear. This necessity and the adoption of predictive point of view quickly lead to the interpretation of the log likelihood of a model as an estimate of the quality of the model, evaluated in terms of the Kullback-Leibler information. The information criterion AIC was thus introduced. Later research clarified the close relation between the concept of information and the work of L. Boltzmann on entropy. The relativistic nature of the informational criterion was then recognized clearly. This subject is philosophically deeply related to the formulation and use of the concept of truth.

3. Contribution of informatioal approach to the generation of hypotheses

The introduction of AIC encouraged researchers to generate hypotheses on their own through the fitting of models of various types, each representing particular point of view. This seems to be the real reason of the wide spread use of AIC in various fields of application. AIC helps people in their endeavor to develop hypotheses. In particular, the informational interpretation of log likelihood allowed the development of the idea of comparing Bayesian models by their likelihoods. The idea produced an effect that emancipated Bayesian modeling activity from the tightly personalistic view of subjective Bayesians. The interpretation also provided a basis for the use of the information provided by the data distribution for the construction of prior distribution. Thus the informational approach contributed to the extension of the possibility of proposing various hypotheses through the proposal and comparison of Bayesian models.

4. Enhancement of the importance of statistics in society

Whith the advancement of informational technology the accumulation of observational data is accelerating. Under this circumstance the potential of statistics is increasing rapidly as the science and technology for the effective use of information supplied by data. However, it is only through the demonstration of significantly successful results of applications that statistics earns the respect of society. Thus we are forced to pay our attention to the development and maintenance of systematic contact with real world. The need for innovative ideas and concepts is recognized only through the contact with the expanding world of data and problems. Thus it looks that the successful operation of cross-disciplinary researches is the key to the full realization of the potential of statistics. The wide spread use of AIC shows that the informational point of view is preparing a basis for the development of statistics in this direction.

FROM COMPARISON DENSITY TO TWO SAMPLE ANALYSIS

Emanuel Parzen
Department of Statistics
Texas A&M University
College Station, Texas 77843-3143 USA

SUMMARY

This paper presents the main methods of our theory of functional statistical inference, based on quantile domain functions, information measures, and *comparison density* estimation. It seeks an approach to data analysis of a sample of n observations by functional and graphical representations of its probability models using various functions on $0 < u < 1$.

Contents of the paper are: 1. Introduction; 2. Quantile domain functions; F, Q; 3. Mid distribution and quantile functions: F^{mid}, Q^{mid}; 4. Sample distribution and quantile: $F^{\sim}$, $Q^{\sim}$; 5. Comparison distribution and comparison density for continuous F, G; 6. Comparison distribution and comparison density for discrete F, G; 7. Information measures: Renyi, Chi-square; 8. Information for comparison density functions; 9. Information measures and entropy tests of fit; 10. Continuous versions of discrete distribution functions: F^c, Q^c; 11. Comparison distributions of one sample (continuous data); 12. One sample parameter estimation; 13. Comparison distributions of two samples (continuous data); 14. Two sample comparison density estimation.

Research supported by the U. S. Army Research Office

STATISTICAL IDENTIFICATION AND OPTIMAL CONTROL OF THERMAL POWER PLANTS

H. NAKAMURA
Bailey Japan Co., Ltd.
511 Baraki, Nirayama-cho
Tagata-gun, Shizuoka-ken
Japan 410-21

SUMMARY

Implementation of an optimal regulator for steam temperature control of thermal power plants is described. In the system, identification of the system dynamics is performed by fitting a multivariate AR model to the plant data time series obtained by a data acquisition experiment. Based on the AR model, system analysis and controller design are carried out mostly by TIMSAC (Time Series Analysis and Control) Library which was developed by Akaike and his group of The Institute of Statistical Mathematics. This library includes programs for analysing mutual dependence among process variables, AR model building, derivation of the state equation, optimal controller design, digital simulation for control performance evaluation, on-line control algorithm, etc.

The study was performed from 1974 through 1977 as a joint research project of Kyushu Electric Power Company, The Institute of Statistical Mathematics, and The Central Research Institute of Electric Power Industry in Japan (CRIEPI), with the help of the boiler, plant computer, and control system manufacturers. After a series of simulation studies and experiments on a large-scale power plant simulator, the first optimal regulator designed by the statistical method was put into commercial operation in February, 1978 at Buzen No.1 Unit, a 500-MW supercritical plant owned by Kyushu Electric Power Company of Japan. Since then, the same or slightly modified types of optimal regulators have been implemented in Japan and abroad.

In the presentation, motivation of the investigation, control system configuration, practical procedures for statistical identification and analysis of the plant dynamics, controller design, etc., are introduced with some data obtained in the field tests or during the routine operation of 500-MW and 600-MW supercritical fossil-fired power plants.

In the introduction, the situation that necessitated the research is briefly explained. Then, the general concept of steam temperature control of a boiler is illustrated. The difficulty in tuning the conventional PID control system for the Multi-Input Multi-Output (MIMO) process, as well as the need of an optimal control system as a means to overcome such difficulty, are described. The advantage of the statistical approach over the conventional method for the derivation of the state equation is stressed.

Then, the description of the procedure for control system design follows. The items discussed are (1) control system configuration to realize the optimal regulator, (2) selection of system variables to compose the optimal regulator, (3) acquisition of time series data for system identification by injecting pseudo-random signals into the plant, (4) system analysis by TIMSAC Library, such as the AR model building, determination of the model order by means of AIC, computation of power spectra and their use for the relative-noise contribution analysis. The way how to use the results of the relative-noise contribution analysis in the selection of system variables, and how to reflect them on the optimal controller design is also discussed.

Next, the theory for deriving a state equation from the AR model and computation of the optimal feedback gain matrix through DP (Dynamic Programming) procedure is briefly introduced.

Prior to the implementation of the optimal regulator, a series of digital simulation studies were performed by using the state equation and the state-feedback gain matrix. The purposes of the simulation are (1) verification of the appropriateness of the state equation, (2) estimation of the control performance by the optimal regulator, (3) determination of the proper values of elements in the weighting matrices for DP computation.

Usually, two kinds of digital simulations are used: the simulation for the rough estimation of control performance, and the one for checking the steam temperature deviations and the amplitudes of manipulated variables when some specified patterns of load changes are imposed on the plant.

Some examples of the simulation results are shown to explain the simulation-based parameter tuning procedure. Compensation for process nonlinearity by means of gain scheduling is briefly referred to. Control algorithm of the implemented optimal regulator is illustrated.

Then, control performance of the superheated and reheated steam temperatures by the proposed optimal regulator is shown for two types of boilers. The examples illustrated are those obtained in the field test or during the routine operation of 500MW and 600 MW supercritical power plants.

Besides the fine performance in the normal operating conditions, the robustness of the proposed system during plant contingency was confirmed by applying a large amount of disturbance to the plant. The record of the state and manipulated variables under such abnormal condition is shown.

The above-mentioned fine performance is produced from the prediction of the plant behavior, which is inherent in the proposed optimal control system. A result of correlation analysis between the boiler load demand and manipulated variables is given to confirm the effectiveness of the prediction.

Today, more than 20 optimal regulators, designed by the statistical method for power plant steam temperature control, are in operation not only in Japan but in China and Canada. In the USA, an experiment using a model of 700MW supercritical sliding pressure plant has been going on under the sponsorship of EPRI (Electric Power Research Institute).

The outstanding feature of the statistical method developed by Akaike and his group is that it provides control engineers with practical means of implementing modern control theory for large industrial processes. In other words, the optimal regulator implementation can be carried out by a well-trained process control engineer, if he uses the TIMSAC Library in a proper way according to the instruction manual.

Simplicity both in system configuration and maintenance is important from the practical viewpoint. In the above-mentioned optimal regulators, main frame computers have been mostly used. Recent progress of microprocessor technique, however, has made it possible to compose the optimal regulator with a few control modules. In this system, system analysis and controller design are carried out in and EWS (Engineering Work Station), and the state equation and state-feedback gain matrices, which are necessary for on-line control, are transferred from the EWS to the control modules of a DCS (Distributed Control System).

APPLICATIONS OF AUTOREGRESSIVE MODEL TO CONTROL SHIP'S MOTIONS AND MARINE ENGINE

K. OHTSU
Tokyo University of Mercantile Marine
2-1-6 Ecchujima
Koto-ku
Tokyo, Japan 135

G. KITAGAWA
The Institute of Statistical Mathematics
4-6-7 Minami-Azabu, Minatoku
Tokyo, Japan 106

SUMMARY

This paper is a survey of applications of the optimal control theory based on a multivariate autoregressive model to control various marine systems. The paper is set out in the following two parts. The first application concerns with a ship's auto pilot system. In this application, a digital type of optimal autopilot system with the fixed gain is discussed at first. The autopilot system is extended also to a noise adaptive controller (NADCON). Furthermore, a rudder roll control system (RRCS) aiming at reducing not only the ship's course deviation and but also rolling motion is also dealt with. The second one is application to a main engine regulating system. In this application, two types of governor system is proposed. The Type I governor aims at regulating propeller revolution and the Type II is one, controlling propeller torque as well as the propeller revolution. All systems proposed in this paper are tested in full scale experiments.

STATISTICAL MODELS FOR EARTHQUAKE OCCURRENCE: CLUSTERS, CYCLES AND CHARACTERISTIC EARTHQUAKES

D. VERE-JONES
Institute of Statistics and Operations Research
Victoria University of Wellington
PO Box 600, Wellington, New Zealand

SUMMARY

This paper reviews different approaches to modelling two key but opposing features of earthquake data, their tendency towards clustering and their tendency towards regular or periodic behaviour. The discussion includes descriptions of earthquake occurrence data, different modelling frameworks (point processes, fractals, self-similar processes) and methods of statistical analysis. The general viewpoint taken is that earthquake occurrence is the product of complex, near-critical dynamic processes in the earth's crust, exhibiting such characteristic features as clustering, long range correlations, distributions with power law tails. Within this general picture many specific types of behaviour, such as foreshock and aftershock activity, periodic or quasiperiodic stress release patterns, fractal-type spatial distribution may be accommodated locally or in particular circumstances. However, given that the processes are complex and comprehensive models are not available, existing models provide partial descriptions only, and apply in limited circumstances and only to a certain degree of approximation.

Clusters are a ubiquitous feature of earthquake occurrence, but their origin is not fully understood and a quantitative theory which would account for their occurrence is lacking. At the same time they are a considerable nuisance factor in applying statistical techniques which have been developed primarily for handling independent (and hence unclustered) observations. The following comments are amplified in the paper.

(i) From a quantitative point of view clustering is probably best thought of as an indication of a complex interacting system close to some critical point, that is analogous to a statistical mechanical system approaching a phase change. It is linked to such features as self-similarity, power-law distribution, and long-range correlations.

(ii) The observed clustering appears to include both local and long-range features; it is not clear whether there are distinct physical processes at work.

(iii) The descriptive models for clustering which appear to be most successful in practice involve self-exciting features and branching models.

(iv) Contour maps of seismicity and earthquake risk may be seriously flawed because of the problems caused by clusters. Clusters which repeat in time in similar locations have very different risk implications to clusters which are randomly located in both time and space.

Cyclic effects have an equally long history in the literature on earthquakes, but the observational evidence to support them is sparse and controversial. Fixed cyclic effects in a strict sense can be modeled by a Poisson process with a periodic rate function, but lack a plausible physical explanation. Periodic waves of seismic activity, operating across a broad front and progressing in space as well as time, are a more recent variation of this theme.

Statistically it is particularly difficult to assess the significance of long term cycles against a background of long term correlation. More plausible are relaxation type models such as the earthquake rebound or stress release model first put forward after the 1906 San Francisco earthquake. Recent studies which attempt to fit the stress release model to historical catalogues will be summarized in the paper. They show remarkable differences in the success with which a stochastic version of this model fits data from different seismic regions. The physical interpretation of these differences is not yet clear.

The last theme, characteristic earthquakes, derives from recent geological studies on repeated earthquakes along the same fault. Records of such earthquakes are "fossilized" in geological or topographical features such as fault scarps, uplifted river or beach terraces, evidence of displacement in peat bogs. The data consists not of points but of lengths of activated fault segments. Most of the records are too short (only a handful of events) for meaningful statistical analyses, and there are many unresolved questions such as the extent of sympathetic movements along parallel or neighbouring faults, which increase the difficulty of interpretation. Some attempt will be made to suggest possible models, and to discuss the statistical evidence in favor of truly "characteristic" (that is, repeating) earthquakes, and such close relatives as the "time predictable" and "slip-predictable" models.

In all of these aspects key statistical questions concern the comparison of and selection between competing models. Use in some form of the principles underlying Akaike's Information Criterion is an almost indispensable component of this work. The paper contains numerous illustrations of this point.

SEISMOLOGICAL APPLICATIONS OF STATISTICAL METHODS FOR POINT-PROCESS MODELLING

Yosihiko OGATA
The Institute of Statistical Mathematics
4-6-7 Minami-Azabu, Minatoku
Tokyo, Japan 106

SUMMARY

This paper is concerned with some statistical models in application to seismology where the occurrence times of earthquakes can be considered to be a point process. A set of systematic methods for the analysis of point-process data is put forward whose key stone is parametric model-building of the conditional intensity function. That is to say, efficient simulation method by the thinning algorithm, the model comparison and parameter estimation by the AIC and the maximum likelihood method, and graphical diagnostic checks of models and data by the residual point process.

The applications to earthquake occurrence data in this paper include causal relationship between seismic activities in distinctive areas, detection of seasonality for shallow earthquakes, aftershock analyses, identification of the standard temporal seismic activities by a point process model, and the detection of an anomalous precursor which is expected to be useful for long-term prediction of large earthquakes.

The computer programs for the analyses discussed in this paper are being prepared for publication in the Computer Monographs of the Institute of Statistical Mathematics, Tokyo, and also in the Software Library of the International Association of Seismology and Physics of the Earth's Interior (IASPEI) in collaboration with Seismological Society of America, accepting the invitation of the editorial committee.

ONE CHANNEL AT-A-TIME MULTICHANNEL AUTOREGRESSIVE MODELING OF STATIONARY AND NONSTATIONARY TIME SERIES

Will GERSCH and David STONE
Department of Information & Computer Sciences
University of Hawaii
Honolulu, HI 96822 U.S.A.

SUMMARY

Multichannel autoregressive (MCAR) modeling of stationary and nonstationary time series data is achieved doing things one channel at-a-time using only scalar computations on instantaneous data. The one channel at-a-time modeling is achieved as an instantaneous response autoregressive (AR) model with orthogonal innovations variance. Conventional MCAR models are expressed as linear algebraic transformations of the instantaneous response-orthogonal innovations models.

Three longstanding time series modeling problems are addressed using this paradigm. In the first, modeling multichannel stationary covariance time series data as one channel at-a-time AR models and subsequently using subset selection achieves a relatively parsimonious representation as compared to conventional MCAR model fitting. This procedure results in reduced one-step-ahead prediction variance. Secondly, the one channel at-a-time paradigm used in conjunction with the quasi-Bayesian "smoothness priors long AR" modeling of the individual channels achieves enhanced power spectral density estimation for the multichannel stationary time series. Thirdly, multichannel time varying AR modeling of nonstationary covariance time series data is achieved using the one channel at-a-time AR models in conjunction with a smoothness priors time-varying stochastic linear regression modeling of the partial correlation coefficients in each AR channel. The d-channel time varying AR model, a model with many more parameters than data, is fitted with only $2d$ explicitly computed hyperparameters. Examples are shown.

SEPARATION OF SPIN SYNCHRONIZED SIGNALS USING A BAYESIAN APPROACH

T. HIGUCHI
The Institute of Statistical Mathematics
4–6–7 Minami–Azabu, Minato–ku
Tokyo 106, Japan

SUMMARY

1. Motivation

Data taken aboard a spinning spacecraft frequently suffers from an unexpected modulation synchronized with the rotation of the spacecraft. If there is a strong noise source along some view direction, a resulting strong periodic noise appears in the measurements. We frequently face situations such that we cannot study the observed data without removing this periodic noise from the original data. It is therefore important to develop methods that can sufficiently separate the noise closely associated with spacecraft spin. In VLF electric filed measurements from the Pioneer Venus spacecraft, spin–synchronized signals often dominate over naturally generated emissions. Consequently, the observed natural emissions are masked by sun–synchronous spacecraft noise sources.

This paper addresses the problem of separating spin–synchronized signals in the Pioneer Venus wave data. A specific problem we are considering is the extraction of the natural emission signal $x(n)$ from given data of the form

$$(y(n), \theta(n)) \quad (n = 1, \ldots, N),$$

where the $y(n)$ are the observed electric fields and $\theta(n)$ is the angle between the sensor and the solar direction. Our goal is to develop a method to automatically and objectively remove any effects caused by the background noise which is closely associated with the spin phase θ.

2. Observational model

Since measurements made with the Pioneer Venus Orbiter (PVO) electric field instrument are masked by sun-synchronous spacecraft noise, the observed data represents a superposition of background noise and natural emissions. We therefore decompose the observed raw data $y(n)$ $(n = 1, \ldots N)$ into $B(n)$ (background noise) and $x(n)$ (real signal), as follows:

$$y(n) = B(n) + x(n). \tag{1}$$

Here it should be noticed that both $B(n)$ and $x(n)$ are positive valued, since they represent electric field spectral intensity measurements $((V/m)^2/Hz)$. Examination of the data shows that the background noise is quite sensitive to the angle between the sensor and the solar direction, $\theta(n)$, so we assume that the background noise, $B(n)$, can be described as a function of time, n, and phase, $\theta(n)$: $B(n) = B(n, \theta(n))$.

A further complication is that the phase response of $B(n)$ is time-varying. However, this can be assumed to be gradually changing with time. Accordingly we model the observed data, $y(n)$, as

$$y(n) = I(n)f(\theta(n)) + x(n), \tag{2}$$

where $f(\theta)$ is the phase response (stationary part of the background noise) and $I(n)$ is the envelope of the background noise at each time n. Namely we assume that the background noise is modulated by the time-varying background noise intensity. Since our data shows that the mean of $I(n)$ sometimes drifts with time, we furthermore decompose $I(n)$ into two factors as $I(n) = t(n)w(n)$, where $t(n)$ and $w(n)$ are a long-term trend and stationary component, respectively. By stationary component, we mean that the distribution can be assumed to be independent of time. Using this decomposition, Eq. (2) can be rewritten as follows:

$$y(n) = t(n)w(n)f(\theta(n)) + x(n). \tag{3}$$

3. Bayesian approach

We first define the background noise component, $B(n, \theta(n)) = t(n)w(n)f(\theta(n))$, at each time n, in order to estimate the natural emission signal $x(n)$. We assume that the long-term trend of the background noise intensity changes slowly with time. In addition, $\log f(\theta)$ is also assumed to show a smooth behavior in the phase domain. $f(\theta)$ is a periodic function, hence $f(360°)$ will be connected to $f(0°)$ continuously. By using this *prior information*, we estimate $t(n)$ and $f(\theta)$ iteratively until the distribution function of $w(n)$ can be satisfactorily assumed to obey the time-independent noise distribution. Based on the final estimate for the background noise component, the determination of the real signal begins by dividing the observations into two groups: candidates for pure background noise (group–B) and that for the background noise + real signal (group–$B+x$). As for group–$B+x$, the real signal $x(n)$ is estimated through Eq. (3).

4. Other models

We will summarize several useful models to extract the background noise which is closely associated with the spin of the satellite.

THE LOCAL LINEARIZATION FILTER WITH APPLICATION TO NONLINEAR SYSTEM IDENTIFICATIONS

Tohru OZAKI
Institute of Statistical Mathematics
4-6-7 Minami Azabu, Minato-ku,
Tokyo, Japan-106

SUMMARY

We consider, in this paper, the identification problem for continuous stochastic dynamical system models with discrete time observation data. Continuous time stochastic dynamical system models include chaos (nonlinear deterministic dynamical system) models, where the variance of the driving system noise is infinitesimal. It is unrealistic to assume that the observed time series data are exactly the same as the true state variable and free from observation noise. Then the system parameter estimation problem for continuous time systems can be reduced to the nonlinear filtering problem. By using a nonlinear filter the observed time series can be transformed into innovation sequences, which are prediction errors. Then the model parameters are estimated by the maximum likelihood method using the innovations and Gaussian likelihood.

The nonlinear filtering problem, although theoretically solved by Kushner, has been well-known to be a difficult problem. This is because the exact solution of the nonlinear filtering problem leads to an infinite dimensional system which is impossible to implement in practice. It has been said, however, that " the non-existence of finite dimensional filters does not address the issue of non exact but high-performance suboptimal filters, which is much more important from a practical point of view." Thus the problem of obtaining good approximations to the exact solutions of nonlinear filtering problems has been studied by various approaches and empirical studies have been published. A common difficulty seen in these empirical studies is the computational instability of nonlinear filtering. Even though the nonlinear filtering is properly defined and is supposed to have a finite valued solution, we often see a numerical explosion in mechanizing the prediction or filtering of the state estimate in discrete time steps. To overcome the computational instability problem we try to get an approximate nonlinear filter, not by approximate continuous time filter equations, but by discrete time filtering equations for an approximate discrete time model of the original continuous time dynamical system. The efficiency of this approach entirely depends on how good the discrete time model approximation is.

The discretization scheme(the local linearization scheme) is introduced from the original continuous time scheme by employing the local linear approximation, i.e. the system is approximated by a linear system on each short interval [t, Δt). The discrete model thus obtained is known to preserve the structural characteristics of the original system; i.e. the discrete model is stable and stationary if the original system is stable and stationary.

Using this discrete model, the nonlinear filter(local linearization filter) is introduced. Since our discretization(the local linearization scheme) preserves the structural characteristics of the original system, the obtained nonlinear filter is free from the computational instability problem. An example of the application of our filtering method to the van der Pol osccillation data, well known for its difficulty, is shown with numerical results.

A maximum likelihood method is given with the likelihood function, explicitly written down by using the local linearization filter. The effectiveness of the maximum likelihood method is checked by applying the method to simulation data generated from stochastic nonlinear vibration models, one is from a stochastic van der Pol equation type model and another is from a stochastic hard spring type osccillation model.

The asymptotic equivalence of the least squares estimation and the maximum likelihood estimation has been shown. We compared the performance of the two methods numerically by applying the two methods to ten sets of simulation data generated from each model. The results show that for our examples the maximum likelihood method is more reliable than the least squares method.

INFERENCE OF EVOLUTIONARY TREES FROM DNA AND PROTEIN SEQUENCE DATA

Masami HASEGAWA
Institute of Statistical Mathematics
4-6-7 Minami-Azabu, Minatoku
Tokyo, Japan 106

SUMMARY

The evolutionary history of life is inscribed in the DNA and protein sequences of living organisms, and we can infer evolutionary trees from these sequences. DNAs and proteins are sequences of four types of bases and of twenty types of amino acids, respectively. At each site of a sequence, one type of base or amino acid may change to another one during evolution, and the substitutions occur independently between lineages after they separate in an evolutionary tree. These substitutions can be best regarded as stochastic, and therefore, it is appropriate to infer evolutionary trees in the framework of maximum likelihood (ML) based on a statistical model for base or amino acid substitutions.

Recent developments of the ML method in molecular phylogenetics and its applications in biological problems are reviewed, and the importance of AIC in improving the model is stressed.

NEW STRUCTURE CRITERIA IN GROUP METHOD OF DATA HANDLING

T. LANGE
Department of Electrical Engineering
College of Technology and Economics
Berlin Lichtenberg / Germany

SUMMARY

The contribution deals with new structure criteria for algorithms of the automatic selection of optimal model structure based on Group Method of Data Handling. These structure criteria were proposed and tested for model selection on the condition that the amount of experimental data is quite limited. Several effects occurring at the simulation of typical Non-Gaussian disturbances are described. A new criterion for indication of data's uncertainty and a new criterion for evaluation of quality of structure criteria are also proposed.

1. Introduction

The task of statistical modeling can be divided into two parts - the parameter estimation (a*) and the selection of the model structure (M*, k*).
The task of modeling, especially the automatic selection of the model structure based on the Group Method of Data Handling GMDH of Ivakhnenko will be discussed in connection with the condition that the amount of experimental data is quite limited.
The availability of a limited amount only of experimental data points often leads to a considerable uncertainty in the task of complex system modeling because the basic assumptions for regression analysis in this case are questionable. Under these conditions it is especially difficult to choose the structure criteria for the selection of a model.

2. Special problems in case of uncertainty

Several investigationsshowed that the origin of uncertainty is a very diverse problem which depends not only on the absolute amount of primary data but also on a lot of correlated factors, such as the amount of measuring points, the length and position of interpolation region, the value of coefficients of the true model, the proportion of correlation rank of columns of the input matrix, the nature of the output disturbance and the completeness of the input vector.

3. Typical effects during object simulation with Non-Gaussian disturbances as symptoms of condition of uncertainty

Often it is impossible to exclude the influence of the following three kinds of disturbances with extreme Non-Gaussian distribution:

- multiplicative-additive disturbance
- Outlier-disturbances
- Disturbance due to "nonobservable variables"

During the investigation of non-standard disturbances on the basis of simulation experiments the following effects could be detected several times :

1. If non-standard disturbances influenced the output of the system, or if in case of Gaussian disturbances the amount of measured data was quite limited, a shift of the RSS functional and its optimum away from the origin was detected.
2. In case of the above effect the tendency could be detected that the growth's curvature of the vector norm of estimated parameters depends on the complexity of the model (the higher the complexity, the steeper the curvature of the growth of the vector norm).
3. The growth of the coefficient's vector norm means that the estimation is biased. This fact led to the idea to observe how the coefficient's vector norm will be changed depending on the value of the Ridge-parameter in case of several simulated disturbances. When carrying out this investigation it was detected that the growth of disturbances led to a reduction of the curvature of the dependence of the coefficient's norm on the Ridge-parameter.

These detected effects were the starting point for asking the following questions:

1. How is it possible to use the data set to decide whether Least Squares is applicable for a given data set ?"
2. If the answer is "No", then what should be used instead of Least Squares ?
3. How should the quality of a model be assessed numerically in the above case ?

To solve the above problems several new criteria were proposed and investigated.These investigations can be divided into three points:

1. the identification of uncertainty by means of the available data set only;
2. the choice of a suitable structure criterion under uncertainty;
3. the evaluation of the quality of the structure criteria under uncertainty.

4. The identification of the existence of uncertainty by means of the available data set

To identify the existence of uncertainty it is proposed to use the common Bootstrap method. The Bootstrap characteristics can therefore be used for the visual recognition of uncertainty .

However, the Residual Sum of Bootstrap square of the model's coefficients on the first layer of GMDH algorithms is a degree of uncertainty and can be used for the automatic selection of the model structure.

5. The choice of the structure criterion

As an answer to the question, which structure criterion has to be chosen if non-standard disturbances influence the system output, two alternatives are possible. These alternatives are based on two kinds of model comparison:

i) the analytic comparison of the models,
ii) the heuristic comparison of the models.

i) The analytic approach to compare models:
One of the models will be named the true one and the other the real one. The true model should provide output values which are identical with the output values of the investigated system if there are no disturbances. In consequence, the true model must be an unbiased one. Naturally, in this case the aim of modeling is to find a real model which is as close to the true model as possible. The real model is searched under the condition that the disturbance was considered as $\delta = 0$ concerning the true model. Thus the quality of modeling depends on the power of the real disturbance. In addition, this quality also depends on the amount of available data.

ii) The heuristic approach to compare models:
The heuristic comparison of models corresponds to verifying the stability of the real model in case of certain changes in the conditions. As well, this resistance will be the criterion of quality. This version demands a minimum of information only.

In principle, the "behavior"-criteria use the information from different data samples, which have, however, the same size in most cases.

The following new structure criteria were proposed or modified and investigated:

- the Gnostic Structure Criteria GSC
- the Criterion of Nonlinear Sum
- the Disturbance Criterion DC2
- the Local Data Uncertainty Criterion LDUC

6. The evaluation of quality of the structure criterion used in case of uncertainty

To evaluate the quality of decision rules, the Tchebyshev inequality can be used. In addition, to estimate both the quality of the model and the quality of measured data the numeric Criterion of Data Reliability has been suggested.

THE USE OF THE KULLBACK-LEIBLER DISTANCE FOR LEARNING IN NEURAL BINARY CLASSIFIERS

D. PADOS, P. PAPANTONI-KAZAKOS,
D. KAZAKOS, and A. KOYIANTIS
Department of Electrical Engineering
Thornton Hall
University of Virginia
Charlottesville, VA 22903

SUMMARY

We consider neural network structures whose objective is binary hypothesis testing and we focus on the development of unsupervised learning techniques. When the operational environment is stochastically described, unsupervised learning becomes a statistical recursive estimation problem. Stochastic approximation methods can be applied towards its solution.

In the neural network literature, unsupervised learning has been basically limited to techniques arising from the Robbins-Monro method and its extensions. Traditionally, the performance criterion adopted is the least squares error.

In this paper we deviate from the classical approach by assuming that the performance measure at the operational stage, for the recognition of the acting hypothesis, is the Neyman-Pearson criterion. Then, we use the Kullback-Leibler distance between the observed and the desired Type-1 error probability as the performance criterion, to derive adaptive learning algorithms for pertinent network parameters.

Let us consider a single network element which at time instant n receives an input x_n and produces the output u_n. With u_n binary, the element can be viewed as providing a hypothesis testing decision (0 for hypothesis H_0 and 1 for hypothesis H_1). If we define

$$U(x) = \begin{cases} 1 \; ; \; x \geq 0 \\ 0 \; ; \; x < 0 \end{cases}$$

then we set

$$u_n = U(g(x_n) - \lambda) \tag{1}$$

where g(x) may be any strictly monotonically increasing and bounded function. The induced Type-1 error $h(\lambda)$ depends on the value of the threshold λ and the distribution of the H_0 hypothesis which generates the input data:

$$h(\lambda) = \Pr(x \geq g^{-1}(\lambda) / H_0) \tag{2}$$

Feeding the network element with a data sequence $\{ x_n \}$ from the H_0 hypothesis, whose statistics are unknown, and observing the binary outputs $\{ u_n \}$, the objective is to recursively adapt the parameter λ to match the induced Type-1 error $h(\lambda)$ with a given desired value α. We use as the "matching" criterion the Kullback-Leibler distance $K(\alpha,h(\lambda))$, where

$$K(\alpha, h(\lambda)) \triangleq \alpha \log \frac{\alpha}{h(\lambda)} + (1-\alpha)\log\frac{1-\alpha}{1-h(\lambda)} \tag{3}$$

Under some general conditions on H_0 and $g(x)$ and due to the analytical and strict convexity properties of the $K(\cdot)$ function, $K(\alpha,h(\lambda))$ has a unique global minimum at $\lambda=\lambda^*$ where

$$\lambda^* : \Pr(x \geq g^{-1}(\lambda^*)/H_0) = \alpha \tag{4}$$

Assuming, in addition, that $h(\lambda)$ is an analytic function and applying Newton's numerical method on equation (3) we obtain:

$$\hat{\lambda}_{n+1} = \hat{\lambda}_n - \frac{1}{\dot{h}(\hat{\lambda}_n)}\left[\frac{\ddot{h}(\hat{\lambda}_n)}{[\dot{h}(\hat{\lambda}_n)]^2} + \frac{(h(\hat{\lambda}_n)-\alpha)^2 + \alpha(1-\alpha)}{h(\hat{\lambda}_n)(1-h(\hat{\lambda}_n))(h(\hat{\lambda}_n)-\alpha)}\right]^{-1} \tag{5}$$

The recursive equation (5) requires explicit knowledge of the $h(\cdot)$ function, which of course cannot be provided. Instead, we propose the use of an estimate $\hat{p}_n$ of $h(\hat{\lambda}_n)$, for each $\hat{\lambda}_n$ encountered. We also ignore the $\frac{\ddot{h}(\hat{\lambda}_n)}{[\dot{h}(\hat{\lambda}_n)]^2}$ term in expression (5), which now becomes

$$\hat{\lambda}_{n+1} = \hat{\lambda}_n - G(\hat{\lambda}_n)\frac{\hat{p}_n(1-\hat{p}_n)(\hat{p}_n-\alpha)}{(\hat{p}_n-\alpha)^2 + \alpha(1-\alpha)} \tag{6}$$

Different procedures for a recursive adaptation of the estimates $\hat{p}_n$ are proposed and stochastic convergence of the random sequences $\{\hat{\lambda}_n\}$, $\{\hat{p}_n\}$ is established. Specific robust selections for the function $G(\cdot)$ (ideally to be $[\dot{h}(\cdot)]^{-1}$) are developed for cases where the probability density function of H_0 is assumed to belong to a given compact class of density functions. In addition, numerical evaluations of the performance of the algorithm are presented.

LIKELIHOOD ESTIMATION OF DIRECTIONAL INTERACTION

M.TANEMURA
The Institute of Statistical Mathematics
4-6-7 Minami-Azabu, Minato-ku, Tokyo 106, JAPAN

SUMMARY

1. Introduction

The spatial pattern of objects in the space has attracted statisticians and researchers in application fields, especially when the objects are 'points'. There appear some new interesting features in the case when each object possesses some characteristics in addition to its locational information. We then suppose the object has a direction or an orientation as its nature.

The main concern here is the spatial pattern which is exhibited by the directional data of each point and the purpose of this paper is the characterization and the model fitting to such data.

2. Directional Interaction Model and Likelihood

Suppose that, in a finite region V of space, a set of coordinates $X \equiv (x_1, x_2, \ldots, x_N)$ of N points is given, and suppose that a set of directional vectors $S \equiv (s_1, s_2, \ldots, s_N)$ is attached to each spatial point ($s_i \in \Omega, |s_i| = 1, i = 1, \cdots, N$; Ω is the angular space). Then, we assume that an interaction potential of form $f_\xi(s_i, s_j)$ works between two points i and j, and that the information of X is assumed to be included implicitly as the lattice model, for instance. Then we call f_ξ a directional interaction model and we suppose that a given S obeys the so-called 'Gibbs canonical distribution' under the directional interaction potential.

Thus, the explicit form of log-likelihood for our directional interaction model is

$$\log L(\xi; S) = -\sum_{i=1}^{N-1} \sum_{j=i+1}^{N} f_\xi(s_i, s_j) - \log Z(f_\xi; N).$$

The second term of the above equation is the normalizing constant and is written in a complicated and highly multiple integration form.

3. Exact Likelihoods

By restricting the interaction model to $f_\xi(s_i, s_j) \equiv f_\xi(s_i \cdot s_j)$, and by assuming the interaction works only between the 'contiguous pair', we can evaluate the normalizing constant in a simple form. If all vectors are located on the points of a regular lattice, the contiguous pair will corresponds to the nearest neighbour pair. Among such lattice models, the so-called 'Ising model' is well known in statistical physics. In this model, the exact log-likelihood is given for 1-D and 2-D models.

One of the other interesting classes of models is the linear chain of unit vectors. We considered two cases of the linear chain where unit vectors themselves are 2-D and 3-D, respectively. For these models, the explicit expressions of the exact log-likelihood are given. Moreover, the serial correlation coefficient is also given in an analytic form for these models.

4. Simulation Study

In order to show the feasibility of the procedure of estimation, some computer simulations for the linear chains of 2-D and 3-D unit vectors were done. For both cases, the model $f_\xi = -\xi s_i \cdot s_{i+1}$ was considered. For the length of the chain of $N = 200$, computer simulation of this model was performed for nine cases of parameter values. For each case, we obtained respectively fifty independent samples of equilibrium configuration. The resulting values indicate good coincidence both between the true parameter values and the mean of estimates and between the experimental and theoretical standard deviations.

Furthermore, we can detect whether the true model is selected or not for each sample data by using the model selection procedure by the value of AIC, $\text{AIC}(k) = (-2)\text{max log likelihood} + 2k$, where k is the number of adjustable parameters. In our case, the value of AIC for true model is AIC(0) and that of AIC for the estimate is AIC(1), because the number of adjustable parameter is ξ only in our case. According to the minimum AIC procedure, the true model is selected if AIC(0) − AIC(1) is negative. By applying this procedure, almost of the samples are selected as being true respectively for all cases of parameter values.

Similar simulation study was done for the linear chain of 3-D unit vectors of the above model. The result of analysis showed that the coincidence between the true values and the means of estimates is excellent as well as the coincidence between the experimental and theoretical standard deviations. In this case, the same order of number of samples are selected to be true as in the 2-D vector case. Thus the feasibility of our procedure and the validity of our simulation was found.

5. Analysis of some real data

In order to show the applicability of our procedure, several real data were analyzed. Firstly, two sets of data of the directions traveled by sea stars were considered. The result of the analysis indicates that, in both cases, the directional interactions are highly attractive, that means, sea stars had a tendency to move the similar direction during the observed period.

As the second example, two sets of data of magnetic remanence of rocks were analyzed. They were assumed as if they consist of one dimensional array of unit vectors. As we have no information on the way of arragement of each vector, the randomization of vectors was also considered. As for the red-bed data ($N = 17$), the values of $\hat{\xi}$ and AIC indicated that almost all vectors are pointing to similar directions. As for the magnetic remanence data of 52 measurements, the values of $\hat{\xi}$ and AIC indicated that there is no interaction among unit vectors. This was also true for respective arrangements of randomized permutation.

The detection of similarity of amino acid sequences is known to be important in the study of protein evolution or other molecular biological studies. In this section a new method of characterizing the pattern of amino acid sequences was presented by applying the likelihood procedure of directional interaction. For that purpose, the similarity table among twenty elemental amino acids was used. In this table, the value 1000 represents the biggest similarity, while the value 0 the most dissimilarity. By using this table, amino acid sequences can be translated into one-dimensional series of unit vectors whose angular information is determined by the similarity among amino acids. Accordingly, we assigned an angle ϕ_{ij} between two amino acids i and j by the relation

$$\phi_{ij} = \pi \left(1 - \frac{d_{ij}}{1000}\right),$$

where d_{ij} represents the similarity value between i and j.

Two types of analysis were then considered, that is, the analysis within a sequence and that between a pair of sequences. Firstly, for a single amino acid sequence, angles were assigned for a contiguous pair of amino acids by the method just discussed. Hence, we obtain a sequence of directions from the original amino acid sequence. Twelve amino acid sequences of length ninety were analyzed. These data are the part of the amino acid sequences translated from the DNA or RNA sequences of retroviral reverse transcriptases of viruses, putative polymerases or copia-like element. The result of our analysis within a sequence showed that, for all sequences, the directional interaction is repulsive. This indicates the dissimilar amino acids are liable to be contiguous, and therefore it means that the specific amino acid sequence is stable.

As the next analysis of amino acid sequences, we compared two sequences in pairs for the above twelve data. Toh *et al.* (1983) and Miyata *et al.* (1986) found by their "homology matrix" comparisons that the above twelve amino acid sequences are similar with each other. In our analysis, the assignment of angle is done such that the angle ϕ_{ij} is given for the two amino acid elements i and j at the same corresponding site of a pair of sequences. Hence, we obtain a sequence of directions for a pair of amino acid sequences. The result of parameter estimation for 66 pairs of comparison indicated, for several pairs of amino acid sequences, the estimates for ξ are large which suggests the strong similarity for these pairs. The result of our analysis was coincident with the conclusions obtained by Toh *et al.* (1983) and Miyata *et al.* (1986).

Thus, the likelihood method proposed in this section is proved to be applicable to the analysis of amino acid sequences.

NEW APPROACH TO OPTIMIZATION OF STABLE ESTIMATION

Alexander M. SHURYGIN
Department of Mechanics and Mathematics
Moscow State University
Moscow 119899, RUSSIA

SUMMARY

Optimality of Fisherian statistics is not still evident in many applications in statistical science. For thirty years the theory of robustness showed that the usual maximum likelihood estimator (MLE) is unstable to the variations of the model used. It casts doubt in the value of classical statistical methods for applications.

This paper reviews the robust estimators and presents a series of results based on the extensive research work of this author to optimization of stable estimation. These results are not very readily available and well-kown beyond the Russian community. It is hoped that this paper can serve to provide a broader exposure for the new Stoyki (in Russian it means steady) estimation techniques in many engineering and scientific applications.

IMPLICATIONS OF INFORMATIONAL POINT OF VIEW ON THE DEVELOPMENT OF STATISTICAL SCIENCE

H. AKAIKE
The Institute of Statistical Mathematics
4-6-7 Minami-Azabu, Minato-ku
Tokyo 106, Japan

0. Introduction

Informational approach represents a new trend in the development of statistical science. This paper is intended for the discussion of the informational outlook in relation to the development of statistics or statistical science.

R. A. Fisher's great success in laying down the foundation of mathematical statistics was largely based on his use of mathematical likelihood as the basic concept for the construction of the theory of estimation. However, from the point of view of the informational approach to be discussed in this paper, it appears that the assumption of a known functional form of the distribution as the starting point of the theory blocked the use of log likelihood as a general criterion of fit of models with possibly different functional forms of the distribution.

The informational approach to statistics was realized by the recognition of the relation between the Kullback-Leibler information and the expected log likelihood of a model. The information criterion AIC was developed based on this observation and found various practical applications before its general acceptance by theoretical statisticians.

In this paper an explanation of this popularity is traced back to the philosophical work of C. S. Peirce on the logic of abduction which is concerned with the first starting of a hypothesis and entertaining it (Peirce, 1955, p.151). Fisher emphasized the use of likelihood in the inductive phase of inference under the assumption of a hypothesis, while Peirce insisted that the most original part of scientific work was related to the abductive phase, or the phase of the selection of proper hypotheses. The existence of examples of successful use of AIC in practical applications shows that the informational approach opened up the possibility of developing a systematic procedure for abductive inference with the aid of the informational interpretation of log likelihood.

Since the development of hypotheses depends on the available knowledge and purpose of application, cross-disciplinary cooperative work is required for successful development

H. Bozdogan (ed.), Proceedings of the First US/Japan Conference on the Frontiers of Statistical Modeling: An Informational Approach, 27–38.

of the science of statistics in the future. The informational approach has prepared a basis for the successful development of such cross-disciplinary activity.

1. Review of Fisherian statistics

In this section the contribution of R. A. Fisher to statistics will be reviewed briefly to clarify its relation with the informational approach to statistics. The distinctive characteristic of Fisher's contribution is the reflection of strong influence of the experience of handling statistical problems in real world. However, his first paper on statistics was not directly related to any particular practical application (Fisher, 1912). In the paper he introduced a criterion of fit of a statistical model which was later to be called likelihood.

Fisher did not describe how he was lead to the use of the criterion. However, taking into account the fact that just before that time the method of fitting frequency curves and the test of the goodness of fit by chi-square statistic had been developed by K. Pearson, one possible guess is that the introduction of the criterion was motivated through the need of calculating theoretical frequencies for the chi-square test. It is known that there is a close relation between the chi-square test and the probabilistic interpretation of entropy as developed by L. Boltzmann and that Boltzmann's final work on the subject is based on a representation identical to negative Kullback-Leibler information (Akaike, 1985). Thus it seems that there is an inherent historical connection of Fisher's contribution to statistics with the probabilistic concept of entropy, or information.

As is observed by Stigler(1976) the 1922 paper of Fisher (Fisher, 1922) on the mathematical foundations of theoretical statistics is characterized by its heavy dependence on the concept of parameter. Fisher defined the purpose of statistical methods as the reduction of data and divided the related problems into three types : problems of specification, estimation and distribution. The importance of the problem of specification or the proper choice of the functional form of the distribution for each particular application was clearly recognized by Fisher. He highly evaluated the contribution of K. Pearson on this aspect of statistics by introducing the Pearsonian system of frequency curves and an objective criterion of goodness of fit, the chi-square test statistic. However, in spite of the insightful observation of the importance of the problems of specification, Fisher considered these as entirely a matter for practical statistician and regarded the discussions of theoretical statistics as alternating between problems of estimation and problems of distribution (Fisher, 1922, p.315).

Fisher was keen to clarify the contribution of mathematical theory of statistics to the logic of inductive reasoning. In an expository paper on the advances of statistical methods he tried to demonstrate the adequacy of the concept of likelihood for inductive reasoning in the particular logical situation for which it had been introduced (Fisher, 1935). This was the use of likelihood in the problem of estimation. He considered that the concept of likelihood had been introduced for the handling of the problem of parameter estimation and that the justification for its use was provided by the successful development of the theory of estimation.

In his presidential address to the First Indian Statistical Conference (Fisher, 1938), Fisher pointed out the importance of sufficiently prolonged experience of practical research and of responsibility for drawing conclusions from actual data for teachers of statistics at universities. However, of the three types of problems mentioned in the 1922 paper, only the importance of the problems of distribution was mentioned in relation to the development of the theory of estimation. Further, stressing the successful development of the estimation theory, he even mentioned that statistician's job was only to produce what the given data set contained. This statement makes a peculiar contrast with the statement made in an earlier paragraph of the same paper. It tells that all extensive bodies of data were liable to contain information on points which were not in view when they were collected, and that to recognize such information, and to find the means of eliciting it, was most stimulating part of statistician's task. It seems that this peculiarity reflects the desire to confirm the contribution of mathematical statistics within the framework of inductive logic.

In spite of the heavy reliance of Fisher's theory of estimation on the concept of likelihood his understanding of the nature of the likelihood remained somewhat rudimentary. This can be seen by the explanation of the concept of mathematical likelihood given in his last book as follows :

> *It is, like Mathematical Probability, a well-defined quantitative feature of the logical situations in which it occurs, and like Mathematical Probability can serve in a well-defined sense as a " measure of rational belief ";* .... (Fisher, 1973, p.72).

Similar explanation appears in several other writings. However, no convincing argument was developed for the characterization of the likelihood as a "measure of rational belief ".

The informational point of view to be discussed in the next section shows that this lack of the characterization restricted Fisher's use of likelihood to the local theory of parameter estimation under the assumption of a given form of the distribution. A framework was then established to view the test of significance as the basic procedure for the solution of the problems of specification and restrict the estimation to the parameter of a given model. Thus the test and estimation formed a paradigm to make statistics into what was called a normal science by Kuhn (1970). However, it seems that the use of test procedures advocated by this paradigm eventually produced a very restricted image of statistics in applications which was conditioned by the availability of proper test procedures.

2. Informational view of statistics

In spite of the suggestive title of the 1912 paper to view likelihood as an absolute criterion of fit of frequency curves Fisher actually defined it as a rational measure for the comparison of parameters within a model. This eliminated the possibility of recognizing the potential of log likelihood as a general criterion of fit of a model and restricted its use to the estimation of parameters within a fixed model. Thus the later development of the theory of estimation was concerned with the refinement under the assumption of a known form of the true distribution.

In a real situation, it is certainly quite rare that the exact form of the distribution is known. The success of the analysis of real data thus depends essentially on the choice of the basic model, or the form of the distribution. Only through proper modeling of the structure of the distribution one can get useful information out of data by statistical procedures. A poor model cannot produce useful information with the refinement of the parameter estimation procedure.

According to the present author's personal experience almost each new problem demanded significant effort for the choice of the basic model. One trivial example which clearly explains this was obtained in the analysis of data for the experimental comparison of chemotherapeutics for the treatment of tuberculosis. The medical doctors conducting the experimental study were bothered by the smallness of the number of complete time series of observations of patients and asked the author to provide a proper procedure to fill in the missing observations. A careful examination of the data showed that the clustering of the time series by the initial condition of each patient, represented by the number of cavities in the x-ray chest image, the time series could produce sufficient information to differentiate the effects of the medicines. The choice of the number of cavities as the concomitant variable was the key to the success of the analysis and the simple recovery rate at each month demonstrated sufficient accuracy in each cluster. However, the detection of this concomitant variable required much work. This simple example sufficiently shows that it is the solution of the problem of specification that is really decisive for the successful handling of a real problem.

If the attention had been limited to the empirical choice of the basic model there would have been no chance to get the idea of the use of information criterion. The analysis of engineering problems showed the existence of a vast area of application of a particular type of models, the constant, or time invariant, linear system models. Their frequent use suggested that the identification of the structure of a constant linear system model, based on the time series of observations of a real system, was a practically very important problem. The limitation of the empirical approach to the handling of time series for the identification became clear when the fitting of a constant linear system model, the autoregressive (AR) model, was considered. Conventional procedure of power spectrum estimation was realized by smoothing the periodgram and this was highly empirical and required much experience for proper choice of the final estimate. The estimation could be realized through AR model fitting and thus eliminate the difficulty if a proper procedure for the determination of the order of an autoregression was known. However, when the choice of AR order was contemplated, it became quickly clear that curiously enough this problem had been handled not as a problem of estimation but of a test.

It is interesting to see that Fisher also faced with the problem of order determination in his first work with the long sequence of yields of wheat from Broadbalk when he tried to extract time trends by fitting orthogonal polynomials (Fisher, 1921). Incidentally, in this connection, J. F. Box (1978, p.104) mentions that, fifty years later, the problem of how best to determine the adequacy of fit of a polynomial of any degree has not been satisfactorily solved. Apparently Fisher tried to confirm the adequacy of the chosen order by checking the autocorrelation coefficients of the residuals.

In the case of the order determination of AR models, after the trial of various criteria, the concept of final prediction error (FPE) was introduced which was defined by the expected mean-square one-step ahead prediction error when the estimated parameters were used for prediction (Akaike, 1969, 1970). However, for the extension of the concept of FPE to the case of multivariate AR model, a measure for the evaluation of the multivariate prediction error was required. A seemingly natural choice was the generalized variance, or the determinant of the variance-covariance matrix of the innovation, or the one-step ahead prediction error. It was then noticed that the log of the sample generalized variance of the residuals was asymptotically related to the maximum log likelihood of the stationary Gaussian multivariate AR model (Whittle, 1953). Thus the attention was turned to the method of maximum likelihood.

Certain similarity was observed between the choice of the order of an autoregression and that of the number of factors in the factor analysis, but, in the case of the factor analysis, it was not easy to identify the equivalent of the prediction error of time series. The recognition of the fact that the fitting of the factor analysis model was realized by maximizing the log likelihood (Jöreskog, 1967) lead to the recognition that the negative log likelihood was playing the role of the mean square prediction error. In the case of AR model fitting the estimated parameters were used to compute the predicted value and the analysis of the expected performance of such a predictor lead to the concept of FPE. By analogy, it was clear that the probability distribution defined by the maximum likelihood estimate of the parameter was providing a predictor of which performance was being evaluated by the expected log likelihood. When the use of log likelihood was considered as the basic criterion it was noticed that the expectation of this quantity was related to the Kullback-Leibler information defined by

$$I(f;g) = E_x \log f(x) - E_x \log g(x) ,$$

where g(x) is an approximation to f(x), the true distribution, and E_x denotes the expectation with respect to f(x). The non-negativity of the K-L information showed that the maximum of the expected log likelihood was attained when the model was true.

The information criterion AIC was derived by considering the typical situation of estimation treated by Fisher. When the parametric model $f(x \mid \theta)$ is fitted to the vector of observations $x(N) = (x_1, x_2, \ldots, x_N)$ under the assumption of independent and identical distribution of observations the asymptotic theory of the maximum likelihood estimate suggests that the difference of the twice the expected log likelihoods

$$2E_x \log f(x \mid \theta_0) - 2E_x \log f(x \mid \theta(x(N)))$$

is, under certain regularity condition, asymptotically distributed as a chi-square statistic with the degrees of freedom k, the dimension of θ, where E_x denotes the expectation with respect to the true distribution $f(x \mid \theta_0)$ of $x(N)$ and $\theta(x(N))$ the maximum likelihood estimate. This quantity is asymptotically approximated by

$$2 \log f(x(N) \mid \theta(x(N))) - 2 \log f(x(N) \mid \theta_0)$$

which is a familiar Neyman-Pearson likelihood ratio test statistic to test the hypothesis $\theta = \theta_0$. This shows that the increase of the log likelihood attained by replacing the true parameter θ_0 by its maximum likelihood estimate $\theta(x(N))$ is asymptotically equivalent to the decrease of the expected log likelihood of the predictive distribution $f(\bullet \mid \theta(x(N)))$ from that of the true distribution $f(\bullet \mid \theta_0)$.

Thus to recover the relative relation between the twice expected log likelihoods from that of the twice log likelihoods it is necessary to reduce the value of twice the maximum log likelihood by twice the corresponding chi-square statistic. However, since this chi-square is unobservable, it is replaced by its expectation k. To establish the direct correspondence between the log likelihood and mean square error the negative of the log likelihood is adopted as the criterion to measure the badness of the model and AIC is defined by

$$\text{AIC} = - 2 \log (\text{maximum likelihood}) + 2 (\text{number of parameters}),$$

where log denotes natural logarithm.

This derivation of AIC is based on the assumption that the parametric model contains the true distribution as its member. This represents the situation where the models being considered are just-or over-parametrized. In actual comparison of models the crucial situation is where some of the models are under-parametrized but the corresponding increase of the badness of fit, as measured by the criterion, is of comparable order to the reduction of the number of parameters. This situation is discussed in detail in the original paper of Akaike (1973) by using the result of the large sample theory of the test of hypotheses of multiple parameters developed by Wald (1943).

3. Contribution of informational approach to the generation of hypotheses

By reading ordinary introductory text books of mathematical statistics one often gets an impression that the main use of the theory of statistics is the evaluation of the level of significance of some test statistics or of the accuracy of estimated parameters. Since the level of mathematics required for the solution of the problems of distribution related to the test and estimation procedures demands professional skill, people who are mainly interested in the application of statistical procedures usually have to seek for a model of which handling procedure is well-developed. Thus it was often felt impossible for the people in applied area to develop the use of new models for their own problems. However, from the point of view of application, it is the production and confirmation of interesting and useful hypotheses through proper use of observational data that is really expected of statistical methods. Thus the emphasis of the mathematical aspect of conventional statistical procedures unexpectedly produced an undesirable effect to impede the participation in the development of statistics of the people with ample knowledge of particular area of application but with less professional skill of mathematics.

In this connection it is interesting to refer to the philosophical works of C. S. Peirce on abduction and induction around 1890 (Peirce, 1955). Peirce discusses the process of making use of observed facts, or data, to extract practical knowledge for application to other circumstances than those under which they were observed. For this the necessary additive ingredients are hypotheses and he calls the inferential step of starting and entertaining a hypothesis abduction, or retroduction, and calls the operation of testing a hypothesis by experiment induction. The testing in this case is performed by checking the validity of a prediction based on the hypotheses by observations and the result is used for the forming of confidence on the hypothesis. It can be seen that the statistical process of the test of significance falls into this category of induction. Fisher (1935) considered the essential effect of the general body of researches in mathematical statistics at his time to be a reconstruction of logical rather than mathematical ideas and the work of making sense of figures was considered to be an attempt of inductive logical process for the use of data in making an inference from the particular to the general. It seems that what Fisher highly evaluated of mathematical statistics was the contribution to the logic of induction as defined by Peirce.

It is interesting to note that Peirce placed more importance on the inferential step of abduction or retroduction than on induction. He considered that originality lies in the suggestion of hypotheses and that typical induction only tests a suggestion already made. He also exemplified the process of abduction by Kepler's work on the analysis of the motion of the planet Mars which was realized through a succession of refinements of hypotheses until a complete fit to observations was attained. By this example the importance of the existence of sound and rational motive for the modification of hypothesis is stressed instead of the use of capricious or random modification. The existence of an inherent relation of this concept of abduction with the scientific treatment of the problems of knowledge can be observed by the recently revived interest in the abductive inference in the area of artificial intelligence; see, for example, Josephson et al. (1987).

Within the context of mathematical statistics the abductive reasoning is realized by the refinement of a hypothesis through successive process of modification and testing. Fisher also considered that the free use of empirical frequency formulae was justified by the use of the tests of goodness of fit. However, this process is essentially concerned with the problem of specification which was considered by Fisher to be outside of the concern of theoretical statisticians. Thus the view of statistics developed by Fisher placed the abductive aspect of inference outside the attention of theoretical statisticians. This was a situation which was not quite favorable for the development of statistics as a science of knowledge.

The introduction of AIC significantly changed the situation. The example of Kepler's work shows that the process of abduction as contemplated by Peirce is realized by successive test or evaluation of hypotheses. Thus for the implementation of an abductive inference procedure it is necessary to label the result of the test or evaluation of each hypothesis by a common measure useful for the purpose of comparison. In this connection it must be remembered that Akaike (1985, p.11) has pointed out the importance of the fact that the log likelihood $\log f(x \mid \theta)$ provides an unbiased estimate of $E_y \log f(y \mid \theta)$, irrespective of the form of the true distribution $f(y)$ of x, where E_y denotes the expectation with respect to the distribution $f(y)$. This fact represents the most significant aspect of the informational approach in relation to the generation of hypotheses. It provides a justification for the choice of a model with larger log likelihood as a better model, whatever it may be the form of the true distribution. This observation essentially frees the statistical modeling activity from the unproductive concept of " fixed but unknown true distribution " and encourages the proposal of new models.

When the use of a statistical model for the handling of the information supplied by data is considered, it becomes obvious that the definition of true model depends on the objective of the modeling. In particular, the concept of true model depends on the available resources, such as tools and knowledge, for the construction and application of the model. In this sense the concept of true model is highly dependent on the circumstances. Further, the basic choice of a model is realized only through the mental activity of the researcher. In this sense, the concept could be quite subjective or personal.

The emphasis of the subjective nature of statistical modeling activity seemingly makes a significant contrast to the emphasis of the objectivity of statistical procedures by Fisher as represented by his careful handling of the Bayes procedure or the inverse probability. However, Fisher (1973, p.81) also clearly recognized the role of statistician's imagination in a test of significance. Since all the great scientific discoveries or inventions are always quite unexpected from the point of view of contemporary observers and the generation of an innovative hypothesis is always highly dependent on personal activity, it is obvious that blind adherence to the concept of objectivity must be eliminated to regain the creative power of statistical methods in scientific activities. M. Polanyi who stresses the personal aspect of science also points out the importance of the talent to select good hypotheses for investigation and mentions that any theory of inductive inference in which this talent plays no role is a *Hamlet* without prince (Polanyi, 1962, p.30). The importance of the role of initiative in producing human intelligence has been noticed earlier by A. M. Turing in a report written in 1947 (Turing, 1969).

Fisher (1973, p.46) considered that the objectivity of a statistical procedure is represented by the communicability and verifiability of the the result. The above mentioned informational interpretation provides a justification of log likelihood as an objective or inter-subjective criterion of the goodness of statistical models. Thus log likelihood attains the position of a general criterion that can lead the unending search for "truth" through successive improvement of hypotheses, represented by statistical models, within a scientific community. AIC is a typical example that has shown the general applicability of this criterion for the realization of an abductive inference procedure, and encouraged the activity related to the production and comparison of hypotheses by both theoretical statisticians and researchers in the area of application. Since the method of maximum likelihood provides a definite procedure for the production of the estimates of unknown parameters and since the computers allow easy implementation of necessary numerical procedures an efficient procedure for the development of a new hypothesis is realized through successive proposal and comparison of statistical models with the aid of AIC.

This procedure provides a typical example of the realization of the abductive inference procedure discussed by Josephson et al. (1987) for the search of a best hypothesis.

However, the limitation of the above procedure becomes quickly apparent. This is the difficulty of handling the situation where extremely large number of hypotheses can be generated mechanically. Typical example is the polynomial fitting to a temporal series of observations. By equating some of the coefficients equal to zero the number of possible models becomes equal to 2^{k+1} for the k-th degree polynomial. The number of models can easily exceed the number of observations and there is no hope of the simplistic minimization procedure of AIC to lead to any reasonable choice of a best hypothesis.

Again it is quite remarkable that Peirce already considered this type of problem in relation to the search for better hypotheses. In his work Peirce remarks, based on his own experience, that it is not the logical simplicity but the simplicity in terms of facility and naturalness that should be preferred. This remark is based on the conviction that human mind is with a natural bent in accordance with nature's. It seems that the notion of simplicity implies the possession of wider applicability for the generation and explanation of other hypotheses. These observations suggest the need of introducing a hypothesis of higher level that properly represents the natural bent of human mind and governs the collection of original hypotheses. It is interesting to note that, in the case of statistical hypotheses, the role of the governing hypothesis is played exactly by the prior distribution over the models that represent the original hypotheses. Thus Peirce's observation provides an idea on the use of Bayesian models for the elimination of the difficulty of handling a large number of models. However, the observation does not directly lend support to the personalistic view of subjective Bayesians. Any behavior that is only based on rigid subjectivism cannot be differentiated from the behavior of a lunatic person and the result can not be communicated as useful information to others.

The observations made in this section suggest the necessity of clarifying the relation between data and hypotheses. In the inductive logic data are used to check or evaluate hypotheses. In the abductive phase of inference the data prompt the generation of hypotheses. It can be seen that inductive logic is concerned with the decisional use of data, while the abduction is concerned with the analysis or interpretation of data. If the process of the analysis of data is viewed as a process of abduction then it becomes natural and necessary that the hypotheses being generated are data dependent, as the mental search for possible hypotheses will be conditioned by the knowledge of the data. This is the actual state of the analysis of data being practiced in scientific activities and the implicit dependence of the hypotheses on the data poses a problem for the strict application of inductive reasoning. However, if the dependence is realized only through the rejection of definitely inadequate models with very low likelihood this would not seriously invalidate the reasoning. The final validation of the hypotheses will only be realized through the application to real problems.

A drastic view of data is obtained by viewing the knowledge represented by, or used in the forming of, the hypothesis also as a kind of data. Actually the prior knowledge used in the representation of a hypothesis may be considered as a reduced representation of the information supplied by a particular set of past observational data. By this view statistical model construction is concerned with the proper use of the information provided by the whole set of data of the present and past. The past and present data are uniquely determined and in that sense certain, but the knowledge based on the past data is not necessarily perfect or certain. Its validity must always be questioned. This simple observation shows that the rigid Bayesian framework which assumes the prior knowledge represented by the prior distribution as certain is useful only in the inductive or decisional framework. In the abductive phase of the analysis of data the validity of the assumption of the prior distribution must be discussed.

The informational approach suggests that if the process of generating a hypothesis is viewed as the process of forming an expectation represented by a probability distribution then a generally acceptable criterion for the comparison of models could be defined by proper definition of the of log likelihood. Thus a Bayesian model is viewed only as a particular type of statistical model and its relative goodness can be evaluated by the likelihood which is defined by the integral of the likelihood of the data distribution with respect to the prior distribution of the parameter. Some early examples given by the present author on the use of Bayesian models based on this observation are the seasonal adjustment procedure BAYSEA, and the control of improper solutions in the factor analysis (Akaike 1980, 1987).

4. Enhancement of the importance of statistics in society

The history of statistics shows the rises and falls of various schools each focused on particular aspect of statistics and demonstrates the difficulty inherent in the definition of the word statistics. K. Pearson (1973) describes vividly how the word "statistics" was first introduced into English. The word has been traditionally used to denote numerical characteristics of a mass of observations, in particular of a society. Another and more recent use is to denote the science and technology of the effective use of information provided by observational data for the construction and use of hypotheses. The use of the word statistical science for this latter purpose would be more appropriate to eliminate the common confusion of the concept of statistics.

Recent development of computer technology is rapidly increasing the number and variety of observations recorded in media directly accessible from computers. The huge accumulation of data necessitates the use of statistical procedures for the extraction of useful information for particular application. Thus this is the time that society is really in need of statistical science. The discussions in the preceding sections show that the contribution of a statistician would be highly evaluated only when he had contributed significantly to the generation of a new effective hypothesis. For the construction and formulation of such a hypothesis, besides the skill of handling statistical models, deep background knowledge of the area is indispensable. It is the ability to provide a reasonable representation of the background knowledge in the form of a statistical model that is really required of a statistician. Thus, basically it is only through cross-disciplinary cooperative researches that statisticians can make significant direct contributions to society.

The present author found several researchers in unexpected areas of application who used AIC, or its variant for the evaluation of Bayesian models, to produce very interesting results. This shows that the informational approach certainly contributed significantly to the enhancement of abductive use of statistical inference and increased the number of successful examples of application of statistical methods. However, when the complexity of the required model is further increased, the cooperation of professional statisticians will be necessary to make fuller use of the information supplied by data. It is also certain that only through this type of contact with real problems statisticians can gain the idea of the future direction of the development of statistics, as statistics is the science of a particular type of activity of human brain and the function of human brain develops with the accumulation of experience and knowledge in human society. Without the contact with outside world, statisticians' research effort will inevitably be restricted to the analysis of those problems that were modeled by the brain activity based only on the remote past experience.

These observations show the need of formulating statistical science as a science for the development of the concept of statistics through the contact with problems and researchers of various disciplines. The reason of existence of the science can be found in the fact that a statistical model developed in a particular area of application can often be profitably transplanted to other areas. Thus the accumulation of statistical models and experiences of their use is particularly important. Also, as the development of the informational approach shows, it would be safe to say that only the theoretical or logical analyses of the accumulated experiences of application can lead to the development of new concepts. These observations show that only the proper implementation of cross-disciplinary activities can enhance the importance of statistics in society. The informational approach prepared a technical and philosophical basis for the development of such cross-disciplinary activities.

5. Conclusion

The informational view of statistics has clarified that the lack of the characterization of log likelihood as a general measure of rational belief limited Fisher's attention to the use of log likelihood in the inductive logic of estimation and this lead to the unlucky dichotomy of statistical methodology into test and estimation. In contrast with this, the informational point of view has been developed based on the recognition that the choice of the basic model or the solution of the problem of specification constitutes the crucial phase of statistical handling of observational data and log likelihood is useful as a general criterion for the evaluation and comparison of models with different distributional structures.

The review of the philosophical work of Peirce on abduction and induction has shown that the abductive inference, or the generation of hypotheses, plays more significant role in the science of knowledge than the inductive inference. It has been confirmed that AIC produced a realization of the abductive inference procedure. It has further been observed that the informational point of view provides a basis for the free use of the originality based on personal knowledge for the construction of hypotheses represented by statistical models, including Bayesian models.

Thus the informational view allows the development of statistical methods for the important scientific activity of hypothesis generation with proper use of observational data. It may be concluded that informational view has prepared a basis for the enhancement of the recognition of importance of statistical science in society.

References

Akaike, H. (1973). Information theory and an extension of the maximum likelihood principle. *2nd International Symposium on Information Theory*. (eds. B.N.Petrov and F.Csaki), 268-281, Akademiai Kiado, Budapest. Reproduced in *Breakthroughs in Statistics*, Vol. 1, (eds. S.Kotz and N.L.Johnson), Springer-Verlag, New York, (1992), 610-624.

Akaike, H. (1980). Seasonal adjustment by a Bayesian modeling. *Journal of Time Series Analysis*, Vol. 1, 1-13.

Akaike, H. (1985). Prediction and entropy. *A celebration of Statistics*, (eds. A.C.Atkinson and S.E.Fienberg), 1-16, Springer-Verlag, New York.

Akaike, H. (1987). Factor analysis and AIC. *Psychometrika*, Vol. 52, 317-332.

Box, J. F. (1978). *R. A. Fisher, the life of a scientist.* Wiley, New York.

Fisher, R. A. (1912). On an absolute criterion for fitting frequency curves. *Messenger of Mathematics*, Vol. 41,155-160.

Fisher, R. A. (1915). Frequency distribution of the values of the correlation coefficient in samples from an infinitely large population. *Biometrika*, Vol. 10, 507-521.

Fisher, R. A. (1921). Studies in crop variation. I. An examination of dressed grain from Broadbalk. *Journal of Agricultural Science*, Vol.11, 107-135.

Fisher, R. A. (1922). On the mathematical foundations of theoretical statistics. *Philosophical Transactions of the Royal Society of London*, A, Vol.222, 309-368.

Fisher, R. A. (1935). The logic of inductive inference. *Journal of the Royal Statistical Society,* Vol. 98, 39-54.

Fisher, R. A. (1938). Presidential address, First Indian Statistical Conference, 1938. *Sankhyá*, Vol.4, 14-17.

Fisher, R. A. (1973). *Statistical Methods and Scientific Inference.* Third edition. Hafner Press, New York.

Jöreskog, K. G. (1967). Some contributions to maximum likelihood factor analysis. *Psychometrika,* Vol.32, 443-482.

Josephson, J. R., Chandrasekarn, B., Smith Jr, J. W. and Tanner, M. (1987). A mechanism for forming composite explanatory hypotheses. *IEEE Transactions on Systems, Man, and Cybernetics,* Vol.17, 445-454.

Kuhn, T. S. (1970). *The Structure of Scientific Revolutions* Second Edition, enlarged. The University of Chicago.

Pearson, K. (1973). The early history of statistics. *The History of Statistics in the 17th & 18 Centuries,* (ed. E.S.Pearson), 1-9, Charles Griffin, London.

Peirce, C. S. (1955). Abduction and induction. *Philosophical Writings of Peirce,* (ed. J. Buchler), 150-156, Dover, New York.

Polanyi, M. (1962). *Personal Knowledge,* Corrected edition. University of Chicago Press.

Stigler, S. M. (1976). Discussion of a paper by L. J. Savage. *The Annals of Statistics,* Vol. 4, 498-500.

Turing, A. M. (1969). Intelligent machinery. *Machine Intelligence 5,* (eds. B. Meltzer and D. Michie), 3-23, Edinburgh University Press.

Wald, A. (1943). Tests of statistical hypotheses concerning several parameters when the number of observations is large. *Transactions of the American Mathematical Society*, Vol.
54, 426-482.

Whittle, P. (1953). The analysis of multiple stationary time series. *Journal of Royal Statistical Society, Series B*, Vol.15,125-139.

FROM COMPARISON DENSITY TO TWO SAMPLE ANALYSIS

Emanuel Parzen
Department of Statistics
Texas A&M University
College Station, Texas 77843-3143 USA

1. Introduction

This paper on statistical data modeling is written to express my esteem for Professor Hirotugu Akaike in celebration of his 65th birthday by a U.S./Japan Conference on *The Frontiers of Statistical Modeling: An Informational Approach* (held May, 1992 at the University of Tennessee). I like to play the game "how long have you known Professor Akaike" because I have the good fortune of knowing him since 1965.

Parzen (1979) argues that statistical data analysis should be defined as fitting probability models to data. This paper presents typical concepts and recent results of our modeling theory which emphasizes quantile domain functions, information measures, and comparison density estimation. Ultimate goals include: unify parametric and nonparametric inference for continuous and discrete data; demonstrate that mathematical statistical and data analytic approaches are both needed for statistical inference; stimulate exoteric methods (applicable by applied researchers) rather than esoteric methods (known only to a small group of mathematical statisticians); combine mathematical statistical and data analytic views to develop methods of statistical analysis which are based on assumptions (known model) which are tested in ways that provide insight how to model deviations of the data from the assumed model (and thus identify a "true" model as an "iterated" model).

Contents of the paper are: 1. Introduction; 2. Quantile domain functions; F, Q; 3. Mid distribution and quantile functions: F^{mid}, Q^{mid}; 4. Sample distribution and quantile: $F^{\tilde{}}$, $Q^{\tilde{}}$; 5. Comparison distribution and comparison density for continuous F, G; 6. Comparison distribution and comparison density for discrete F, G; 7. Information measures: Renyi, Chi-square; 8. Information for comparison density functions; 9. Information measures and entropy tests of fit; 10. Continuous versions of discrete distribution functions: F^c, Q^c; 11. Comparison distributions of one sample (continuous data); 12. One sample parameter estimation; 13. Comparison distributions of two samples (continuous data); 14. Two sample comparison density estimation.

Research supported by the U. S. Army Research Office

H. Bozdogan (ed.), Proceedings of the First US/Japan Conference on the Frontiers of Statistical Modeling: An Informational Approach, 39–56.

2. Quantile domain functions: F, Q

The probability law of a random variable Y is described by its true *distribution* function

$$F(y) = \text{Prob}[Y \le y], \quad -\infty < y < \infty,$$

and true *quantile* function

$$Q(u) = F^{-1}(u) = \inf[y : F(y) \ge u].$$

The most famous parametric model $F(y;\theta)$ is the Normal Distribution:

$$\begin{aligned} F(y;\mu,\sigma) &= \Phi((y-\mu)/\sigma), \\ \Phi(y) &= \int_{-\infty}^{y} \phi(x)dx, \\ \phi(x) &= (2\pi)^{-.5}\exp(-.5x^2), \\ Q(u;\mu,\sigma) &= \mu + \sigma\Phi^{-1}(u). \end{aligned}$$

Every random variable Y has a *probability mass* function, defined

$$p(y) = \text{Prob}[Y = y].$$

One can define $p(y)$ analytically in terms of the distribution function $F(y)$ as the jump in F at y. Similarly the *spacing* function of Y, denoted $sp(u)$, is defined as the jump in the quantile function Q at u; the jump at u is the difference between the right hand and left hand limits at u.

Continuous random variables obey $p(y) = 0$ for all y. Discrete random variables obey

$$\sum_{\text{all } y} p(y) = 1.$$

We call a random variable *bi-continuous* if both p and sp are identically zero. One of our goals is to unify data analysis for continuous and discrete data.

An (absolutely) continuous random variable Y has a distribution function which is determined by its probability density function $f(y) = F'(y)$. It is bi-continuous if $f(y) > 0$ for all y satisfying $0 < F(y) < 1$.

For a continuous random variable the *density quantile* function is defined by

$$fQ(u) = f(Q(u)).$$

Then $Q(u)$ has a *quantile density* $q(u) = Q'(u)$ satisfying

$$q(u) = 1/fQ(u), fQ(u)q(u) = 1.$$

To prove this important formula verify that the indefinite integral of $1/fQ(u)$ equals $Q(u)$, or differentiate the identity $FQ(u) = u$ which holds for continuous F.

Quantile data analysis implements methods of Data Analysis in the distribution F and the quantile Q domains. Parallel functions in the two domains are: F, Q; p, sp; f, q. Important interpretations are given by $\log f$ and $\log q$. The inverse of f is not used but the inverse of q is important and is given by fQ.

Three general properties of quantile functions are:

$F(Q(u)) = u$ if F is continuous at $y = Q(u)$;

$F(y) \geq u$ if and only if $Q(u) \leq y$;

$F^{-1}_{g(Y)}(u) = g(F^{-1}_Y(u))$ if g is a function with the mathematical properties of a quantile function: non-decreasing and left-continuous.

Two important applications of quantile functions are concerned with transforming Y to and from a random variable U which is Uniform[0,1]:

Y and $Q(U)$ are identically distributed (since $Q_{Q(U)}(u) = Q(Q_U(u)) = Q(u)$).

$F(Y)$ and U are identically distributed if F is continuous (since $Q_{F(Y)}(u) = F(Q(u)) = u$).

3. Mid-distribution and quantile functions: F^{mid}, Q^{mid}

We define several versions of the distribution function which we believe should play important roles in statistical data analysis. Versions of F and Q which we believe should be used routinely in the theory and practice of statistics are the mid-distribution function $F^{mid}(y)$, defined by

$$F^{mid}(y) = F(y) - .5p(y),$$

and the mid-quantile function defined by

$$Q^{mid}(u) = Q(u) + .5sp(u).$$

Note that $F(y)$ is right continuous, and $Q(u)$ is left continuous. Note that F^{mid} and Q^{mid} are not inverses of each other.

We recommend as the definition of the *probability integral* transformation (or *rank transform*) of a continuous or discrete random variable Y

$$W = F^{mid}(Y);$$

it has mean $E[W] = .5$ and variance

$$\text{VAR}[Y] = (1/12)(1 - \sum_{\text{all } y} p^3(y)).$$

It is easy to verify this for a continuous Y (then W is Uniform[0,1]) and for a Bernoulli random variable Y taking values 0 and 1 with probabilities q and p; $F^{mid}(0) = .5q$ and $F^{mid}(1) = 1 - .5p$.

4. Sample distribution and quantile: $F^{\sim}$, $Q^{\sim}$

Our approach to data analysis of a data set $Y_1, \ldots, Y_n$ involves defining sample versions of F and Q. The initial way to represent a sample is to form its sample distribution function

$$F^{\sim}(y) = \text{ fraction of sample } \leq y$$

and its sample quantile function

$$Q^{\sim}(u) = F^{\sim -1}(u)$$

Explicit formulas for $F^{\sim}$ and $Q^{\sim}$ are expressed in terms of the distinct values in the sample, denoted $v_1 < \ldots < v_k$, their relative frequencies

$$p_j^{\sim} = \text{ fraction of sample } = v_j,$$

and their cumulative relative frequencies

$$u_j^{\sim} = p_1^{\sim} + \ldots + p_j^{\sim}, j = 1, \ldots, k.$$

Define $u_0^{\sim} = 0, v_0 = -\infty$, $v_{k+1} = \infty$.

The sample distribution function is discrete (piecewise constant) satisfying (for $j = 0, 1, \ldots, k$)

$$F^{\sim}(y) = u_j^{\sim} \text{ for } v_j \leq y < v_{j+1}.$$

The sample quantile function is discrete (piecewise constant) satisfying (for $j = 1, \ldots, k$)

$$Q^{\sim}(u) = v_j \text{ for } u_{j-1}^{\sim} < u \leq u_j^{\sim}.$$

We summarize these formulas by saying $F^{\sim}$ and $Q^{\sim}$ are piecewise constant between their values

$$F^{\sim}(v_j) = u_j^{\sim},$$
$$Q^{\sim}(u_j^{\sim}) = v_j.$$

5. Comparison distribution and comparison density for continuous F, G

Let F_θ be a specified continuous distribution (satisfying $Q_\theta F_\theta(y) = y$), which is a model for F, the unknown true distribution function of a continuous random variable Y. An important conceptual tool in statistical data analysis is transforming Y to the random variable $W = F_\theta(Y)$ which has quantile function

$$Q_W(u) = F_\theta(Q_Y(u))$$

and distribution function

$$F_W(y) = F_Y(Q_\theta(u))$$

How does one benefit by transforming probability law estimation problems to probability law estimation for a variable W on the unit interval? One could form an estimator of the probability density of Y from an estimator of the probability density of W by sample analogies of the formulas relating their probability densities

$$f_W(y) = f_Y(Q_\theta(u))/f_0Q_\theta(u),$$
$$f_Y(y) = f_0(y)f_W(F_\theta(y)).$$

Estimation of $f_W(u)$ provides estimation of $f_Y(y)$. One could form an estimator of the quantile and quantile density function of Y by sample analogues of the formulas

$$Q_Y(u) = Q_\theta(Q_W(u)),$$
$$q_Y(u) = q_0(Q_W(u))q_W(u)$$

which require estimation of $Q_W(u)$ and $q_W(u)$. In our view the conceptual importance of the transformation to W comes from interpreting q_W as a comparison density function $d(u; F_Y, F_\theta)$.

To two distribution functions F and G we associate a comparison distribution function on $0 < u < 1$, denoted $D(u) = D(u; G, F)$. We consider three cases: both continuous; both discrete; one continuous and one discrete.

When F and G are continuous we define

$$D(u; G, F) = F(G^{-1}(u))$$

with comparison density function $d(u) = D'(u)$ given by

$$d(u) = d(u; G, F) = f(G^{-1}(u))/g(G^{-1}(u)).$$

We assume that $f(y) > 0$ implies $g(y) > 0$. Then $D(0) = 0$, $D(1) = 1$. We call F and G equivalent if $f(y) > 0$ if and only if $g(y) > 0$.

In terms of comparison distribution functions we express the quantile and distribution functions of $W = F_\theta(Y)$:

$$Q_W(u) = D(u; F_Y, F_\theta) = D(u; \text{ data,model})$$
$$F_W(u) = D(u; F_\theta, F_Y) = D(u: \text{ model,data}).$$

6. Comparison distribution and comparison density for discrete F, G.

When F and G are discrete we assume that their respective probability mass functions $p_F(y)$ and $p_G(y)$ satisfy

$$p_F(y) > 0 \text{ implies } p_G(y) > 0.$$

We call F and G equivalent if $p_F(y) > 0$ if and only if $p_G(y) > 0$. In the discrete case we define first the comparison density

$$d(u) = d(u; G, F) = p_F(G^{-1}(u))/p_G(G^{-1}(u))$$

and define its integral $D(u) = D(u; G, F)$ by

$$D(u) = D(u; G, F) = \int_0^u d(u')du'.$$

Our assumptions guarantee that $D(1) = 1$.

Analogues of this definition will be given in section 11 for F continuous and G discrete based on the following characterization of $D(u)$; it is a $P-P$ plot obtained by joining linearly the points

$$(0,0); (G(v_j), F(v_j)), j = 1, ..., k; (1,1)$$

where $v_1 < \ldots, v_k$ are the distinct values at which G jumps (which we have assumed to include all values at which F jumps).

We call our approach to data analysis "functional" because it emphasizes forming and smoothing functions on the interval $0 \leq u \leq 1$; raw estimators $\tilde{d}(u; F_Y, F_\theta)$ and $\tilde{d}(F_\theta, F_Y)$ are smoothed to form estimators $\hat{d}(u; F_Y, F_\theta)$ and $\hat{d}(u; F_\theta, F_Y)$. The graphs it provides for graphical data analysis are pictures of functions.

7. Information measures: Renyi, Chi-square

Comparison densities provide insight into information methods because information measures of univariate distributions can be expressed in terms of $d(u; F, G)$. Information measures play a central role in statistical data analysis because they provide tools to measure the "distance" between two probability distributions F and G. The (Kullback-Liebler) information divergence is defined (Kullback (1959)) by (our definitions differ from usual definitions by a factor of 2)

$$I(F; G) = (-2) \int_{-\infty}^{\infty} \log(g(x)/f(x)) f(x) dx$$

when F and G are continuous with probability density functions $f(x)$ and $g(x)$. When F and G are discrete, with probability mass functions $p_F(x)$ and $p_G(x)$, information divergence has an analogous definition:

$$I(F; G) = (-2) \sum \log\{p_G(x)/p_F(x)\} p_F(x).$$

An information decomposition of information divergence is

$$I(F; G) = H(F; G) - H(F),$$

in terms of entropy $H(F)$ and cross-entropy $H(F; G)$:

$$H(F) = (-2) \int_{-\infty}^{\infty} \{\log f(x)\} f(x) dx,$$

$$H(F; G) = (-2) \int_{-\infty}^{\infty} \{\log g(x)\} f(x) dx.$$

Adapting the fundamental work of Renyi (1961) we define Renyi information of index λ. For continuous F and G: for $\lambda \neq 0$, -1

$$IR_\lambda(F;G) = \frac{2}{\lambda(1+\lambda)} \log \int \left\{ \frac{g(y)}{f(y)} \right\}^{1+\lambda} f(y)dy$$
$$= \frac{2}{\lambda(1+\lambda)} \log \int \left\{ \frac{g(y)}{f(y)} \right\}^{\lambda} g(y)dy$$
$$IR_0(F;G) = 2 \int \left\{ \frac{g(y)}{f(y)} \log \frac{g(y)}{f(y)} \right\} f(y)dy$$
$$= 2 \int \left\{ \frac{g(y)}{f(y)} \log \frac{g(y)}{f(y)} - \frac{g(y)}{f(y)} + 1 \right\} f(y)dy$$
$$IR_{-1}(F;G) = -2 \int \left\{ \log \frac{g(y)}{f(y)} \right\} f(y)dy$$
$$= -2 \int \left\{ \log \frac{g(y)}{f(y)} - \frac{g(y)}{f(y)} + 1 \right\} f(y)dy$$

An analogous definition holds for discrete F and G.

The second definition provides: (1) extensions to non-negative functions which are not densities, and also (2) a non-negative integrand which can provide diagnostic measures at each value of y.

Renyi information, for $-1 < \lambda < 0$, is equivalent to Bhattacharyya distance. Hellinger distance corresponds to $\lambda = -.5$.

In addition to Renyi information divergence (an extension of information statistics) one uses as information divergence between two non-negative functions an extension of chi-square statistics which has been developed by Read and Cressie (1988). For $\lambda \neq 0$, -1, Chi-square divergence of index λ is defined for continuous F and G by

$$C_\lambda(F;G) = \int B_\lambda \left(\frac{g(y)}{f(y)} \right) f(y)dy$$

where

$$B_\lambda(d) = \frac{2}{(1+\lambda)} \left\{ d \left(\frac{d^\lambda - 1}{\lambda} \right) - d + 1 \right\}$$
$$B_0(d) = 2\{d \log d - d + 1\}$$
$$B_{-1}(d) = -2\{\log d - d + 1\}$$

Important properties of $B_\lambda(d)$ are:

$$
\begin{aligned}
&B_\lambda(d) \geq 0, B_\lambda(1) = B'_\lambda(1) = 0, \\
&B'_\lambda(d) = \frac{2}{\lambda}\left(d^\lambda - 1\right), B''_\lambda(d) = 2d^{\lambda-1} \\
&B_1(d) = (d-1)^2 \\
&B_0(d) = 2(d\log d - d + 1) \\
&B_{-.5}(d) = 4\left(d^{.5} - 1\right)^2 \\
&B_{-1}(d) = -2(\log d - d + 1) \\
&B_{-2}(d) = d\left(d^{-1} - 1\right)^2
\end{aligned}
$$

An analogous definition holds for discrete F and G. Axiomatic derivations of information measures similar to C_λ are given by Jones and Byrne (1990).

The Renyi information and chi-square divergence measures are related:

$$
\begin{aligned}
IR_0(F;G) &= C_0(F;G) \\
IR_{-1}(F;G) &= C_{-1}(F;G)
\end{aligned}
$$

For $\lambda \neq 0, -1$,

$$
IR_\lambda(F;G) = \frac{2}{\lambda(1+\lambda)}\log\left\{1 + \left(\frac{\lambda(1+\lambda)}{2}\right) C_\lambda(F;G)\right\}
$$

Interchange of F and G is provided by the *Lemma*: when F and G are equivalent,

$$
\begin{aligned}
C_\lambda(F;G) &= C_{-(1+\lambda)}(G;F) \\
IR_\lambda(F;G) &= IR_{-(1+\lambda)}(G;F)
\end{aligned}
$$

Example: Renyi information divergence of two zero mean univariate normal distributions. Let P_j be the distribution on the real line corresponding to Normal $(0, K_j)$ with variance K_j. Let $\kappa = \dfrac{K_2}{K_1}$. Then

$$
\begin{aligned}
&d(u; P_2, P_1) = \kappa^{.5}\exp\left\{-.5(\kappa - 1)|\Phi^{-1}(u)|^2\right\} \\
&IR_{-1}(P_2; P_1) = \kappa - 1 - \log\kappa, \\
&IR_\lambda(P_2; P_1) = (1/\lambda)\{\log\kappa - (1+\lambda)^{-1}\log\{1 + (1+\lambda)(\kappa - 1)\}_+\} \\
&C_\lambda(P_2; P_1) = \{2/\lambda(1+\lambda)\}\,\kappa^{.5(1+\lambda)}\{1 + (1+\lambda)(\kappa - 1)\}_+^{-.5}
\end{aligned}
$$

8. Information for comparison density functions

Information divergence $I(F;G)$ is a concept that works for both multivariate and univariate distributions. In the univariate case we are able to relate $I(F;G)$ to the concept of comparison density $d(u;F,G)$,

For a density $d(u)$, $0 < u < 1$, Renyi information (of index λ), denoted $IR_\lambda(d)$, is non-negative and measures the divergence of $d(u)$ from uniform density $d_0(u) = 1$, $0 < u < 1$. It is defined:

$$IR_0(d) = 2\int_0^1 \{d(u)\log d(u)\}du = 2\int_0^1 \{d(u)\log d(u) - d(u) + 1\}\,du$$

$$IR_{-1}(d) = -2\int_0^1 \{\log d(u)\}du = -2\int_0^1 \{\log d(u) - d(u) + 1\}\,du$$

for $\lambda \neq 0$ or -1

$$IR_\lambda(d) = \{2/\lambda(1+\lambda)\}\log\int_0^1 \{d(u)\}^{1+\lambda}du$$

$$= \{2/\lambda(1+\lambda)\}\log\int_0^1 \Big(\{d(u)\}^{1+\lambda} - (1+\lambda)\,\{d(u)-1\}\Big)\,du.$$

To relate comparison density to information divergence we use the concept of Renyi information IR_λ which yields the important identity (and interpretation of $I(F;G)$!)

$$I(F;G) = (-2)\int_0^1 \log d(u;F,G)du$$

$$= IR_{-1}(d(u;F,G)) = IR_0(d(u;G,F)).$$

For a density $d(u)$, $0 < u < 1$, define

$$C_\lambda(d) = \int_0^1 B_\lambda(d(u))du.$$

The comparison density again unifies the continuous and discrete cases. One can show that for univariate F and G

$$C_\lambda(F,G) = C_\lambda(d(u;F,G))$$

For a random sample of a random variable with unknown probability density f, maximum likelihood estimators $\theta^\frown$ of the parameters of a finite parameter model f_θ of the probability density f can be shown to be equivalent to minimizing

$$IR_{-1}(f^\sim; f_\theta) = IR_{-1}\,(d\,(u; F^\sim, F_\theta))$$

where $f^{\sim}$ is a raw estimator of f (initially, a symbolic sample probability density formed from the sample distribution function $F^{\sim}$).

9. Information measures and entropy tests of fit

To test the goodness of fit hypothesis for a continuous random variable Y

$$H_0 : F_Y(y) = F_\theta(y),$$

many statistics have been proposed which start with the probability integral transformation

$$W = F_\theta(Y)$$

for which the goodnes of fit hypothesis is H_0: W is Uniform[0,1].

An entropy test of fit is Moran's statistic which transforms $Y_1, \dots, Y_n$ to $W_1, \dots, W_n$ and forms the order statistics $W(1;n) < \dots < W(n;n)$ with spacings

$$S_i(\theta) = W(i;n) - W(i-1;n), j = 1, \dots, n+1,$$

defining $W(0,n) = 0$, $W(n+1;n) = 1$ Moran's statistic is often defined as

$$\sum_{i=1}^{n+1} -\log S_i(\theta),$$

We prefer to define it as

$$M^{\sim}(\theta) = (n+1)^{-1} \sum_{i=1}^{n+1} (-2) \log(n+1) S_i(\theta).$$

Under the null hypothesis, it is asymptotically normal with mean 2γ ($\gamma = .57722$, Euler's constant), and variance

$$4(n+1)^{-1}((\pi^2/6) - 1).$$

Small sample asymptotic chi-square and beta distributions (given by Smethurst and Mudholkar (1991)) *are more appropriate for an entropy interpretation.*

In order to understand and extend $M^{\sim}(\theta)$, we regard it as an estimator of a quantity $M(\theta)$ defined by probability theory. The original observation Y has true distribution function F_Y and quantile function $Q_Y(u) = F_Y^{-1}(u)$. The transformation $W = F_\theta(Y)$ has quantile function $Q_W(u) = F_\theta(Q_Y(u))$ and distribution function $F_W(u) = F_Y(Q_\theta(u))$; W has quantile density $q_W(u) = f_\theta(Q_Y(u))/f_Y Q_Y(u)$, and entropy

$$H(W) = \int_0^1 (-2 \log f_W(w)) f_W(w) dw = \int_0^1 2 \log q_W(u) du.$$

Under the null hypothesis $H_0 : F_Y = F_\theta$, $q_W(u)$ is identically 1, and $H(W) = 0$.

Note $-H(W)$, the neg-entropy of W, is non-negative and is the population parameter, denoted $M(\theta)$, which is being non-parametrically estimated by Moran's statistic $M^{\sim}(\theta)$.

How do we benefit from estimating entropy (or neg-entropy) of W? It provides tests of H_0 and can provide (through suitable analogues of Akaike Information Criterion) insight about selection of alternative models to fit when one rejects H_0. Thus understanding and improving Moran's statistic requires us to solve problems of density estimation, especially estimation of the smooth comparison density $d(u; F_Y, F_\theta)$ from raw estimators

$$d^{\sim}(u; F_Y, F_\theta) = d^{\sim}(u; F^{\sim c}, F_\theta),$$

where $F^{\sim c}$ is a continuous version of the discrete distribution function $F^{\sim}$.

Another important interpretation of $M(\theta)$ is $M(\theta) = I(f; f_\theta)$, the Kullback information divergence between the true $F(y)$ and the model $F_\theta(y)$.

10. Continuous versions of discrete distribution functions

To compare a continuous and a discrete distribution we propose forming a continuous distribution function version of a discrete one. To estimate a continuous distribution function F from data we recommend first forming our continuous version $F^{\sim c}$ of the discrete distribution function given by the sample distribution function $F^{\sim}$ formed from the data.

For discrete data we recommend estimating the continuous version F^c of its discrete distribution function F. We conjecture that these recommendations provide a unified theory of discrete and continuous data analysis as well as improved methods of continuous data analysis.

To define the continuous version of a discrete distribution F, we assume that it can be described by a finite number of points (v_j, u_j) such that

$$F(v_j) = u_j \text{ for } j = 1, \ldots, k.$$

Note that $u_k = 1$. Define $u_0 = 0$; then $0 = u_0 < u_1 < \ldots < u_k = 1$.

Its quantile function $Q(u)$ is discrete and satisfies, for $j = 1, \ldots, k$,

$$Q(u_j) = v_j.$$

Define "mid-values" v_j^c, $j = 0, \ldots, k$, by

$$v_0^c = v_1, v_k^c = v_k,$$
$$v_j^c = .5(v_j + v_{j+1}) \text{ for } j = 1, \ldots, k-1.$$

Define F^c and Q^c to be piecewise linear between its values (for $j = 0, 1, \ldots, k$)

$$Q^c(u_j) = v_j^c,$$
$$F^c(v_j^c) = u_j$$

We call F^c and Q^c continuous (piecewise linear) versions of the discrete (piecewise constant) functions F and Q.

It is interesting to compare the continuous version of a discrete distribution to its mid-distribution F^{mid} whose definition we recall; define $p_j = u_j - u_{j-1}$ and

$$F^{mid}(v_j) = u_j - .5p_j \text{ for } j = 1, ..., k.$$

One conjectures that approximately (and exactly when v_j are equi-spaced)

$$F^c(v_j) = u_j - .5p_j,$$

so that approximations to F^c are also approximations to F^{mid}.

To justify our view that these concepts are very natural, we would argue that the continuity correction when one approximates a discrete distribution by a continuous one (say the binomial by the normal) can be explained by regarding the limiting continuous distribution as approximating F^c and F^{mid} rather than F.

Let F be the distribution function of a Binomial(n, p) random variable. The continuity correction says that (for $x = 0, 1, \ldots, n$) approximately

$$F(x) = F^c(x + .5) = \Phi((x + .5 - np)/(np(1-p))^{.5})$$

Note that (for $x = 0, 1, \ldots, n$) approximately

$$F^c(x) = F^{mid}(x) = \Phi((x - np)/(np(1-p)^{.5}).$$

11. Comparison distributions of one sample (continuous data)

Given a sample $Y_1, \ldots, Y_n$ from a continuous distribution F, we recommend as the first step in data analysis to compute and plot $F^{\sim c}$ and $Q^{\sim c}$, the continuous versions of the discrete sample distribution and quantile functions. We regard $Q^{\sim c}$ as a raw estimator of the true quantile Q which provides a minimal amount of smoothing of the observations.

The process of fitting a model to the data can be formulated in terms of a specified continuous distribution function F_θ (whose form may be guessed from a visual examination of $Q^{\sim c}$, normalized at $u = .5$ to equal 0 and to have slope 1 (compare Parzen (1986))). We not only estimate F but also test the goodness of fit hypothesis $H_0 : F = F_\theta$. To motivate our approach let us review some methods of testing a goodness of fit hypothesis H_0 for continuous data.

A graphical diagnostic of H_0 is the $Q - Q$ plot which looks for linearity in the graph of

$$(Q_\theta(F^{\sim mid}(v_j)), v_j), j = 1, \ldots, k;$$

an alternative is the $Q - Q^c$ plot which graphs the points

$$(Q_\theta(u_j), v_j^c) = (Q_\theta(F^{\sim}(v_j)), v_j^c), j = 1, ..., k - 1.$$

Goodness of fit tests of H_0 have traditionally been expressed by statisticians as measures of $F^{\sim}(y) - F_\theta(y)$, the difference of distribution functions. Typical measures are non-linear functionals such as

$$D[\text{KolmogorovSmirnov}] = \text{supremum}_y |F^{\sim}(y) - F_\theta(y)|$$

$$D[\text{CramervonMises}] = \int_{-\infty}^{\infty} |F^{\sim}(y) - F_\theta(y)|^2 dF_\theta(y).$$

Goodness of fit compares probabilities; we believe that probabilities $p^{\sim}$ and $p^{\wedge}$, representing ***data and model, should be compared not by their differences but by their ratio!*** In symbols, measure $(p^{\sim}/p^{\wedge}) - 1$ rather than $p^{\sim} - p^{\wedge}$. Therefore goodness of fit tests should be based on measures of the difference from the identity function $D_0(u) = u, 0 < u < 1$, of comparison distribution functions. Goodness of fit tests for uniformity are traditionally based on

$$D(u; F_\theta, F^{\sim}) = F_W{}^{\sim}(u) = F^{\sim}(Q_\theta(u)), 0 < u < 1,$$

the sample distribution function of the probability integral transformation $W = F_\theta(Y)$. Traditional maximum likelihood estimators of θ are chosen by the criterion that the sample quantile function

$$D(u; F^{\sim}, F_\theta) = Q_W{}^{\sim}(u) = F_\theta(Q^{\sim}(u)), 0 < u < 1,$$

has minimum Kullback information distance from $D_o(u) = u$.

This paper proposes that we need to overcome the problem that $F^{\sim}$ and $Q^{\sim}$ are discrete and are not directly covered by our definitions of comparison distribution functions; we recommend that data analysis be based on the definitions below of continuous raw comparison functions, denoted $D^c(u; F^{\sim}, F_\theta)$ and $D^c(u; F_\theta, F^{\sim})$. Analogously one defines comparison distribution functions, denoted $D^c(u; G, F)$ and $D^c(u; F, G)$ rather than $D(u; G, F)$ and $D(u; F, G)$, when F is continuous and G is discrete.

Recall $F^{\sim}(v_j) = u_j$ for $j = 1, \ldots, k$. Let v_j^c, $j = 0, \ldots, k$ be mid values. Define for $j = 1, \ldots, k-1$

$$w_j = F_\theta(Q^{\sim c}(u_j)) = F_\theta(v_j^c)).$$

Assume $0 < w_1 < \ldots w_{k-1} < 1$.

Define $Q_W^c(u)$ as a piecewise linear curve connecting the values

$$(0,0), (u_j, w_j) \text{ for } j = 1, ..., k-1, (1,1).$$

Define $F_W^c(u)$ as a piecewise linear curve connecting the values

$$(0,0), (w_j, u_j) for j = 1, ..., k-1, (1,1).$$

The derivatives, denoted $q_W^c(u)$ and $f_W^c(u)$ respectively, are sample quantile density and probability density functions. Define $d^c(u; F^{\sim}, F_\theta) = q_W^c(u), d^c(u; F_\theta, F^{\sim}) = f_W^c(u)$.

One smooths raw densities to form smooth estimators, denoted

$$q_W\hat{\ }(u), Q_W\hat{\ }(u), f_W\hat{\ }(u), F_W\hat{\ }(u).$$

The adequacy of the smoothing can be judged by comparing on one graph Q^c_W and $Q_W\hat{\ }$, and comparing on one graph F^c_W and $F_W\hat{\ }$. In this way one can develop $f_Y\hat{\ }(y)$ and $Q_Y\hat{\ }(u)$.

12. One sample parameter estimation.

Regular *maximum likelihood* estimators $\theta\hat{\ }$ are parameter values minimizing

$$\int_0^1 -\log f_\theta(Q\tilde{\ }(u))du$$

or equivalently minimizing the negative of the average log likelihood

$$-L(\theta) = (1/n)\sum_{j=1}^{n} -\log f_\theta(Y(j;n))$$

A *maximum spacings* estimator, denoted $\theta\hat{\ }$, can be obtained (compare Ranneby (1984)) by minimizing (with respect to all possible parameter values θ) the neg-entropy

$$2\int_0^1 -\log d^c(u;F\tilde{\ },F_\theta)du$$

or equivalently minimizing

$$-2\sum_{j=1}^{k}(u_j - u_{j-1})\log((F_\theta(Q^{\sim c}(u_j)) - F_\theta(Q^{\sim c}(u_{j-1}))/(u_j - u_{j-1}))\,.$$

In this expression, logarithm is taken after integration rather than before; consequently it provides estimators in non-regular cases.

Minimum information estimators (more precisely, minimum Renyi information of index λ estimators) $\theta\hat{\ }$, minimize (for $-1 < \lambda < 0$, and especially $\lambda = -.5$ (compare Beran (1977))).

$$IR_\lambda\left(d^c\left(u;F\tilde{\ },F_\theta\right)\right).$$

Regular maximum likelihood estimators (which correspond to $\lambda = -1$) satisfy the estimating equations

$$\int_0^1 S_{\theta_i}\left(Q^{\sim c}(u)\right)du = 0$$

where

$$S_{\theta_i}(y) = \frac{\partial}{\partial\theta_i}\log f_\theta(y)$$

is the score function of component θ_i of the vector parameter θ. Minimum information estimators satisfy the estimating equations

$$\int_0^1 (d^c(u; F^\sim, F_\theta))^{1+\lambda} S_{\theta_i}(Q^{\sim c}(u))\, du = 0.$$

Minimum information estimators provide ***robust*** estimators.

To test if robust minimum information estimators of a given data set are to be preferred to regular maximum likelihood estimators, one could test if the latter satisfy the estimating equations of the former. The theory and practice of this "test for robustness" are open research problems.

13. Comparison distributions of two samples (continuous data)

A central problem of statistics is test

$$H_0 : F_1 = F_2,$$

the equality of two continuous distribution functions F_1 and F_2. The data are assumed to be independent observations of a first sample

$X_1, \ldots, X_{n_1}$ assumed to be distributed as F_1,

and a second sample

$Y_1, \ldots, Y_{n_2}$ assumed to be distributed as F_2.

Let $F_1^\sim$ and $F_2^\sim$ denote the sample distribution functions of the two samples.

The pooled sample consists of all X and Y values; its sample distribution function is denoted $F^\sim$. Let $n = n_1 + n_2$ be the total sample size. Let $p_j = n_j/n$ be the fraction of the pooled sample in the j-th sample. One can represent

$$F^\sim = p_1 F_1^\sim + p_2 F_2^\sim.$$

it is an estimator of the true pooled distribution

$$F = p_1 F_1 + p_2 F_2.$$

The novelty of our approach to testing H_0 is our proposed comparison distribution function

$$D^\sim(u) = D(u; F^\sim, F_1^\sim)$$

which estimates $D(u) = D(u; F, F_1)$. Because $F^\sim$ and $F_1^\sim$ are both discrete, the comparison distribution $D^\sim(u)$ is defined in terms of the comparison density function

$$d^\sim(u) = d(u; F^\sim, F_1^\sim).$$

The asymptotic distribution of $D^\sim(u)$ as an estimator of $D(u) = D(u; F, F_1)$ can be shown to be the same as the Pyke-Shorack two sample process. The mathematical statistics is the same, but the data analysis is greatly improved.

Linear rank statistics to test H_0 can be represented as linear functionals

$$< J(u), d^\sim(u) >= \int_0^1 J(u) d^\sim(u) du$$

for suitable score functions $J(u)$. The Wilcoxon statistic (which tests for a shift in location) corresponds to $J(u) = u$, whose orthonormal version is $J(u) = 12^{.5}(u - .5)$.

14. Two sample comparison density estimation.

We propose that the best summary of two sample data analysis is not a p value of a linear rank statistic but a smooth density estimator $\hat{d}(u)$ of the true comparison density $d(u) = d(u; F, F_1)$.

The asymptotic distribution of estimators $d\hat{}(u)$ which smooth $d\tilde{}(u)$ is best understood by normalizing it to be between 0 and 1 by defining

$$p\tilde{}(u) = p_1 d\tilde{}(u)$$

whose smooth estimators are denoted $p\hat{}(u)$. Their asymptotic distribution theory (outlined below) is developed on the assumption that they are estimators of a true comparison density $d(u) = d(u; F, F_1)$; let $p(u) = p_1 d(u)$. The asymptotic variance of $p\hat{}(u)$ can be shown to be proportional to $p(u)(1 - p(u))$ which means that its asymptotic distribution is similar to that of an estimator $p\hat{}$ of a probabiity p.

In contrast the asymptotic variance in the one sample case of the smooth quantile density estimator $q\hat{}$ is proportional to q^2, and the asymptotic variance of the smooth probability density estimator $f\hat{}$ is proportional to f if $f\hat{}$ is a standard kernel estimator, and is proportional to f^2 if $f\hat{}$ is a nearest neighbor estimator.

One of the joys of a unifed framework for one sample and two sample data analysis is that it can comprehend and explain the different qualitative behavior of estimators of different types of densities. Parzen (1983) states, and outlines proofs of the following results about comparison density estimation.

A *kernel* comparison density estimator has the form

$$d\hat{}(u) = \int_0^1 K_M(u - t) d\tilde{}(t)dt$$

where

$$K_M(t) = \sum_{v=-\infty}^{\infty} e^{-2\pi i t v} k_M(v)$$

$$k_M(v) = k\left(\frac{v}{M}\right)$$

We call M truncation point (or effective number of parameters); it is chosen as a function of n and tends to ∞, as n tends to ∞, at a suitable rate. Let

$$\overline{K^2} = \int_{-\infty}^{\infty} K^2(t)dt = \int_{-\infty}^{\infty} k^2(x)dx$$

One can show that (by letting M tend to ∞ at a suitable rate)

$$\lim_{n\to\infty} \frac{n}{M} \operatorname{Var}\left[p\hat{}(u)\right] = \overline{K^2} p(u)(1 - p(u))$$

The *numerical derivative* density estimator

$$d\hat{}(u) = (2h)^{-1}(D\tilde{}(u+h) - D\tilde{}(u-h))$$

corresponds to $M = 1/h$ and $k(x) = (\sin 2\pi x)/2\pi x$; it has asymptotic variance

$$\lim_{n\to\infty} hn \operatorname{Var}[p\hat{}(u)] = .5p(u)(1-p(u)).$$

Evaluate $\overline{K^2}$ from $K(t) = .5$ for $|t| \leq 1, 0$ otherwise.

An *autoregressive* (of order m) estimator has asymptotic variance

$$\lim_{n\to\infty} \frac{n}{m} \operatorname{Var}[p\hat{}(u)] = 2p(u)(1-p(u))$$

Model order selection techniques can be developed by adapting Akaike (1973), Atilgan and Bozdogan (1990), Sakamoto, Ishiguro, Kitagawa (1983).

To obtain one sample probability density estimator results from the two sample results, let the first sample have unknown distribution F_1, the second sample have known distribution F_2, and let n_2 be very large. Then the pooled distribution $F = F_2$, $d(u)$ equals $d(u; F_2, F_1)$, and p_1 tends to 0. The kernel probability density estimator of f_1 has asymptotic variance

$$\lim_{n_1\to\infty} \frac{n_1}{M} \operatorname{Var}[d\hat{}(u; F_2, F_1)] = \overline{K^2} d(u; F_2, F_1)$$

since

$$p(u) = p_1 f_1(F^{-1}(u))/f(F^{-1}(u)) = p_1 d(u; F_2, F_1).$$

References

Akaike, H. (1973). *Information theory and an extension of the maximum likelihood principle.* Proc. of the Second International Symposium on Information Theory, B. N. Petrov and F. Csaki, Akademiai Kiado, Budapest, pp. 267-281. Reprinted in *Breakthroughs in Statistics*, I. edited by S. Kotz and N. L. Johnson, Springer-Verlag: New York, 1992.

Arimoto, S. (1971). "Informational-theoretical considerations on estimation problems," *Inf. and Control*, 19, 181-194.

Atilgan, T. and Bozdogan, H. (1990). "Selecting number knots fitting cardinal B-splines for density estimation using AIC." *J. Japan Statist. Soc*, 20 (2), 179-190.

Beran, R. J. (1977). "Minimum Hellinger Distance Estimates for Parametric Models." *Annals of Statistics*, 5, 445-463.

Jones, L. K. and Byrne, C. L. (1990). "General entropy criteria for inverse problems, with applications to data compression, pattern classification, and cluster analysis," *IEEE Transactions on Information Theory*, vol. 36, no. 1, pp. 23-30.

Kullback, S. (1959). *Information Theory and Statistics.* New York: Wiley.

Parzen, E. (1979). "Nonparametric Statistical Data Modeling", *Journal of the American Statistical Association*, (with discussion), 74. 105-131.

Parzen, E. (1983). "FUN.STAT Quantile Approach to Two Sample Statistical Data Analysis." Technical Report.

Parzen, E. (1986). "Quantile Spectral Analysis and Long Memory Time Series," *Journal of Applied Probability*, Vol. 23A, 41-55.

Ranneby, B. (1984). "The maximum spacing method," *Scand. J. Statist.*, 11, 93-112.

Read, T. R. C. and Cressie, N. A. C. (1988). *Goodness of Fit Statistics for Discrete Multivariate Data*, Springer Verlag, New York.

Renyi, A. (1961). "On measures of entropy and information." *Proc. 4th Berkeley Symp. Math. Statist. Probability, 1960*, 1, 547-561. University of California Press: Berkeley.

Sakamoto, Y., Ishiguro, M., and Kitagawa, G. (1983). *Akaike Information Criterion Statistics*, D. Reidel: Boston.

Smethurst, P. A. and Mudholkar, G. S. (1991). "Two Facets of the Moran Statistics." *J. Statist. Comput. Simul.* 39, 215-220.

STATISTICAL IDENTIFICATION AND OPTIMAL CONTROL OF THERMAL POWER PLANTS

Hideo Nakamura
Bailey Japan Co., Ltd.
511 Baraki, Nirayama-cho
Tagata-gun, Shizuoka-ken
Japan 410-21

1. Introduction

This paper describes the implementation of an optimal regulator for thermal power plants which is designed on the basis of the state space concept. System identification, system analysis, and optimal controller design are performed based on the TIMSAC (Time Series Analysis and Control) Library (Akaike, 1972a, 1979b, 1988c), which was developed and programmed by Prof. Akaike, Miss Arahata, and other members of The Institute of Statistical Mathematics.

First of all, the author will briefly refer to the motivation of the investigation, or in other words, the situation that necessitated the implementation of an optimal regulator for thermal power plants.

1.1 REQUIREMENT FROM ELECTRIC POWER SYSTEM OPERATION

In an electric power pool, it is required to accurately maintain both the power system's frequency and the contract power between neighboring electric utilities. For this purpose, computers in the power system's dispatch center perform the so-called active-power control, i.e., the load-frequency control and the economic load dispatch control. It is essential for fine active power control that power generation at thermal power plants can be adjusted quickly and smoothly in response to the commands from the power dispatch center, because in most power systems of Japan, thermal power represents a large portion of adjustable power generation.

1.2 PROBLEMS ON THE THERMAL POWER PLANT CONTROL

For thermal power generation, large-capacity boilers producing high-pressure, high-temperature steam, are widely used. In such boilers, deviations of steam temperatures from their set-points must be kept within one or two percent of their rated values in order to maintain the nominal operating efficiency and ensure the safety and the maximum equipment life of the plant. As described above, main purpose of the boiler control is to allow the increase or decrease of steam generation as fast as possible in response to the load command from the power system's dispatch center, while satisfying the above-mentioned operating conditions.

H. Bozdogan (ed.), Proceedings of the First US/Japan Conference on the Frontiers of Statistical Modeling: An Informational Approach, 57–79.

However, since a modern thermal power plant usually includes many control loops with significant mutual interactions within the boiler process, it is not easy for the conventional PID controller to fully compensate for these interactions and satisfy the required steam conditions for large and fast changes in plant load. This difficulty of controlling a mutually interacting multivariable system had been one of the principal factors that set the limit to the response of a thermal power plant to the load changes required for the load-frequency control (LFC) of an electric power system.

1.3 THE HISTORY OF THE RESEARCH & DEVELOPMENT

More than twenty-five years ago, the author started an investigation in Kyushu Electric Power Company for improving load following capability of thermal power plants. In the study, the author collaborated with Kyushu University. Through theoretical and experimental works the study group completed several transfer- function-based plant models describing the dynamics of once-through boilers. Figure 1 shows an example of the block diagram worked out in the study group.

Based on these results we composed a simulation models with a large-scale analog computer, and later with a digital computer and using them studied the problem of PID controller tuning. Although we tried hard to find how to improve the control performance of the integrated control system we could not reach a definite conclusion.

This was due to the fact that parameters in the conventional PID controller have different influences on each of the process output variables, and accordingly feedback control loops, connecting process-output variables with process-input variables, mutually interact within the boiler process. This experience made the author realize the difficulty of tuning a multivariable system, i.e., a multi-input and multi-output (MIMO) system, and consequently to distrust the term "optimal tuning" of conventional PID controllers in MIMO systems.

After some while, the author learned the theory of optimal control designed on the state-space concept. Impressed deeply by its prominent design concept, he tried to apply the theory to boiler control. However, the trial which continued for a couple of years finally failed, because of the difficulty in deriving a state space representation of the complex boiler process as depicted in the block diagram in Figure 1. This taught the author that, so long as we stuck to the conventional approach based on the simultaneous differential equations representing the energy and mass balances in the boiler process, the derivation of a state equation for practical use would be almost impossible.

In 1972, Otomo, Nakagawa, and Akaike (1972), reported a successful implementation of the optimal control system for a cement kiln process, in which the method of statistical system identification and analysis for controller design had been adopted. The author was greatly animated by this success and immediately understood the usefulness of the statistical approach developed by Prof. Akaike. and believed that this method must be the best and the most promising one for solving the difficulties he was confronting with and then leading to the implementation of the optimal control system for thermal power plants. This made the author re-start the investigation.

The investigation actually began in 1974 as a joint project of Kyushu Electric Power Company, The Institute of Statistical Mathematics, The Central Research Institute of Electric Power Industry of Japan (CRIEPI). and Toshiba Corporation. with the collaboration of Ishikawajima-Harima Heavy Industries Co., Ltd. About four years later, the first optimal regulator was implemented at Buzen No. 1 unit. a 500-MW supercritical constant pressure plant owned by Kyushu Electric Power Company.

Before implementing the optimal regulator. experimental studies were repeated with plant models built on digital and analog-digital hybrid computers. and also with an exact power plant model established on a power plant simulator at CRIEPI.

As had been anticipated from the results of simulation studies, the optimal regulator implemented at Buzen No.1 plant demonstrated excellent control performance. As of 1987. five supercritical plants (including a sliding-pressure plant) with a total output power of 2700-MW. were in commercial operation in Kyushu Electric Power Company under optimal regulators. With their high load-following capability. these plants have been contributing to the load frequency control of the company.

The successful implementation of the optimal regulator (Nakamura and Akaike. 1981) had a great impact on boiler control. Today, no boiler manufacturer in Japan can propose its control system to electric utilities without referring to optimal or advanced control. The implementation of optimal regulators designed by Japanese boiler and control system manufacturers on the basis of statistical approach, has extended throughout Japan and abroad. Microprocessor-based optimal regulators that are installed with the distributed control system have also been prevailing today.

In this paper, the author will explain the concept of power plant control, the control system configuration, and how to apply the TIMSAC Library for implementing an optimal regulator at thermal power plants. Then, the excellent control performance of the optimal regulator will be introduced with some records obtained in the field tests and during routine operation.

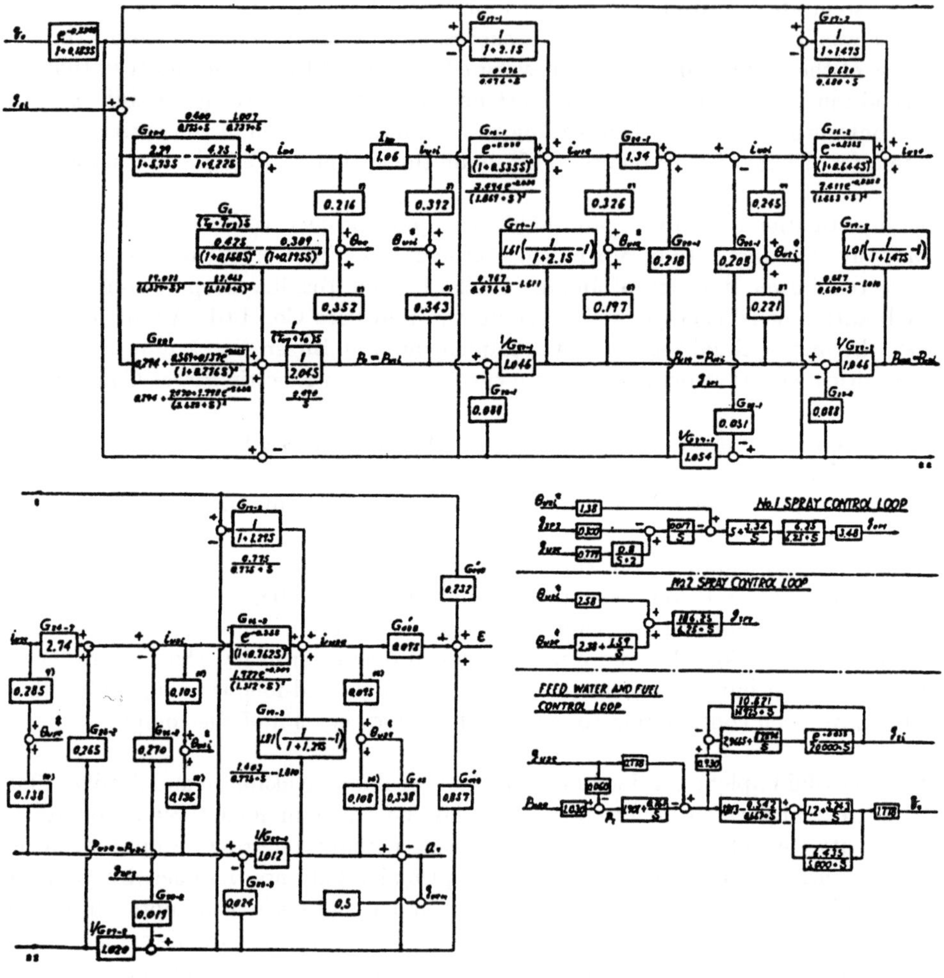

Figure 1: Block disgram expressing once-through boiler dynamics

2. Control of A Power Plant

At first, we will discuss the control problems of a power plant equipped with a supercritical once-through boiler.

Figure 2 shows conceptual view of an F-W type supercritical boiler we implemented an optimal regulator for the first time.

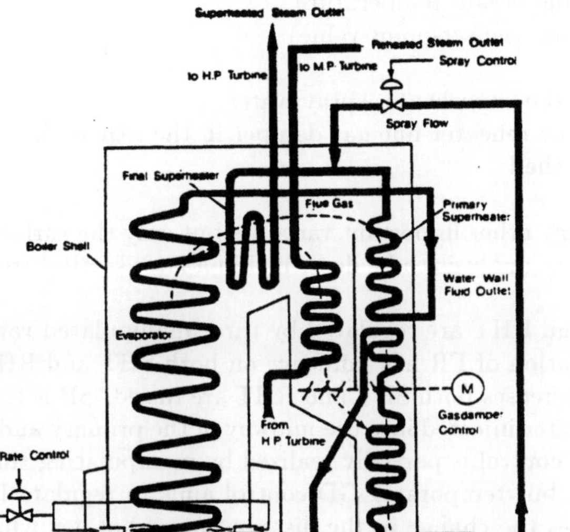

Figure 2: Conceptual view of F-W type supercritical boiler

In the boiler process, steam temperature control is regarded as the most difficult one. Therefore, we will hereafter put our attention on the control of the steam temperatures at the outlets of the superheater and the reheater from where the steam produced in the boiler supplied to the high-pressure turbine and the medium-pressure turbine respectively.

For convenience sake, we provide below a list of abbreviated notations of the process input and output variables relevant to steam temperature control.

Nomenclature

MWD	Megawatt demand or load command measured at the output of the rate limiter
WWT	Waterwall or evaporator outlet fluid temperature (transient deviation from the exponentially smoothed mean value)
SHT	Superheater outlet steam temperature (deviation from the set-point value)
RHT	Reheater outlet steam temperature (deviation from the set-point value)
FR	Fuel flow rate
SP	Flow rate of the superheater spray water
GD	Opening of the reheater flue gas damper in the rear path of the boiler shell

Certainly there are many other important variables but only the variables which were selected by a preliminary analysis to be described in a later section are retained here.

Plant variables SHT and RHT are regulated by three manipulated variables FR, SP, and GD. The manipulation of FR has influence on both SHT and RHT in the same direction. When FR increases both SHT and RHT are raised. SP is the portion (3 or 4 percent) of the feedwater injected into the midway of the primary and the secondary superheaters and SHT control is partially realized by manipulating this quantity. Its effect on SHT is quick but temporary. GD control aims to regulate RHT. The manipulation of GD causes the change in the distribution of hot gas which flows along the heat conduction surfaces of the reheater and superheater. The increase in GD increases the hot gas flow along the reheater and raises RHT, but it lowers SHT by reducing the gas flow along the superheater. Consequently, the GD control acts on SHT and RHT in an opposite direction.

Figure 3 shows a configuration of the conventional PID (Proportional, Integral, Derivative) control system for the boiler shown in Figure 2.

The steam temperature control system works in the following manner. When the change in load command (MWD) takes place, boiler input variables are manipulated through feedforward and feedback control loops. Of these control loops, the feedforward loops work to adjust the input variables, such as the governing valve opening, the flue gas damper opening (GD), fuel flow rate (FR), feedwater flow rate, etc., to the values corresponding to the new required load (MWD). Since the manipulation of each input variable has influence on more than one output variables in different manner, controlled variables, such as main steam pressure (TP), superheater outlet steam temperature (SHT), reheater outlet steam temperature (RHT), etc., deviate from their set-points.

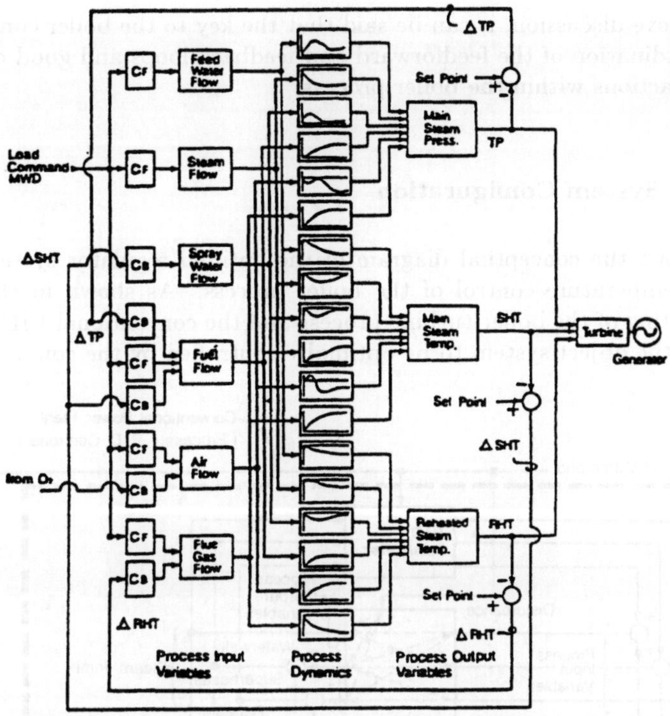

Figure 3: PID control system for a once-through boiler

To cancel such deviations. feedback loops connecting controlled variables with manipulated variables adjust boiler inputs so as to achieve final thermal-hydraulic balance in the boiler process.

However, as these feedback control loops interact each other within the boiler process, they form a typical mutually interacting multivariable system. This is the principal factor that makes steam temperature control quite difficult and consequently limits the rate of load change of power plants.

In the case of a supercritical variable-pressure boiler, which has been widely adopted recently, the above-mentioned situation becomes more serious, because in a variable-pressure boiler, its main steam pressure is controlled to the specified value in proportion to the boiler load. Such pressure manipulation causes the changes in the temperature of the working fluid within the evaporator tubes, which is approximately at the saturation temperature of the pressure, and hence enhances the nonlinearity of the process. The above fact also means that for the load change of a variable-pressure boiler a larger amount of feedforward fuel is necessary to complement the energy variation within the boiler.

From the above discussion, it can be said that the key to the boiler control is to find optimal coordination of the feedforward and feedback loops and good compensation for the interactions within the boiler process.

3. Control System Configuration

Figure 4 shows the conceptual diagram of the optimal regulator system applied to the steam temperature control of the boiler process. As shown in the figure, the plant, consisting of the boiler-turbine process and the conventional PID controller, is regarded as the object system to be optimally controlled by the computer.

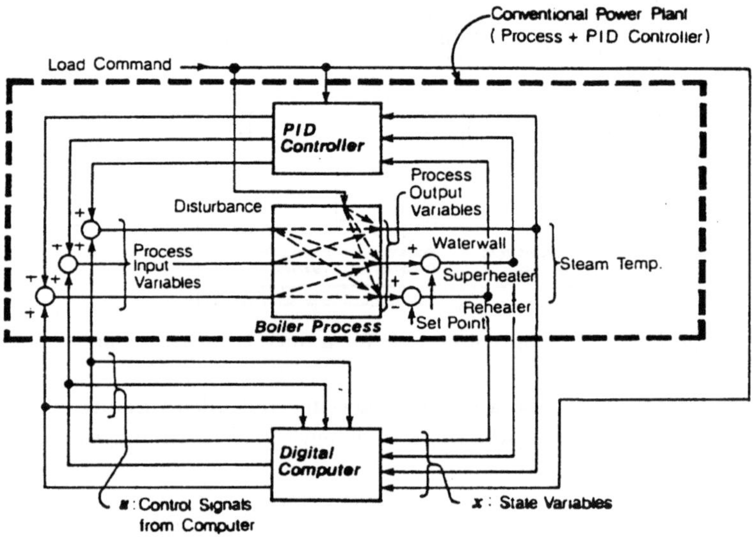

Figure 4: Configuration of proposed optimal control system

Process variables such as steam temperatures along the boiler tube, and other necessary process variables compose the process output vector $x(n)$. The vector $u(n)$ consisting of control signals to the manipulated variables, such as the fuel flow rate, spray flow rate, RH flue gas damper opening, is computed by the algorithm prepared in the computer and added via A to D converters to the control signals from the PID controller at summing amplifiers at every control time. It should be noted that in order to realize a better control performance the exogenous variable MWD is included in $x(n)$ as a pseudo-state variable.

4. Procedure for Control System Design

System analysis and controller design are performed based on the TIMSAC Library shown in Figure 5, the details of which will be described below.

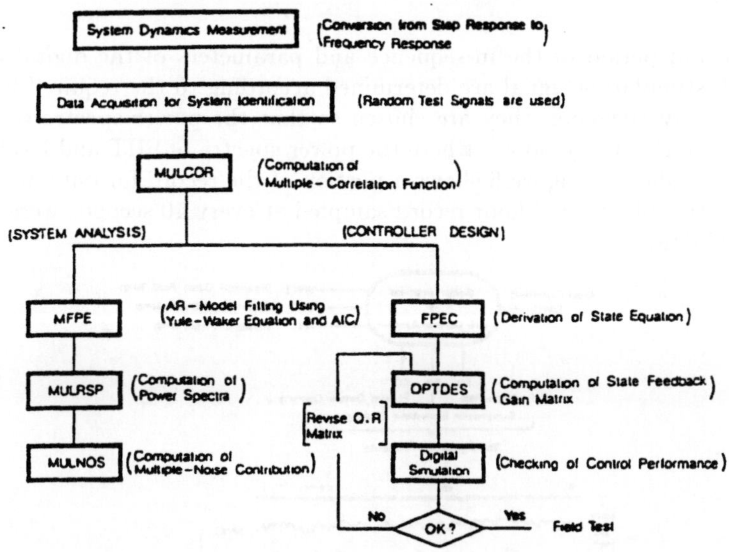

Figure 5: Procedure of roptimal regulator implementation based on TIMSAC Library

4.1 SELECTION OF SYSTEM VARIABLES

At first, system variables are selected for model building. The fundamental policy for the selection of system variables is as follows:

For state variables, controlled variables such as SHT, RHT, etc., are selected together with other process output variables that seem relevant to the dynamics of the controlled variables. For manipulated variables, process-input variables are chosen, such as FR, SP, GD (control signals issued from the computer to the plant), and other input variables that have influence on the controlled variables and are available for the control purposes.

In addition to these variables, MWD is usually included in the state-variable vector $x(n)$ as a pseudo-state variable in order to utilize the information of MWD changes which are the largest measurable disturbances to the plant.

4.2 DATA ACQUISITION FOR AR MODEL BUILDING

The vector time series data for AR model building are obtained by applying stimulating signals to MWD and other manipulated variables from the computer. For each stimulating signal, a maximum period sequence (M-sequence, a pseudo-random binary time series) produced in the computer is used after modification by a digital low-pass filter.

The fundamental period of the m-sequence and parameters of the digital low-pass filter for each stimulating signal are determined according to the required frequency ranges. Generally speaking, they are chosen so that the power spectra of the test signals cover the frequency ranges where the power spectra of SHT and RHT are assumed to be significant. Figure 6 shows a portion of the record for data acquisition. In the case of this plant, an 8-hour record sampled at every 40 seconds were used for AR model building.

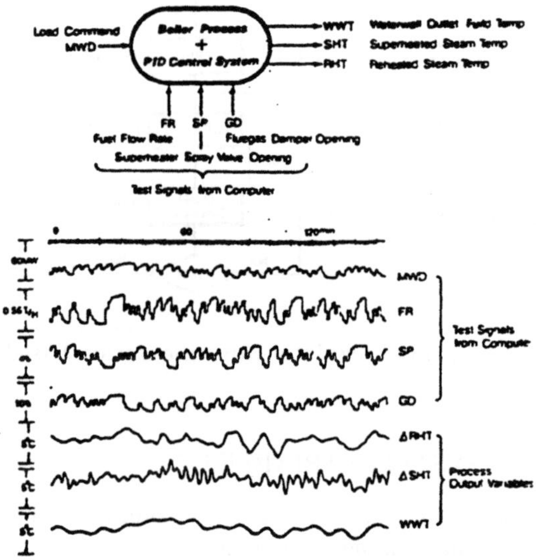

Figure 6: Record of data acquisition for system identification

4.3 SYSTEM ANALYSIS

Using the data obtained in the data acquisition procedure, the TIMSAC Library calculates the covariance matrices among system variables. The covariance functions are used in the Yule-Walker equation which is derived through the least square estimation procedure. By solving the Yule-Walker equation the coefficient matrices of the multi-variate AR model are obtained: Levinson-Durbin's recursive algorithm considerably reduces the computer memory area and computation time. The order of the AR model is automatically determined by the information criterion AIC, or FPE (Akaike, 1976a, 1978b).

As the result of this computation, a description of the system by a multivariate AR model is obtained as

$$X(n) = \sum_{m=1}^{M} A(m)X(n-m) + W(n), \tag{1}$$

where $X(n), X(n-m)$ are the system variable vectors. $A(m)$'s are the coefficient matrices of AR model. M is the model order. and $W(n)$ is the innovation vector.

Use of the coefficient matrices and the innovation vector of the AR model yields power spectral density functions of the system variables and the cross-spectral density functions among them. These spectral density functions provide useful information on the dynamic properties of the controlled system.

Assuming orthogonality among the elements in the innovation vector $U(n)$ in Equation (1), the contribution of the elements in $W(n)$ to each of the system variables can be computed. Figure 7 shows an example of the relative noise contribution (solid lines) of each element in $U(n)$ to the variance of the controlled variables, SHT and RHT. For reference sake, power spectral density functions of SHT and RHT are shown by dotted lines. in the figure.

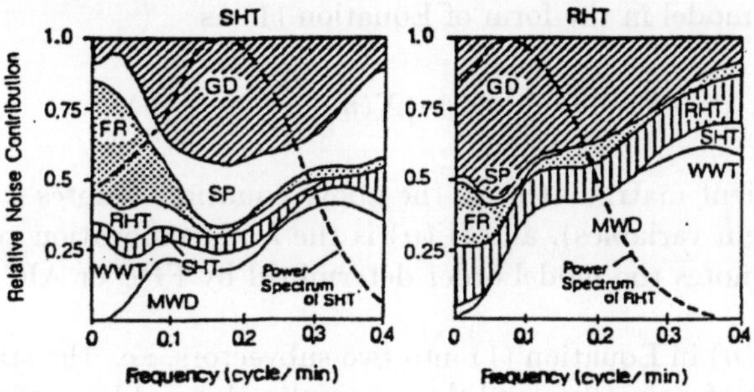

Figure 7: Relative noise contribution to the power spectra of SHT and RHT

Figure 7 shows that FR dominates in the variance of SHT in the low-frequency range, while in the frequency range above 0.1 cycle/min. the effects of SP. GD, and MWD on SHT are significant. Also, it is found that the manipulation of GD and MWD have quite a large influence on the variance of RHT.

As for the state variables. inclusion of WWT in the state variable vector appears to be effective to improve the expression of SHT. because the contribution of WWT to SHT is clearly recognizable in the lower-frequency range.

From the above analysis, we included WWT in the system variables for model building. MWD was also included in the state variable vector as a pseudo-state variable, in order to reflect its influence on the future behavior of SHT and RHT. As a matter of fact, it was found from the result of the digital simulation study that, if we remove MWD from the state variables, control performance of the optimal regulator considerably deteriorates.

AS explained above, the result of the relative noise contribution analysis provides useful information in determining the state variables and omitting unnecessary variables for model building. Usually, a noise contribution analysis is performed using process variables that appear to be relevant to SHT and RHT. The variables with small contributions to these controlled variables are omitted one by one until the final state variables are determined. This is desirable because expressing system dynamics with the smaller number of system variables is important to obtain a good model from a limited length of data time series.

4.4 DERIVATION OF STATE EQUATION

Once the system variables have been determined through the analysis described above, the state equation, which represents the system dynamics in the state space, is derived in the following way.

At first, a multivariate AR model is fit to vector data time series $X(n), n = 1, 2, \cdots, N$, to obtain the AR model in the form of Equation (1) as

$$X(n) = \sum_{m=1}^{M} A(m)X(n-m) + W(n),$$

where each coefficient matrix $A(m)$ in the above equation denotes $k \times k$ matrices (k = number of system variables), and $W(n)$ is the $k \times 1$ innovation vector. M in the above equation denotes the model order determined by FPE or AIC. [5]

Now, we divide $X(n)$ in Equation (1) into two subvectors, i.e., the state-variable vector $x(n)$ consisting of r elements and the manipulated variable vector $u(n)$ consisting of l elements. At the same time we divide the matrix $A(m)$ into subvector as shown below:

$$X(n) = \begin{bmatrix} x(n) \\ u(n) \end{bmatrix}, \quad A(m) = \begin{bmatrix} A_m & B_m \\ * & * \end{bmatrix} \quad , \tag{2}$$

where $*$ means the submatrices irrelevant to the prediction of the state vector $x(n)$. By using Equation (2), Equation (1) can be written as

$$x(n) = \sum_{m=1}^{M} A_m x(n-m) + \sum_{m=1}^{M} B_m u(n-m) + w(n). \tag{3}$$

If we introduce an operator z, providing a time delay $m\Delta t$ (Δt is the sampling period), and denote $X(n-m)$ by $z^{-m}X(n)$, then Equation (3) can be expressed as

$$(I-A_1z^{-1}-A_2z^{-2}-\cdots-A_Mz^{-M})x(n)=(B_1z^{-1}+B_2z^{-2}+\cdots+B_Mz^{-M})u(n)+w(n)$$

or

$$A(z^{-1})x(n)=B(z^{-1})u(n)+w(n), \tag{4}$$

where

$$A(z^{-1})\triangleq I-\sum_{m=1}^{M}A_mz^{-m}, \quad B(z^{-1})\triangleq\sum_{m=1}^{M}B_mz^{-m}$$

Equation (4) is a well-known expression of system dynamics.

In Equation (3), if we denote

$$\begin{aligned} x_0(n) &= x(n) \\ x_k(n) &= \sum_{m=k+1}^{M}[A_mx(n+k-m)+B_mu(n+k-m)] \\ & \qquad k=1,2,\cdots,M-1 \end{aligned} \tag{5}$$

then we get

$$\begin{aligned} x_0(n) &= A_1x_0(n-1)+x_1(n-1)+B_1u(n-1)+w(n) \\ x_1(n) &= A_2x_0(n-1)+x_2(n-1)+B_2u(n-1) \\ &\ \ \vdots \qquad\qquad \vdots \\ x_{M-2}(n) &= A_{M-1}x_0(n-1)+x_{M-1}(n-1)+B_{M-1}u(n-1) \\ x_{M-1}(n) &= A_Mx_0(n-1)+B_Mu(n-1) \end{aligned}$$

which will be expressed in the matrix form as

$$\begin{aligned} Z(n) &= AZ(n-1)+Bu(n-1)+W(n) \\ Y(n) &= CZ(n) \end{aligned} \tag{6}$$

where

$$Z(n)=\begin{bmatrix} x_0(n) \\ x_1(n) \\ \vdots \\ x_{M-2}(n) \\ x_{M-1}(n) \end{bmatrix}, \quad A=\begin{bmatrix} A_1 & I & 0 & \cdots & 0 \\ A_2 & 0 & I & \cdots & 0 \\ \vdots & \vdots & & & \\ A_{M-1} & 0 & 0 & \cdots & I \\ A_M & 0 & 0 & \cdots & 0 \end{bmatrix}, \quad B=\begin{bmatrix} B_1 \\ B_2 \\ \vdots \\ B_{M-1} \\ B_M \end{bmatrix},$$

$$W(n)=\begin{bmatrix} w(n) \\ 0 \\ \vdots \\ 0 \\ 0 \end{bmatrix}, \quad C=[I\ 0\ \cdots\ 0]\ .$$

Equation (6) is the state space representation or the state equation, which is commonly called "observable canonical form", or "observable companion form".

4.5 COMPUTATION OF OPTIMAL FEEDBACK GAIN MATRIX

Based on the state Equation (6) and a quadratic criterion function defined by Equation (7) below, an optimal state-feedback gain matrix G is obtained through the Dynamic Programming (DP) procedure.

$$J_1 = E\sum_{n=1}^{I}[Z'(n)QZ(n) + u'(n-1)Ru(n-1)] \tag{7}$$

where E is the symbol expressing expectation, I is a properly chosen integer, matrix Q is a non-negative definite, matrix R is positive definite, and $'$ denotes transpose.

In the Dynamic Programming, the computation of the state-feedback gain matrix G goes on backward from the final stage of the control span to its first stage. If we take the control span I in Equation (7) large enough, say 40 or so in this case, then the elements in the feedback gain matrix G converge to a certain steady values. In our control system, assuming a linear stationary system, we use the converged G matrix, and compute the control signal $U(n)$ by $U(n) = -G \cdot Z(n)$ at every control time.

4.6 DIGITAL SIMULATION

Validity of the state equation and appropriateness of the state-feedback gain matrix can be checked by digital simulation in the following way.

To simulate the response curves of the state variables under PID control to a specified change in MWD, we perform state transition using the state equation by replacing the MWD element at each step with the specified value of the MWD, while keeping the other system inputs to zero. Then, the record of the state variables obtained in the state transition process represents the response of the state variables to the specified MWD change. The same procedure applies to the other manipulated variables. By comparing the simulation results with the responses of the actual plant, the validity of the state equation can be verified.

The above-mentioned digital simulation is also useful for the adjustment of the state-feedback gain matrix. In this case in addition to the MWD changes, the control signals computed from $u(n) = -G \cdot Z(n)$ are provided to the manipulated variable vector in the state equation at each step of the state transition. The weighting matrices Q and R are adjusted by reviewing both the behavior of the state variables and the amplitudes of the control signals.

In the optimal regulator design procedure shown in Figure 5, DP computation and digital simulation are repeated with revised Q and R matrices at each interation step, until candidate gain matrices suitable for the field test are finally obtained.

Figure 8 shows an example of the digital simulation described above which was obtained during the tuning process of a 500-MW supercritical variable-pressure plant. Figure 8-(a) is an example of checking of the state equation, while Figure 8-(b) is the simulation results during the tuning process of the state-feedback gain matrix. In the simulation, the load-adaptive adjustment of control parameters to be described in the following paragraph was used.

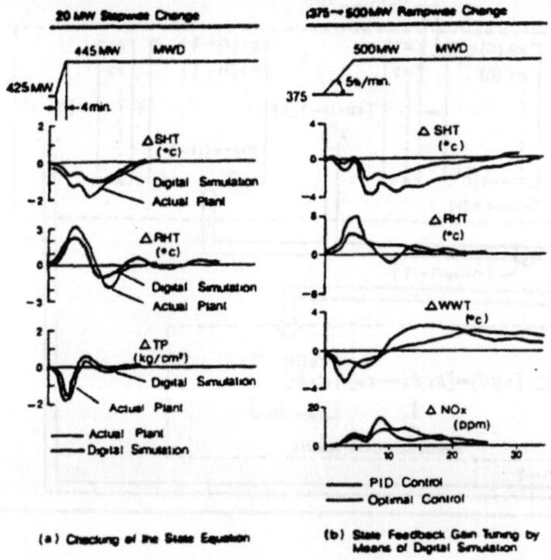

Figure 8: An example of digital simulation

4.7 COMPENSATION FOR PROCESS NONLINEARITY

To compensate for the significant nonlinearity of the boiler process, system identification and determination of the state-feedback gain matrix are performed at two or three load levels. In the actual operation, the parameters in the state equation and the state-feedback gain matrix are adjusted at each control time by a linear interpolation between neighboring matrix pairs according to the magnitude of MWD at each control time. Experiments using an exact power plant simulator at the Central Research Institute revealed that for a large ramp-wise load change, this load-adaptive adjustment of control parameters or gain scheduling strategy considerably reduces the amplitudes of both steam temperatures and manipulated variables compared with the results obtained by the system with parameter fixed.

5. Control algorithm

With the state equation and the feedback gain matrix obtained in the off-line computation, the control algorithm of the optimal regulator is formulated as shown in Figure 9. For on-line control, the state equation (6) is converted into the form as shown in Figure 9, which is equivalent to the original equation (6).

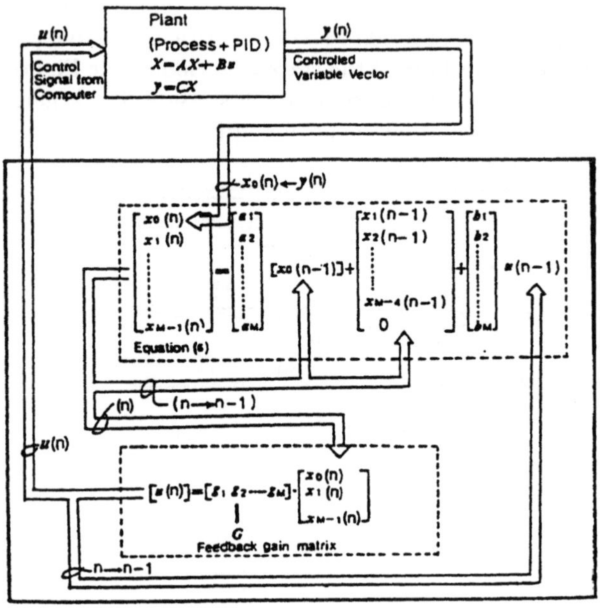

Figure 9: Control Algorithm

The optimal control proceeds in the following way:
At time $n-1$, the control computer calculates the state vector $\hat{Z}(n)$, the prediction of $Z(n)$ at time $n-1$. When the plant data $y(n) = x_0(n)$ become available at time n, the upper-most vector $\hat{x}_0(n)$ of $Z(n)$ (the prediction of $x_0(n)$) is replaced by the actual plant data $x_0(n)$ to obtain $Z(n)$. Then, the control signal $u(n)$ is computed by $u(n) = -G \cdot Z(n)$ and immediately sent to the summing amplifier and added to the control signals from the PID controller. After this, the time index n in the state equation is replaced by $n-1$ and the control computer calculates the state vector $\hat{Z}(n)$, thereby preparing for the computation at the next control time. The same procedure is repeated at every control time.

6. Control performance

Before implementing an optimal regulator, preliminary studies are usually performed using a plant model with the dynamics similar to the target plant. This process is important to get a general insight to the plant in advance, because dynamic property varies from plant to plant. It is helpful to prepare some control strategy in advance, such as the selection of the control interval, selection of system variables, compensation for system nonlinearity, and gain coordination between PID controllers and control computer etc., since the time to be allowed for optimal controller tuning at site is usually quite restricted. Because an optimal regulator is designed on the basis of linear control theory. it is desirable to reduce the nonlinearity of the object system before applying the regulator. From this point of view function generators, included in the PID control system for non-linearity compensation should be tuned in the stage of PID controller tuning so that they produce linearized dynamic property as much as possible over the range of the plant load where the optimal regulator is applied. Figure 10 shows a comparison of the control performance between the conventional PID controller and the optimal regulator obtained in the field test of a 600-MW supercritical constant-pressure plant. In this experiment. randomly-varying signals produced by the control computer were applied to the MWD setter of the plant.

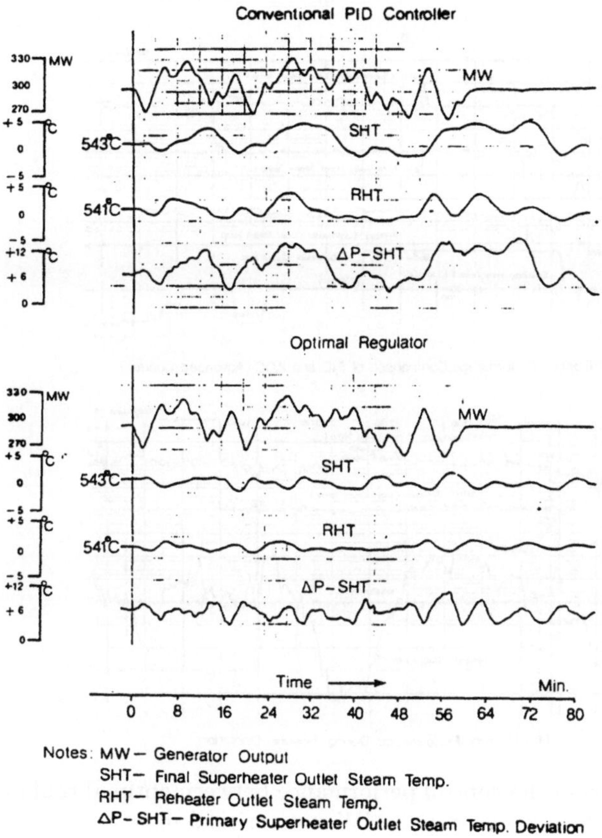

Figure 10: Comparison of the control performance between optimal regulator and PID controller (field test record of 600MW supercritical plant)

As shown in the figure. the fluctuations of the controlled variables, SHT and RHT. are considerably reduced by the application of the optimal regulator. The deviations of the primary superheater outlet steam temperature, which is considered as an index expressing the internal state of the plant, is also smaller in the optimal regulator than in the conventional PID controller. Thus, it is found that the optimal regulator has a favorable effect on the internal conditions of the boiler process.

Figure 11, which provides a similar comparison between PID control and an optimal regulator, was the record obtained during the routine operation of a 500-MW supercritical constant-pressure plant when the plant was put under load change commands from the system's load dispatch center.

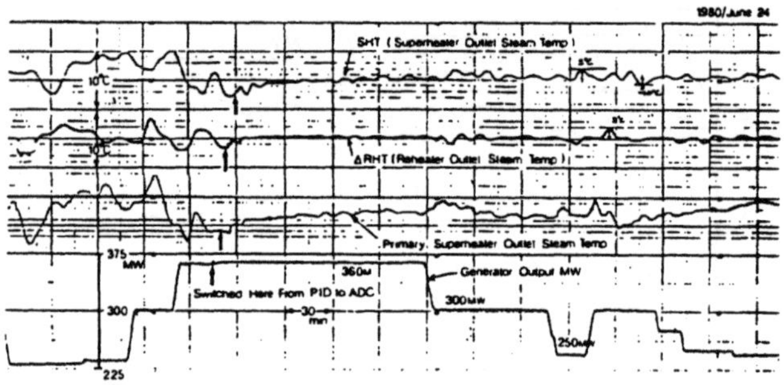

(a) Control Performance Comparison of PID and ADC (Advanced Control)

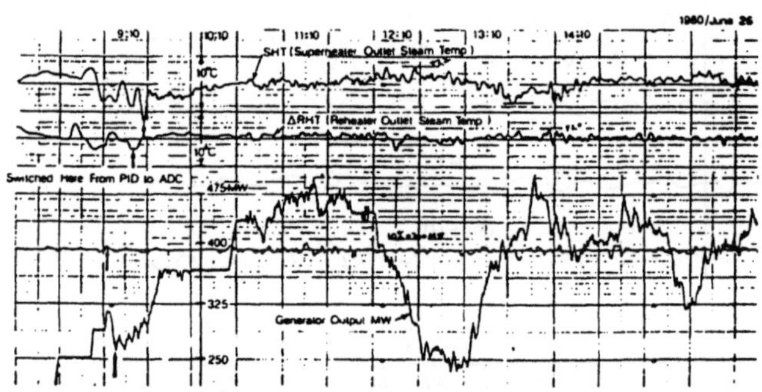

(b) Control Performance During Routine Operation

Figure 11: Comparison of the control performance between optimal rgulator and PID controller during routine operation (500MW supercritical plant)

In these two cases (a) and (b), in order to examine the effectiveness of the optimal regulator, the control was switched from the conventional PID control to the optimal control by adding the signals from the computer to those from the PID controller at the instant shown with the arrow marks in the charts. From these records the improvement of control performance by the optimal control system is also clearly recognized. The record (b) in Figure 11 is the one obtained when the plant operating under the optimal regulator was subjected to fast and large load changes. As observed in this figure, the deviations of SHT and RHT from their set points remain in small ranges even under such demanding operating conditions.

7. Consideration On The Improvement of Control Performance

An optimal regulator basically works to cancel the deviations of state variables from their values at steady state. From this point of view, it is important to include MWD, the largest measurable disturbance to the plant, in the state variables in order to predict its influence on the steam temperature. In the Dynamic Programming for computing the feedback gain matrix, zero is assigned to the element corresponding to MWD in the weighting matrix Q so as to allow free movement of MWD, because MWD is an exogenous variable that cannot be controlled at the plant.

Figure 12 depicts the cross-correlation functions between MWD and manipulated variables, computed from the data obtained during the routine operation of a 600-MW supercritical constant-pressure plant. In the figure, cross-correlation functions calculated from the plant data operating under PID controller are compared with those of the optimal regulator. In the abscissa of Figure 12, the amount of time shift between zero and the peak point of the cross-correlation function approximately represents the response time of each manipulated variable to the change in MWD.

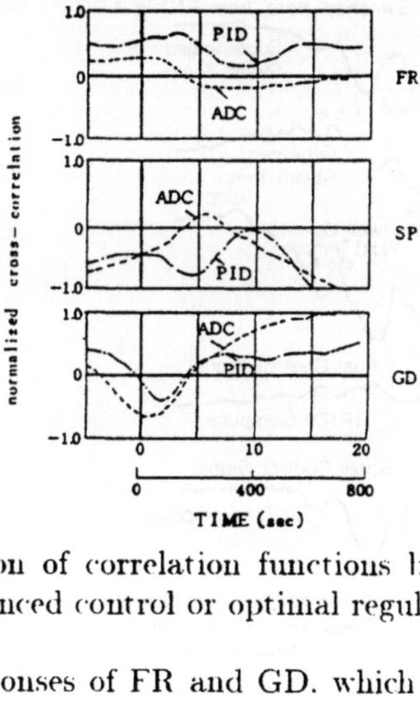

Figure 12: Comparison of correlation functions between MWD and manipulated variables (ADC : advanced control or optimal regulator)

In Figure 12, the responses of FR and GD, which are manipulated by feedforward signals to follow MWD changes, are slightly faster in the optimal regulator than in the PID controller. On the other hand, the response of the SP, which receives the influence of MWD change indirectly through the deviations of SHT, is remarkably faster in the optimal regulator than in the PID controller. Such faster response of the manipulated variables to MWD changes seems to be brought about owing to the prediction of state variables in the optimal regulator and effectively works in reducing the deviations of the controlled variables with less control action.

8. Robustness Of The Optimal Regulator

In addition to the excellent features described above, experience in commercial operation revealed that the optimal regulator had a stabilizing effect on the system under various disturbances.

Figure 13 illustrates an example of such stabilizing effect of the optimal regulator: the chart was recorded when the plant was subjected to an extraordinary disturbance caused by the starting of one of the two forced draft fans. As shown in the figure, the process input and output variables were restored to their normal conditions quite soon after the control system was switched from PID control to optimal control, despite the plant load was increasing.

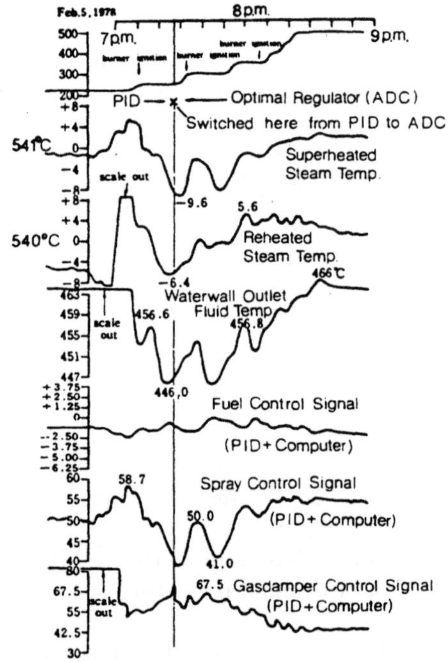

Figure 13: Control performance of the optimal regulator under extraordinary disturbance

Figure 14 shows a comparison of the behavior of the manipulated variables in the optimal regulator and that in the PID controller under identical rampwise change in MWD shown at the top of the figure. As can be seen in the figure, the moves of the manipulated variables FR, SP, GD in the optimal regulator are quite different from those in the PID control system. To be more exact, for the identical change in plant load, the manipulated variables in the optimal regulator move at a larger rate of change than that in the PID control. Such behavior seems to be resulted from the prediction of state variables and efficiently works to suppress the deviations of steam temperatures from their set points and consequently contributes to stabilize the behavior of the plant under disturbances.

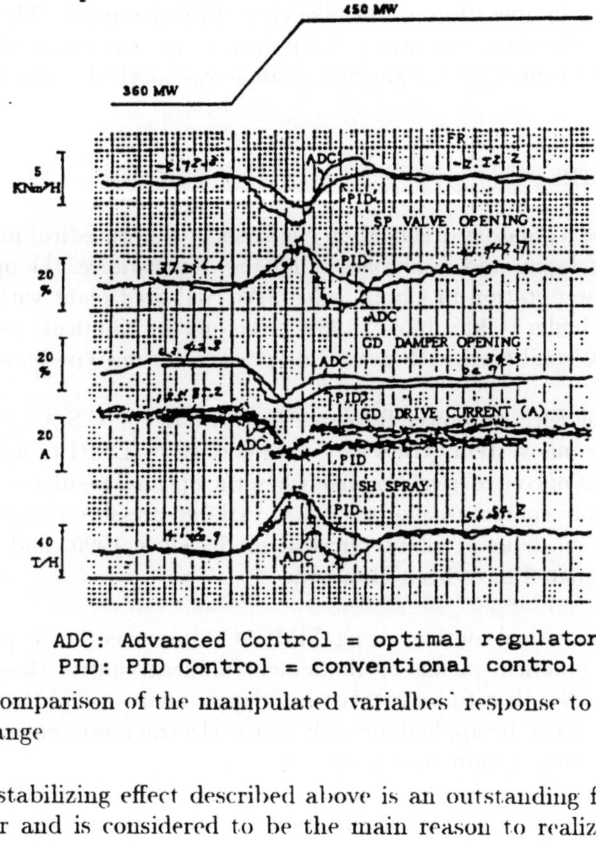

ADC: Advanced Control = optimal regulator
PID: PID Control = conventional control

Figure 14: Comparison of the manipulated varialbes' response to an identical rampwise load change

The system stabilizing effect described above is an outstanding feature of the optimal regulator and is considered to be the main reason to realize excellent control performance. This property can be effectively used to suppress unfavorable steam temperature fluctuation due to predictable disturbances to the furnace. The following is an example of this kind, which was adapted in the implementation of the optimal regulator at a Canadian 500MW plant. In this plant, the superheated steam temperature was considerably affected by the ignition of the burner placed at the upper part of the furnace: a large hump appeared in the steam temperature record after the ignition of the burner which took place during the increase in plant load.

Since the ignition of the burner can be detected in advance by observing its trigger signal, a gasdamper maneuvering called gas tempering was applied simultaneously with the trigger signal, which increases the gas recirculation flow temporarily to cool the upper part of the furnace. Although the gas tempering was known as an effective means of reducing the peak value of steam temperature rise at burner ignition, the extent of maneuvering had been limited to lessen its aftereffect.

In the case of this plant, we applied a larger amount of gas tempering, expecting that the optimal regulator in operation would take care of its aftermath. The experiment proved that, the amplitude of the steam temperature rise was remarkably reduced, while the disturbance caused by the gas flow change was soothed to the level allowed for plant operation.

9. Closing Remarks

Today many optimal regulators designed on the basis of the statistical method are in routine operation in Japan and in abroad. They are, for instance, the optimal regulator for Shajiao power station in China, a 350MW coal-fired plant with a variable-pressure drum-type boiler (Ohki, Inoue, 1989) : the one for Nanticoke power station in Canada, a 500MW coal-fired with a variable-pressure drum-type boiler.

Presently, EPRI, The Electric Power Research Institute in the U.S.A., is considering the adoption of the same system. We were commissioned by EPRI to investigate the possible improvement of control performance when the optimal regulator is applied to the steam temperature control of a supercritical variable-pressure boiler. The study has been continued on a power plant physical model since 1990, and by now the results we had anticipated have been obtained.

The feature of the statistical method using TIMSAC Library is that it provides control engineers with practical means to implement modern control theory for large industrial processes. Needless to say, statistical system analysis and the control system design based on it can be applied not only to the electric power generating plants but also to various kinds of industrial processes.

Some examples of these kinds are the application to the sewage process by an electric manufacture in which the author himself was involved as a consultant, the one to the activated sludge process (Kanaya, Fujita, 1988), to the control of steam flow and NO_x in the incinerator (Hiraoka, Tsumura, 1988) both by Kyoto University, and the application to the control of hydrogen chloride in the exhaust gas from the incinerator by a steel company, etc.

Since the TIMSAC Library includes all the programs required for the design of the optimal control system, implementation of the optimal controller can be made by a well-talented plant engineer, rather than a specialist with expertise on the analysis of the process dynamics and modern control theory. From the practical point of view, this is quite an important aspect for a technique to survive for generations.

References

[1] Akaike, H., and Nakagawa, T. (1972). *Statistical Analysis and Control of Dynamic Systems*, Saiensu-sha, Tokyo (in Japanese, with a computer control package TIMSAC written in FORTRAN IV with English comments.

[2] Akaike, H. (1976). *Canonical correlation analysis of time series and the ues of an information criterion. System Identification: Advances and Case Studies, (eds. R.H. Mehra and D.G. Lainiotis)*, 27-96, Academic Press, New York.

[3] Akaike, H. (1978). *On the identification of state space models and their use in control. Direction in Time Series. (eds. D.R. Brillinger and G.C. Tiao)*, 175-187, The Institute of Mathematical Statistics, Hyward, California.

[4] Akaike, H., Kitagawa, G., Arahata, E., Tada, F. (1979). *TIMSAC-78, Computer Science Monographs*, A Publication of The Institute of Statistical Mathematics

[5] Akaike, H., and Nakagawa, T. (1988). *Statistical Analysis and control of Dynamic Systems*, Kluwer Academic Publisher.

[6] Otomo, T., Nakagawa, T., and Akaike, H. (1972). *Statistical approach to computer control of cement rotary kilns*, Automatica-J, IFAC, 8, 35.

[7] Ohki, T., Inoue, T. (1989). *Advanced Control System for Thermal Power Plants Based on New Technologies. Conference Notebook, Conference: Power Plant Controls and Automation*, IPRI, Feb. 1989

[8] Kanaya, T., Fujita, I. (1988). *Introduction of Statistical Control Technology in the Sewage Process. (in Japanese). Systems and Control*, Vo.32, No.10, PP.559~506, 1988, THE INSTITUTE OF SYSTEMS, CONTROL AND INFORMATION ENGINEERS

[9] Hiraoka, M., Tsumura, K. (1988). *System Identification and Control of Activated Sludge Process By Use of Statistical Model*. ICA Seminar, IAWPRC 14th Biennial Conference 17-23 July, 1988 Brighton ENGLAND.

APPLICATIONS OF AUTO REGRESSIVE MODEL TO CONTROL SHIP'S MOTIONS AND MARINE ENGINE

K.OHTSU
Tokyo University of
Mercantile Marine
2-1-6 Etchujima
Koto-ku Tokyo
Japan 135

G.KITAGAWA
The Institute of
Mathematical Statistics
4-6-7 Minami Azabu
Shinagawa-ku Tokyo
Japan 106

1. Introduction

A marine engine governor and a ship's autopilot system are typical marine control systems. The marine governor was developed by J.Watt for regulating the propeller revolution of the steam engine. This system was named as a centrifugal force type of mechanical governor and is used even today. The ship's autopilot system with which general ocean going ships are now installed, was investigated theoretically by Minorsky(1922) and whose paper became the fundamental of the classical control theory. The salient feature of such marine systems is that those are exposed to strong stochastic sea disturbances such as winds and waves. A marine engineer should always pay large attention to this destiny of marine systems in designing. To withstand this difficulty, it is necessary to provide a correct prediction model for the ship's motions in winds and waves and vigorous control theory.

The leading paper for identification of a multivariate auto regressive model to a stochastic system, proposed by Akaike(1971) yielded powerful procedure to cope with such a tedious designing problem. The worthiest point in this method is that we can design the control system through the data under operating in the real plant and accordingly to get along without the complicated preliminary experiments. The first successful example of this method for the actual plant control was given by Ootomo, Nakagawa and Akaike(1972) for controlling a rotary cement kiln.

The main aim of this paper is put on the control problems of ship's heading and rolling motions by use of rudder and that of propeller revolution of marine engine by use of governor, using the auto regressive model for predictions of their motions. The paper is set out in the following parts. First, we discuss an autopilot system for course keeping control of full scale ships. Furthermore, we extend this system to a noise adaptive autopilot system, which is capable of following to the change of disturbances. Secondly, we show a new autopilot system called rudder roll control system(RRCS), which has ability to reduce not only the ship's course deviation but

H. Bozdogan (ed.), Proceedings of the First US/Japan Conference on the Frontiers of Statistical Modeling: An Informational Approach, 81–104.

also her rolling motion by use of rudder. Lastly, we consider a governor system in a main engine for regulating fluctuations of propeller revolution and also its torque. In this system, we utilize the auto regressive model for predicting fluctuations of the propeller revolution and the torque.

2. Ship's Motion and Marine Engine

In this section, we briefly review the ship's motions, the marine engine and their controllers. Fig.1 shows the nomenclature of ship's motions with six degrees of freedom. In these motions, yaw, roll and sway are called the lateral motion and the remaining ones, the longitudinal one. The former one are under strong influences of rudder motion.

Yaw is also called ship's heading or simply heading in the course keeping motion. The ship's heading is generally controlled with an autopilot system whose control law is based on the classical proportional, integral and differential (PID) feedback control theory (Fig.2).

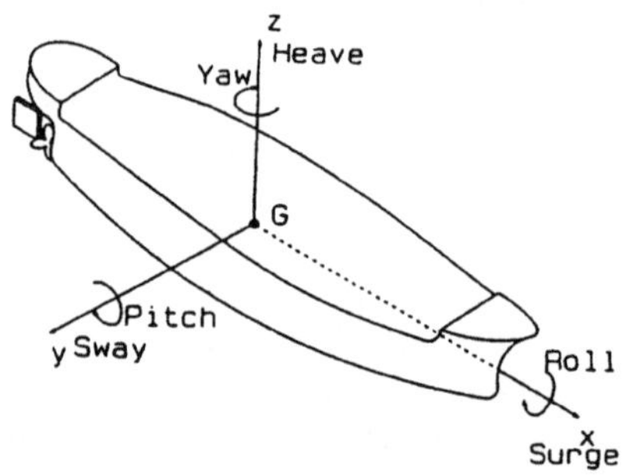

Fig.1 Nomenclature of ship's motion

Fig.3 shows a conceptional diagram of marine governor system. Responding to the signal through the feedback loop, the controller moves a rack bar in the actuator of the governor system to control the injecting quantities of fuel oil into cylinders and regulates fluctuations of propeller revolution. The mechanical governor determines the injecting quantities of fuel oil by centrifugal force brought about with propeller revolution. Recently, the governor system has changed from a mechanical type to an electronic one. The digital PID controller with some filtering facility is also realized in the electronic governor system. The rate of propeller revolution is written by 'rpm' (revolution per minutue) in the figures, hereafter.

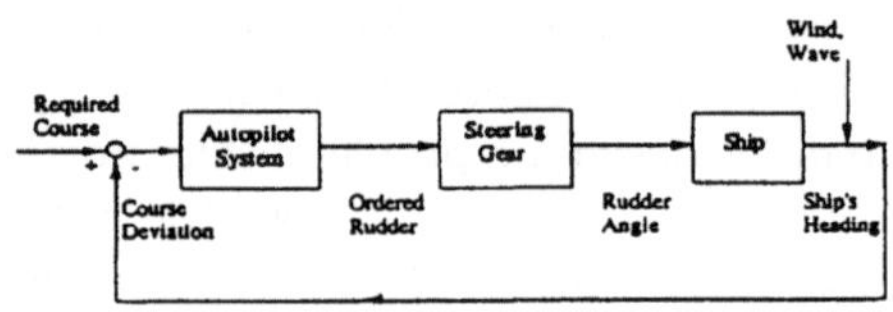

Fig.2 Ship's autopilot system

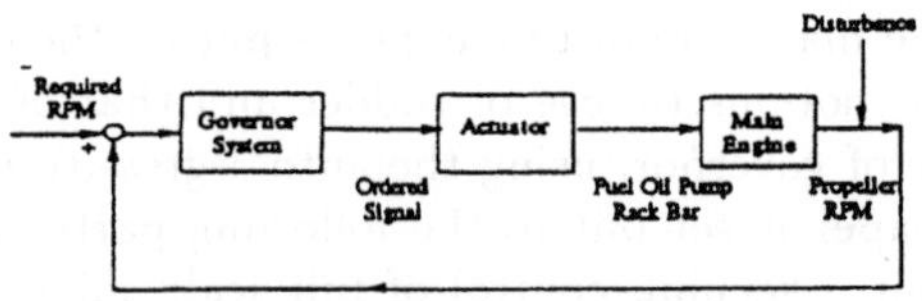

Fig.3 Ship's governor system

3. A Review of Multivariate Auto Regressive Model Fitting for Control

The basic model adopted in this paper for predicting ship's motions is a multivariate auto regressive model with control inputs,

$$X(n) = \sum_{i=1}^{M} A(i)\, X(n-i) + \sum_{i=1}^{M} B(i)\, Y(n-i) + U(n), \tag{1}$$

where $X(n)$ denotes the r dimensional vector of controlled variable such as heading, roll, the rate of propeller revolution and so on and $Y(n)$, the s dimensional vector of control variables such as rudder angle and rack bar's position in the governor.

In fitting this model, the goodness of the model is evaluated by the AIC (Akaike Information Criterion) given by Akaike(1974),

$$AIC(M) = N \log(\det(d_{r,M})) + 2r(r+s)M + r(r+1), \tag{2}$$

where $d_{r,M}$ denotes the covariance matrix of the residual of the fitted auto regressive model. The optimal order in the auto regressive model is determined by minimizing the value of the AIC. We refer to this method as ' minimum AIC method'. In order to formulate the problem in the framework of the modern control theory, it is necessary to transform the model (1) to a state space representation. However, this transformation is not unique. Ootomo *et al.*(1972) adopted the following transformation, by which we can obtain the new control signal in the extremely short elapsed time.

Let us define a state vector and a white noise one by

$$Z(n) = \left(Z_0(n)^t \cdots Z_{M-1}(n)^t\right)^t, \quad W(n) = \left(U(n)^t\, 0\, 0 \cdots 0\right)^t \tag{3}$$

where

$$\begin{cases} Z_0(n) &= X(n) \\ Z_i(n) &= Z_{i+1}(n-1) + A(i+1)Z_0(n-1) + B(i+1)Y(n-1), \\ & \qquad i = 1, 2, \cdots, M-2 \\ Z_{M-1}(n) &= A(M)Z_0(n-1) + B(M)Y(n-1), \end{cases} \tag{4}$$

then (1) is equivalent to the following state space representation

$$\begin{cases} Z(n) &= \Phi\, Z(n-1) + \Gamma\, Y(n-1) + W(n) \\ X(n) &= H\, Z(n), \end{cases} \tag{5}$$

where

$$\Phi = \begin{pmatrix} A(1) & I & 0 & \dots & 0 \\ A(2) & 0 & I & \dots & 0 \\ \vdots & \vdots & \vdots & \ddots & \vdots \\ A(M-1) & 0 & 0 & \dots & I \\ A(M) & 0 & 0 & \dots & 0 \end{pmatrix}, \quad \Gamma = \begin{pmatrix} B(1) \\ B(2) \\ \vdots \\ B(M-1) \\ B(M) \end{pmatrix} \tag{6}$$

$$H = (I \; 0 \; \cdots \; 0)$$

Using this state space representation, we obtain the optimal control law

$$Y^*(n) = G_P \; Z(n) \tag{7}$$

under a quadratic performance function,

$$I_P = E\Big[\sum_{n=1}^{P} [\; X(n)^t \; Q \; X(n) + Y(n-1)^t \; R \; Y(n-1) \;] \Big] \tag{8}$$

where Q and R are the weighting matrices to the controlled and the control variables, respectively. In this feedback law, the optimal gain G_P is obtained by solving a Riccatti equation. And when the optimization interval P is sufficiently long, it approaches a stationary gain G. Hence, the optimal control law can be written by

$$Y^*(n) = G \; Z(n). \tag{9}$$

Thus, we can obtain the optimal gains before actually executing a full scale experiment, basing on the model which has been obtained from the records in the preliminary experiment for identification of (1). However, in order to provide an appropriate model for controlling, it is desirable that the input signal used in fitting the auto regressive model contains abundant frequecy components expanding to wide frequency domain. From this point of view, the steering in the preliminary autopilot experiment is required to be as random as possible.

Lastly, it is important to assess the effects of the designed system by some simulation study. For the sake of it, we can utilize a white noise simulation based on the following model;

$$\begin{cases} Z(n) & = \; \Phi \; Z(n-1) + \Gamma \; Y^*(n-1) + W(n) \\ Y^*(n) & = \; G \; Z(n) \end{cases} \tag{10}$$

where $W(n)$ denotes white noise vector with appropriate covariance.

Summarizing the procedure for designing an autopilot system;

1. Observe the records of yaw, roll and rudder which are driven by steering as randomly as possible.
2. Fit an auto regressive model with control inputs (1) using the minimum AIC procedure.
3. Calculate the optimal gain by solving the Ricatti equation.
4. Perform the digital simulation with white noise using the relation (10) .
5. Repeat from 3. to 5., changing weighting factors in the performance function to various values, until we obtain a satisfactory result .

We refer to Akaike and Nakagawa(1988) regarding to the selection of the weighting matrices Q and R in the step 3.

4. Applications to Autopilot System

4.1 SHIPS FOR THE EXPERIMENTS

In the remainder of this paper, we discuss the applications of the preceding theory to the real marine systems.

Table 1 and Table 2 show the principal dimensions of two ships, Shioji Maru II and III, on which we have mainly performed the full scale experiments. Both these ships belong to the Tokyo University of Mercantile Marine and are mainly utilized for training and studying. The last one is the renewed ship of the former one and is equipped with the sensors for observing the ship's motion, the condition of the main engine and the actuators for driving the control devices. And we can access to the information through the computers.

Length	41.70 m
Breadth	8.00 m
Tonnage	311.37 G.T.
Max.Speed	11.49 knot

Table 1 Principal dimensions of Shioji Maru II

Length	49.93 m
Breadth	10.00 m
Depth	2.80 m
Tonnage	425.00 G.T.
Bow Thruster	2.4 tons
Stern Thruster	1.8 tons

Table 2 Principal dimensions of Shioji Maru III

4.2 AR AUTOPILOT SYSTEM

The first example is application to a ship's autopilot system to keep her course. Let us assume that the controlled variable $X(n)$ in (1) consists of the deviation of ship's heading from the required course (hereafter, we call this the course deviation) and the control one,$Y(n)$, rudder motion. Thus, the model for predicting the course deviation is resulted in a single input single output auto regressive model by the selection of these variables.

Ohtsu, Horigome and Kitagawa(1979) developed a new autopilot system called AR autopilot system, using the optimal control theory based on this model. And full scale experiments were performed by SHIOJI MARU II. Fig.4 shows a digital experimental system for the new autopilot system. The sampling rate was fixed on 1.0 sec..

Fig.5 shows the records of the actual experiments. The left half of it shows the result with the AR autopilot system and the right half, the one with the conventional autopilot system whose control law is designed with the conventional PID theory. The latter experiment was done just after the former one. We notice that the AR autopilot system performs eminent reduction of the course deviation. Moreover, it is remarkable property that the performance is achieved by a small amount of rudder motion. Fig.6 shows the power spectra of the course deviation and the rudder motion in both experiments. We can also see that the AR autopilot realizes a large amount of reduction of the course deviation and the rudder motion, in particular, in the low frequency domain. Up to now, we have carried out various kinds of test with the AR autopilot system, using some types of ship. As a result, we have confirmed that it performs robust property to the changes of direction and forces of winds and waves with the fixed gain and good turning ability.

4.3 NOISE ADAPTIVE AUTOPILOT SYSTEM

Ozaki and Tongs(1975) proposed an effective algorithm for analyzing a locally stationary time series. Using this algorithm, we developed an autopilot system which is called noise adaptive controller(NADCON) (Ohtsu,Horigome and Kitagawa(1979)). The new model for noise adaptive autopilot is seemingly the same as the auto regressive model

$$x(n) = \sum_{i=1}^{M} a(i)\ x(n-i) + \sum_{i=1}^{L} b(i)\ y(n-i) + u(n), \tag{11}$$

where $x(n)$ denotes the course deviation at time n and $y(n)$, rudder motion. But we no longer assume the whiteness of the noise term $u(n)$ and rather consider it as a non stationary colored process. Then, the term $u(n)$ can be regarded as a driving input by the external sea disturbances. And this noise part is assumed to be an auto regressive process represented as follows;

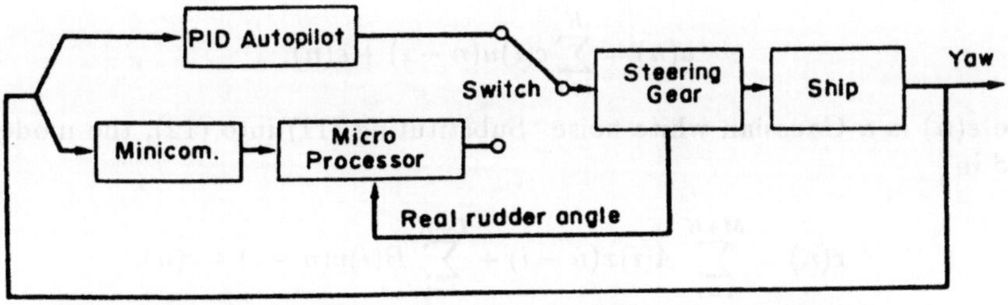

Fig.4 Digital autopilot system for the full scale experiment

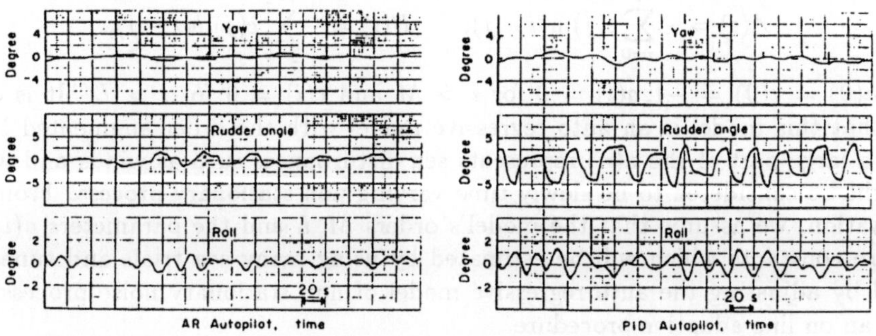

Fig.5 Records of AR and ship's autopilot in the full experiments
(Ohtsu *et al.*(1979))

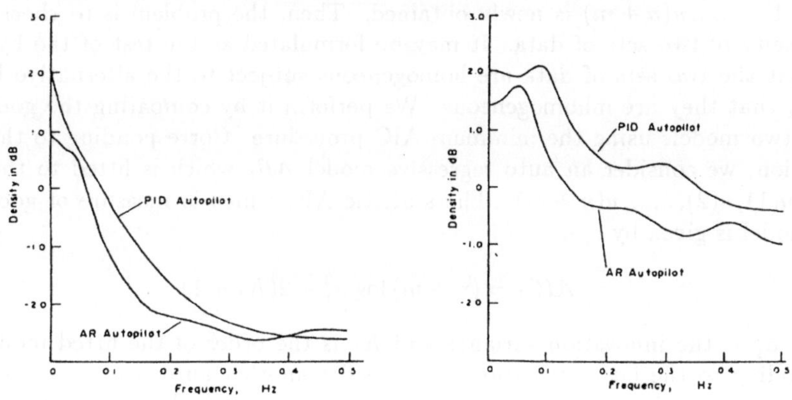

[A] Heading angle [B] Rudder angle

Fig.6 Power spectra of heading angle and rudder angle
(Ohtsu *et al.*(1979))

$$u(n) = \sum_{i=1}^{K} c(i)u(n-i) + \varepsilon(n), \tag{12}$$

where $\varepsilon(n)$ is a Gaussian white noise. Substituting (11) into (12), the model is resulted in

$$x(n) = \sum_{i=1}^{M+K} A(i)x(n-i) + \sum_{i=1}^{L+K} B(i)y(n-i) + \varepsilon(n), \tag{13}$$

where

$$A(i) = -\sum_{j=0}^{i} c(j)\, a(i-j), \qquad B(i) = -\sum_{j=0}^{i-1} c(j)\, b(i-j), \tag{14}$$

with $a(0) = c(0) = -1, a(i) = 0$ for $i > M$ and $c(i) = 0$ for $i > L$. It is easily seen that this model is an auto regressive model with the order augmented by K. Since a large part of the $u(n)$ represents sea disturbances such as winds and waves, it might be considered to be slowly time varying non stationary process. From this observation, we assume that the model's orders M, L and the parameters $a(i)$ and $b(i)$ are constants which will be estimated by using many sea trials and tuned the model by adjusting the auto regressive model of non stationary noise process $u(n)$ using an on line adaptive procedure.

For it, let us consider the situation where an autoregressive model AR_0 has been fitted to the noise process $u(1), u(2), ...u(n)$ and additional set of m observations $u(n+1), ..., u(n+m)$ is newly obtained. Then, the problem is to check the homogeneity of two sets of data. It may be formulated as the test of the hypothesis H_1 that the two sets of data are homogeneous subject to the alternative hypothesis H_2 that they are inhomogeneous. We perform it by comparing the goodness of fit of two models using the minimum AIC procedure. Corresponding to the former situation, we consider an auto regressive model AR_1 which is fitted to the pooled data $u(1), u(2), ..., u(n+m)$. The statistic AIC which is measure of goodness of the model is given by

$$AIC_1 = (n+m)\log\sigma_1^2 + 2(K_1+1) \tag{15}$$

where σ_1^2 is the innovation variance and K_1 is the order of the fitted model. Corresponding to the latter situation, we consider another auto regressive model AR_2 which is fitted to just the new data $u(n+1), ..., u(n+m)$. This means that the data $u(1), u(2), ...u(n+m)$ is divided into two spans and auto regressive models AR_0 and AR_2 are fitted on each data set. The AIC of such jointed model is given by

$$AIC_2 = n\log\sigma_0^2 + m\log\sigma_2^2 + 2(K_0+K_2+2) \tag{16}$$

where σ_0^2 and K_0 are innovation variance and the order of the auto regressive model AR_0, σ_2^2 and K_2 are those of AR_2, respectively. If AIC_2 is less than AIC_1, we are recommended to reject the former model AR_0 and switch to the new model AR_2, since the change of the statistical structure between these two spans are significant. Otherwise, we are recommended to adopt the refined model AR_1 since these two data are judged to be homogeneous (Fig.7).

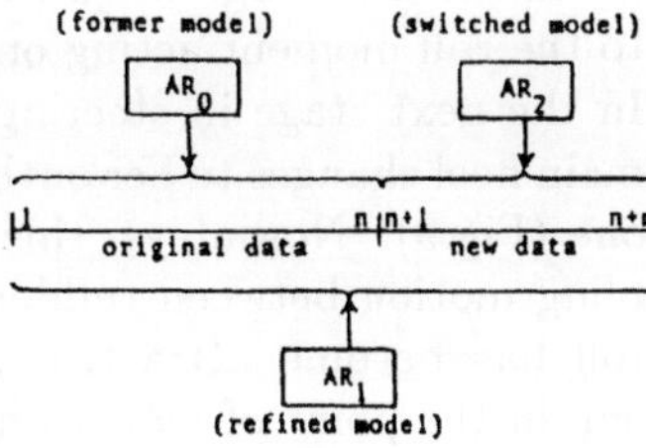

Fig.7 Locally stationary model

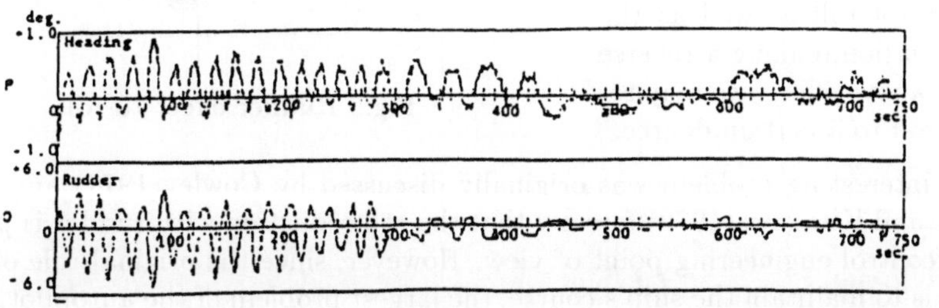

Fig.8 Performance of NADCON system in the full scale experiment (above:Heading, below:Rudder angle)

Fig.8 shows a typical example of full scale experiments using the almost same size of ship with SHIOJI MARU II. In this experiment, $n = m = 200sec.$ and the sampling rate is fixed on 1.0 sec.. We can detect from this figure that until about 300 sec. from the beginning, the movements of the heading and the rudder angle had become larger, because the first AR model had not already fitted well to the actual motions. However, after 300 sec. both the motions became smaller than before it, the model being refined by the procedure mentioned above.

5. Roll Damping Type of Autopilot System

5.1 RUDDER ROLL YAW COUPLING MOTION

For a mariner, it is common experience that steering with rudder generally induces roll, though the original work of rudder keeps ship's heading to the required course. We can comprehend the effects, observing the record of the roll in Fig.5. This figure evidently demonstrates that the steering by large angles of rudder induces roll in the conventional autopilot, comparing with the AR autopilot system. Thus, as stated in Kroukowsky(1961)'s text, we can assert that rudder motion is the seventh ship's motion.

In the first stage when rudder is steered, a ship heels to her inboard direction due to the roll moment acting on the rudder surface.
In the next stage in steering, the main heel changes to her outboard one (Fig.9). Nowadays, this coupling motion between rudder and roll has become attractive problem in the point of view from roll stabilizing with rudder, because it is natural insight that if we steer so skillfully the rudder action to the change of roll as well as the course deviation,making a reverse use of the above effect, we may reduce the roll to a certain degree.

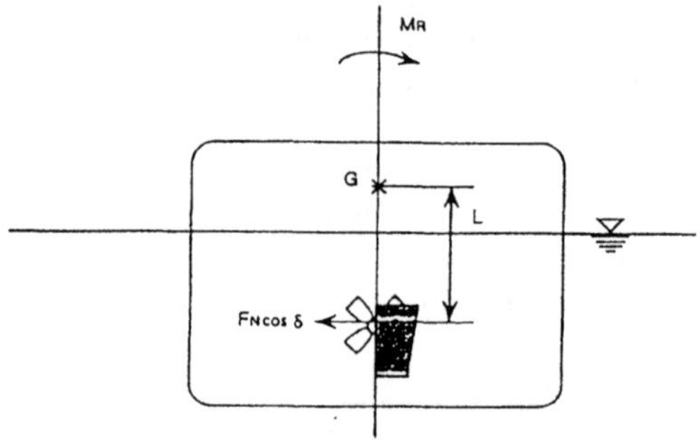

Fig.9 Rudder effect to roll

This interesting problem was originally discussed by Cowley(1974) and Ohtsu, Horigome and Kitagawa(1979) foresaw that the roll reduction with rudder is possible from the control engineering point of view. However, since the original role of ship's autopilot is to maintain the ship's course, the largest problem of the autopilot system having the ability to reduce rolling is how to make both roles of autopilot system compromised.

5.2 DESIGN OF RRCS

Let us consider that the controlled variable $X(n)$ in (1) consists of course deviation and moreover roll rate and the control one $Y(n)$, of rudder angle. In this system, the reasons that we selected not roll angle itself but roll rate, are;

1. the roll rate signal can be more easily measured than the roll angle and
2. the rudder action to roll rate gives stronger roll damping effect than the one to roll angle.

Thus, the problem is resulted in an one input - two outputs control system.

Basing on the optimal control law derived from this model, Oda,Ohtsu *et al.*(1992) designed a new autopilot system called Rudder Roll Control System (RRCS) aiming at course keeping and moreover roll reduction, simultaneously. We have carried out the full scale experiments on 3 middle classes of ships. In here, we shall indicate the results using Shioji Maru III. In this experiments, we measured the yaw signal through a gyro repeater and the roll rate signal through a vertical gyro system in the ship and fixed the sampling period on 0.5sec., basing on the experiences. By the reason described in the last chapter, the input rudder signal for identification of the model (1), given by a seaman, was as random as possible. The fitted auto regressive model (1) by the minimum AIC procedure had 6th order.

Table 3 shows the variances of heading, roll rate, rudder angle and the reduction ratios of roll rate in the full scale experiments. The reduction ratio of roll rate is calculated by

$$Reduction(\%) = \frac{AP - RRCS}{AP} \times 100, \tag{17}$$

where AP denotes the standard deviation of roll rate in the ordinary ship's autopilot, $RRCS$, that in the RRCS, respectively. Fig.10 shows typical examples of the experiments with the RRCS (left) and with the ship's autopilot system (right) which was carried out in very rough sea soon after the first one.

Test No.	Control Mode	Head (deg)	Speed (kts)	Wind Direction (deg)	Standard Deviation				Reduction of Roll Rate(%)
					Yaw (deg)	Roll (deg)	Roll Rate (deg/s)	Rudder (deg)	
S300	AP	220	13.2	SSE	0.42	2.07	1.99	0.60	...
S302	RRCS(27)	205	13.7	SSE	0.73	1.53	1.53	3.20	23
S303	RRCS(29)	210	13.8	SSE	0.83	1.62	1.60	3.70	20
S306	RRCS(29)	15	13.6	SSE	0.91	0.98	0.88	3.44	30
S307	RRCS(27)	15	13.8	SSE	0.86	1.30	1.00	3.37	21
S308	AP	13	13.9	SSE	0.99	1.47	1.26	1.84	...

Table 3 Statistics of the full scale experiments with the RRCS and the ship's autopilot

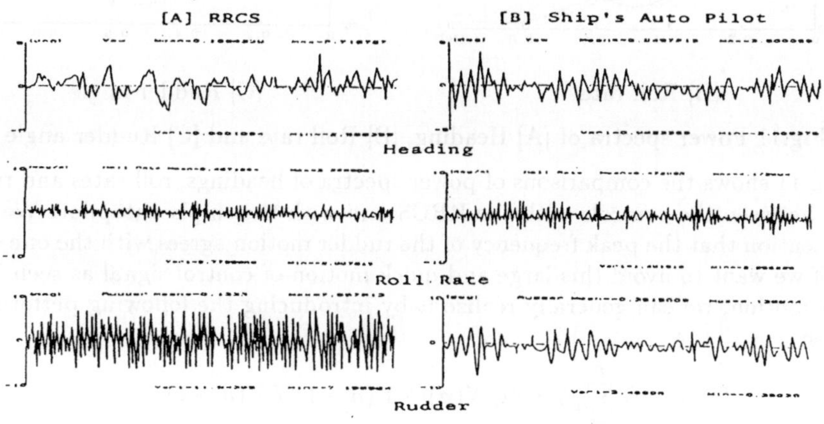

Fig.10 Records of the headings, the roll rates and the rudder angles with the RRCS(left) and the ship's autopilot(right) in the full scale experiments

According to these figures and the table, we can conclude that;

1. The rolling motion with the RRCS decreases than the one with the ship's autopilot.
2. The change of heading with the RRCS maintains the desired course in the allowable limit, though the variance of it becomes larger than the one with the ship's autopilot.
3. The amount of rudder angle becomes rather larger than the one with the ship's autopilot.

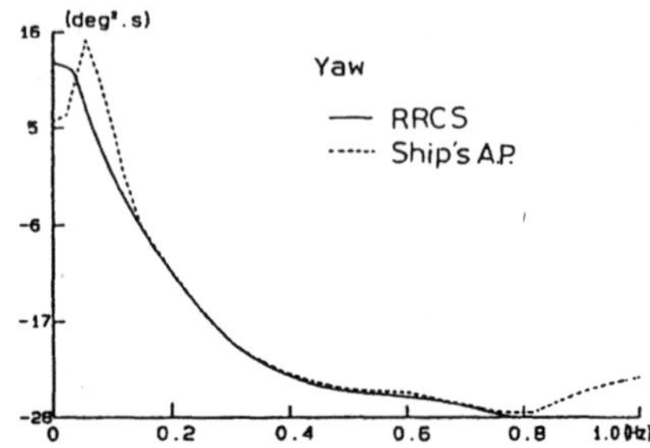

[A] Heading

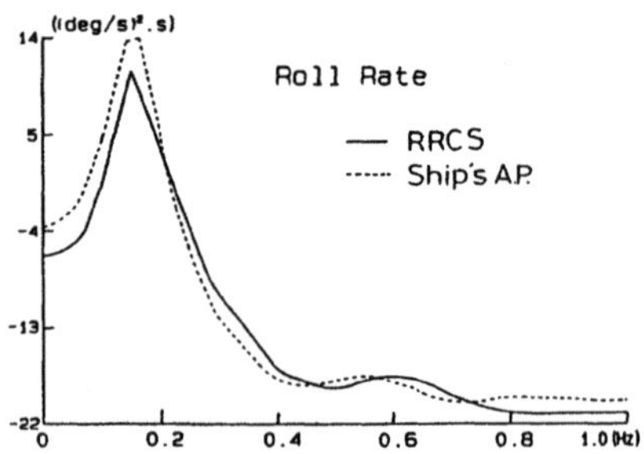

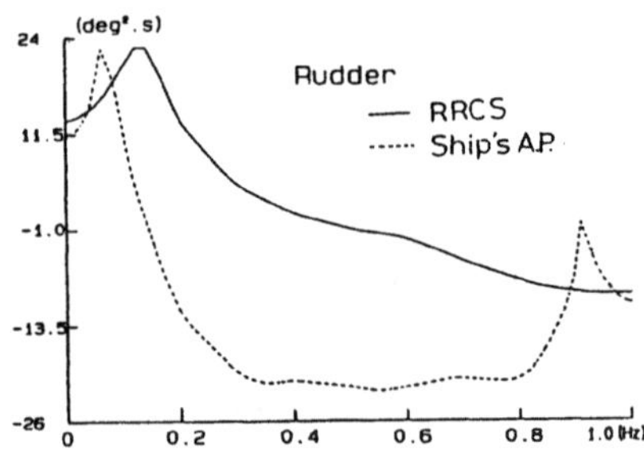

[B] Roll rate

[C] Rudder angle

Fig.11 Power spectra of [A] Heading, [B] Roll rate and [C] Rudder angle

Fig.11 shows the comparisons of power spectra of headings, roll rates and rudder angles between the controls with the RRCS and with the ship's autopilot. We must pay attention that the peak frequency of the rudder motion agrees with the one of roll rate. If we want to avoid this large and quick motion of control signal as seen in the rudder motion, we can generally realize it by introducing the following performance function

$$\begin{aligned} I_P &= E\sum_{n=1}^{P}[X(n)^t Q X(n) + Y(n-1)^t R Y(n-1) \\ &+ (Y(n-1) - Y(n-2))^t T (Y(n-1) - Y(n-2))] \end{aligned} \quad (18)$$

instead of (8). Then, we can obtain a new control law

$$Y(n) = G\ Z(n) + H\ Y(n-1) \quad (19)$$

(Ohtsu, Horigome and Kitagawa(1979)). This performance function plays important role in reducing the mechanical frictional loss in an actuator.

6. Main Engine Control System

6.1 COMPUTER CONTROLLED GOVERNOR SYSTEM

A marine main engine certainly furnishes with some governor system in order to regulate fluctuations of propeller revolution, because it is essentially unstable in the condition without it. Nowadays, an electronic governor system has become popular, instead of the conventional mechanical governor.

Aiming at development of an entirely new governor, we designed a computer controlled system for a direct digital controlling of the main engine in the replaced new training ship SHIOJI MARU III. The principal dimensions of the main engine are shown in Table 4. Especially, we should take account of that this ship equips with a controllable pitch propeller(CPP) . Fig.12 shows the construction of the experimental system. Switching to the computer mode in Fig.12, the right of controlling the main engine completely shifts to the computer and the control signal from the computer is given to the actuator to move the fuel oil pump rack bar.

Engine Type	Middle type 4 cycle Diesel Engine
Normal Output	1,200 PS
Nr.of Cylinder	6
Normal rpm	700 rpm
Propeller	CPP

Table 4 Principal dimensions of the main engine

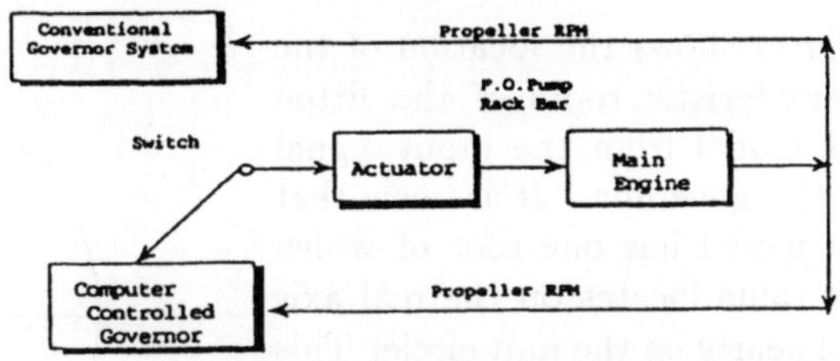

Fig.12 Direct digital system for governor's test

6.2 IDENTIFICATION USING AN AUTO REGRESSIVE MODEL

6.2.1 THE MULTIVARIATE AUTO REGRESSIVE MODEL FOR IDENTIFICATION

At first, we analyze the various dynamical properties of the system, using the auto regressive model of the governor - propeller revolution system in the marine diesel engine of Shioji Maru III, fitted to the records gained at sea (Ishizuka,Ohtsu and Hotta(1991)).

It is necessary that the optimal input for such statistical identification is also as random as possible, as stated before. However, a complete random input inevitably gives large causes to plunge the main engine into dangerous state.

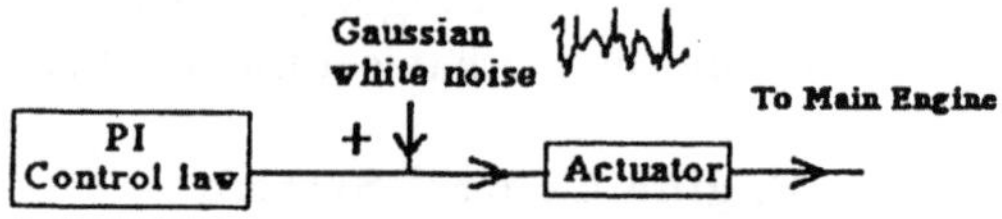

Fig.13 PI plus random signal

For this reason, we input the signal which is added appropriate amount of white noise to the original proportional and integral (PI) control signal to cope with large fluctuations of propeller revolution (Fig.13). We observed the records which were composed of the rate of propeller revolution (rpm) and the ordered signal given by the computer to the governor actuator under the above conditions. The data length were 10,000 points sampled at 0.05 second.

We fitted a multivariate auto regressive model

$$X(n) = \sum_{i=1}^{M} A(i)\ X(n-i) + U(n), \tag{20}$$

to the observed data, using the minimum AIC procedure, in which the state vector X(n) consists of propeller revolution and the ordered signal from the computer to the governor's actuator. The fitted model had 6-th order.

Fig.14 shows the location of the characteristic roots of the fitted AR model from the input signal to the governor. It is seen that the model has one root of which the value locates on the real axis and nearly on the unit circle. This implies that the system has $1/s$ pole (i.e. an integral component) in continuous domain. This is the reason why the main engine governor is unstable and hence it is necessary that the marine engine system furnishes with some governor system.

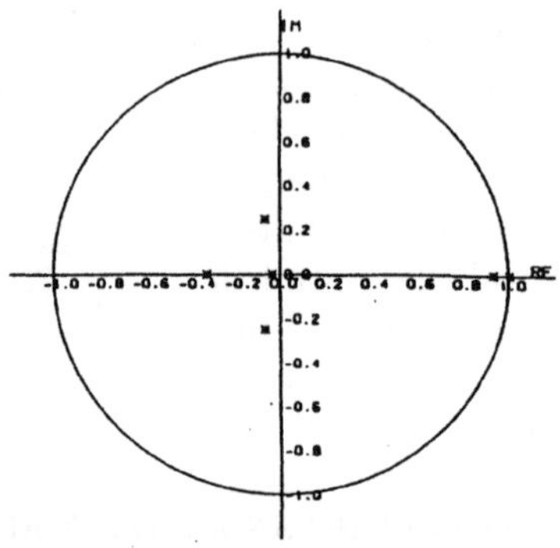

Fig.14 Pole location of the AR model

6.2.2 FREQUENCY RESPONSE FUNCTION OF THE GOVERNOR-PROPELLER REVOLUTION SYSTEM

Using the fitted auto regressive model, we obtain some useful tools for analysis of the system (Akaike(1988)). Fig.15 shows the frequency response function from the input signal in the governor to propeller revolution, illustrated by Bode plot style in which the solid line denotes the amplitude and the dot line, the phase shift. It is easily detected that the system has a low pass characteristic. However in the high frequency domain, the frequency response function associates with vibrations. The last tendency might be due to a nonlinear property such as a time delay element in the system.

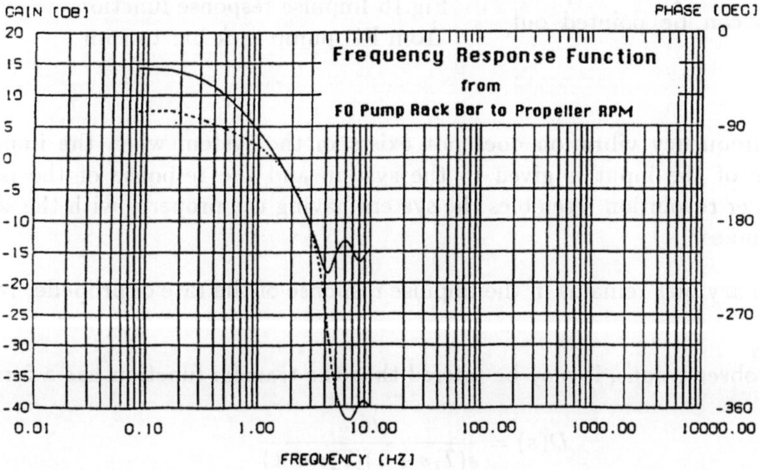

Fig.15 Frequency response function from FO pump rack bar to propeller rpm

6.2.3 IMPULSE RESPONSE FUNCTION OF THE GOVERNOR-PROPELLER REVOLUTION SYSTEM

Let us consider that the variables in the feedback system with k dimensional state vector whose element is $x_i(n)$, connect with each other in the relation,

$$x_i(n) = \sum_{j=1}^{k}\sum_{m=1}^{M} h_{ij}(m)\ x_j(n-m) + u_i(n), i = 1, 2, ..., k \tag{21}$$

where h_{ij} is called impulse response function, we assume that $h_{ii} = 0$, but the $u_i(n)$ is generally a colored noise. This model is called impulse response model. The unbiased estimate of h_{ij} is then given by

$$\begin{aligned} h_{ij}(1) &= A_{ij}(1) \\ h_{ij}(m) &= A_{ij}(m) + \sum_{l=1}^{m-1} A_{ii}(l)h_{ij}(m-l), (m = 2, 3, ...) \end{aligned} \tag{22}$$

using the fitted auto regressive coefficients $A_{ij}(m)$ (Akaike(1968)). Fig.16 shows the impulse response function from the governor to propeller revolution. Observed the shape of this impulse response function, it can be pointed out that;

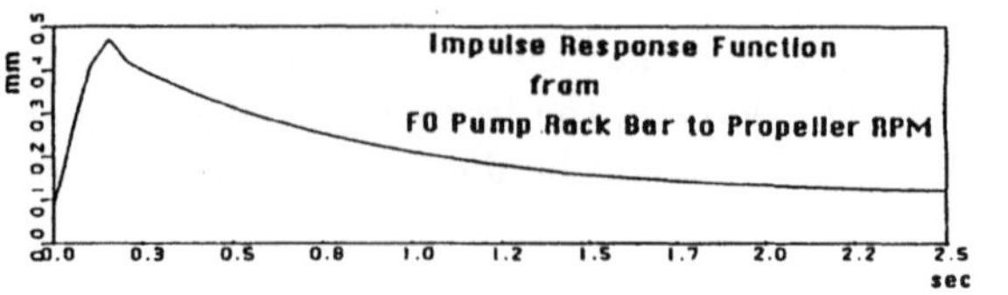

Fig.16 Impulse response function from FO pump rack bar to rpm

1. High frequency vibration does not exists in the system when the impulsive change of the input is given to the system and the response of the rate of propeller revolution resembles the system having the property with the second order poles.

2. Stationary bias remains in the impulse response of the rate of propeller revolution.

From these observations, it may be infered that the transfer function has a form as,

$$D(s) = \frac{K}{s(T_1 s + 1)(T_2 s + 1)} \tag{23}$$

or

$$D(s) = e^{-Ls} \frac{K}{s(T_1 s + 1)(T_2 s + 1)} \tag{24}$$

in continuous domain, where L denotes a time delay element.

6.3 OPTIMAL CONTROL OF MARINE ENGINE

The main aim of the conventional governor has been put on regulating only fluctuations of propeller revolution from the required one. However, in rough sea, marine engineers might require, besides it, that it furnishes with the ability to regulate also propeller torque, taking care of harmful influence such as torque rich, propeller racing and turbo charger's surging of main engine. Considering these requirements, we developed two types of main engine governor system (Ishizuka,Ohtsu and Hotta(1992)), using the designing procedure described in Chapter 3. The first system aims at regulating only fluctuations of propeller revolution from the required value. This will be called Type I governor. The second one aims at regulating not only propeller revolution but also propeller shaft torque. This will be called Type II governor.

6.4 ACTUAL SEA EXPERIMENTS

6.4.1 THE PROCEDURE OF THE ACTUAL EXPERIMENTS

We adopted SHIOJI MARU III quoted in the previous chapter for the actual experiments. And we fixed the sampling period on 0.1 sec, taking account of the speed of computer system. The elements of the records consist of the rate of propeller revolution, propeller torque and the ordered signal to the actuator, whose signals were generated under the condition of the main engine regulated with the PI plus white noise control law. We carried out the actual experiments according to the following procedure;

1. Switch from the ship's governor to a well tuned digital PI control system, for a moment,

2. After the fluctuations of propeller revolution is enough settled, switch to the optimal control system .

6.4.2 TYPE I GOVERNOR

In the model (1), the Type I governor consists of the following variables;

[1] The state vector, X(n): the rate of propeller revolution (rpm).

[2] The control vector, Y(n): the ordered signal to the actuator.

The minimum AIC procedure yielded the auto regressive model having 6-th order to the data for identification.

Table 5 shows the statistics of the records in the full scale experiments with the Type I governor having various gains and that with the ship's one immediately carried out after the former ones. Fig.17 demonstrates the relations between the gains and the variances of both variables. The time length of each experiment has about 5 or 6 minutes, respectively. The wind scale was the 2nd class in Beaufort scale and the ship received it from the port beam direction during the experiment. According to the table, the following results were obtained;

1. The variances of fluctuations of the rate of propeller revolution in the optimal governor are smaller than those in the ship's one except for in the case of the optimal one with the weakest gain,

2. The variances of fluctuations of the fuel oil rack bar position in the optimal governor are also very smaller than those in the case with the ship's one.

Gain's Name	Q/R	Variance of RPM	Variance of Rack Bar
A0001	0.001 / 1.0	7.1952	0.01963
A0005	0.005 / 1.0	2.2512	0.02541
A001	0.010 / 1.0	1.2730	0.07563
Ship's Governor	-	4.0120	0.171

Table 5 Performance of Type I governor and ship's one

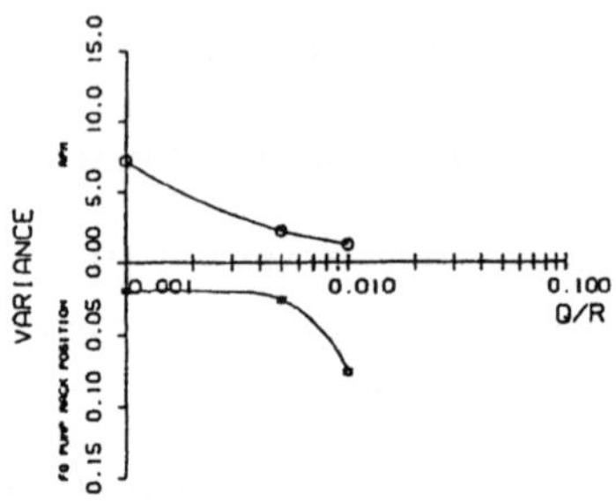

Fig.17 Q/R and the variance of rpm, rack position

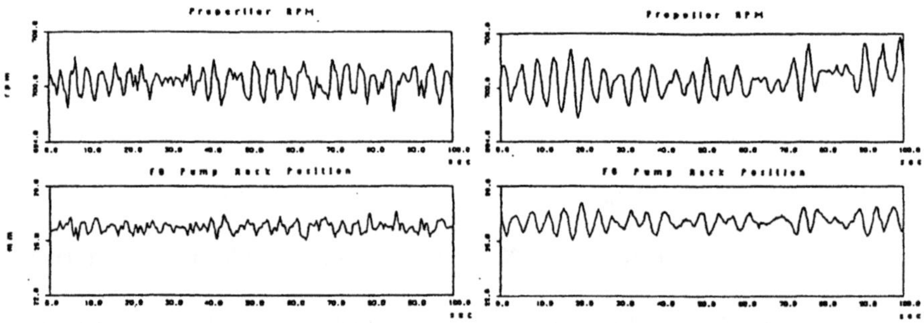

[A] Type I governor [B] Ship's governor

Fig.18 Records of Type I gov.[A] and ship's one[B] in the full scale test (Upper;propeller rpm, lower; rack position)

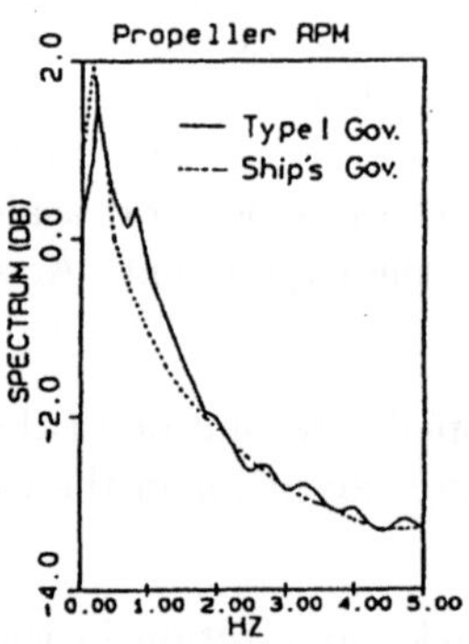

Fig.19 Power spectra of propeller rpm

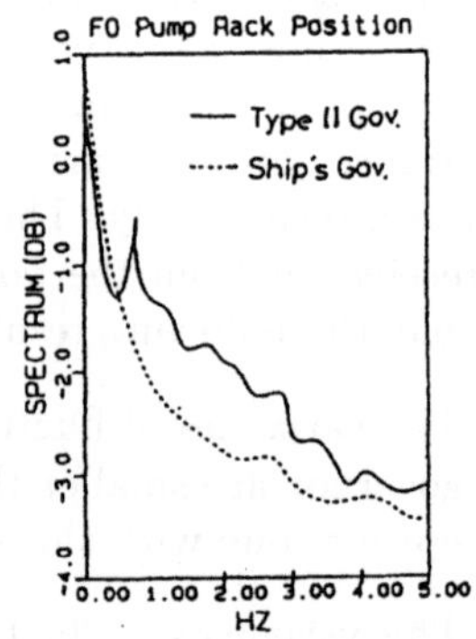

Fig.20 Power spectra of rack position

Fig.18 shows typical examples of the time histories of the rate of propeller revolution and the fuel oil rack position in the optimal governor ($Q/R = 0.001/1$) and those in the ship's one performed soon after the first one, respectively. Fig.19 and Fig.20 show the power spectra of the rate of propeller revolution and the fuel oil pump rack bar position in the both cases, respectively. We can detect that the long period vibration of propeller revolution is eliminated in the Type I governor. However, in the high frequency domain around about 1 Hz, we find a needless peak of spectrum. It is possible to eliminate this peak by introducing the new performance function (18) stated in Chapter 5. Up to now, we have repeated many experiments of this type of governor. And this governor has experienced the trials up to the 4th class of wind force in Beaufort scale.

6.4.3 TYPE II GOVERNOR

The Type II governor consists of the following variables ;

[1] The state vector, X(n) : the rate of propeller revolution and propeller torque.

[2] The control vector, Y(n): the ordered signal to the actuator

The minimum AIC procedure yielded the auto regressive model having 9-th order to the data for identification.

Table 6 shows the statistics of records of the three variables with the type II governor and the ship's one in the full scale experiments. The wind force was the 3rd class in Beaufort scale and the ship navigated in oblique sea condition.

		Type II Governor	Ship's Governor
RPM	maximum	703.0640	702.1480
	minimum	690.8570	690.8570
	mean	697.2564	696.5109
	variance	4.2182	5.1283
TORQ	maximum	2.5880	2.5670
	minimum	2.2170	2.1690
	mean	2.3927	2.4016
	variance	0.0025	0.0032
RACK	maximum	27.3560	26.7610
	minimum	23.8620	24.1060
	mean	25.5485	25.6383
	variance	0.2980	0.2599

Table 6 Statistics of the full scale tests with Type II and ship's governor

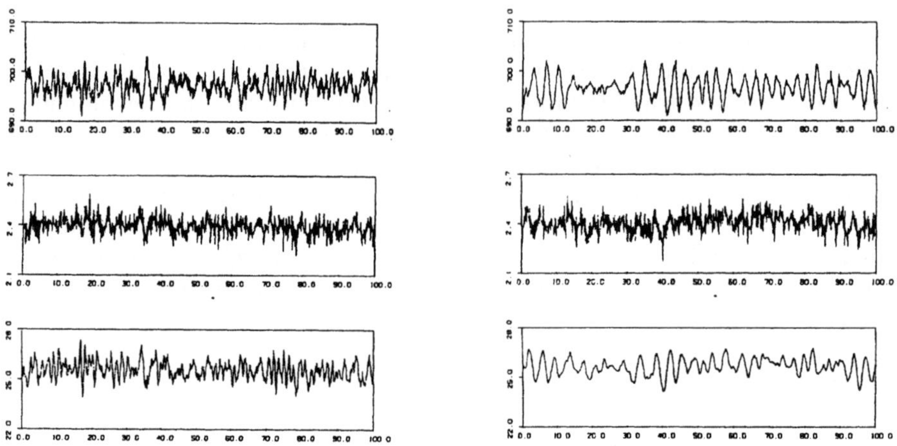

[A] Type II governor [B] Ship's governor

Fig.21 Records of Type II gov.[A] and ship's one[B] in the full scale test (Upper;propeller rpm, middle;propeller torque, lower; rack position)

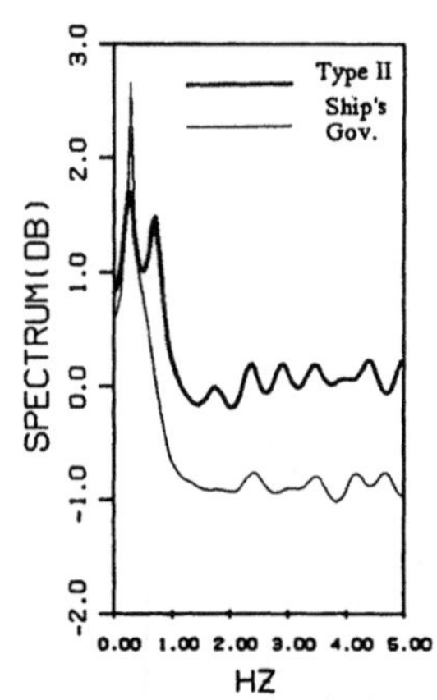

Fig.22 Power spectra of propeller rpm

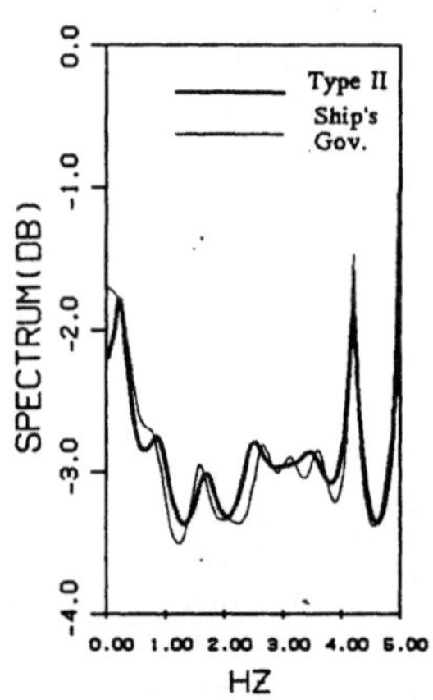

Fig.23 Power spectra of torque

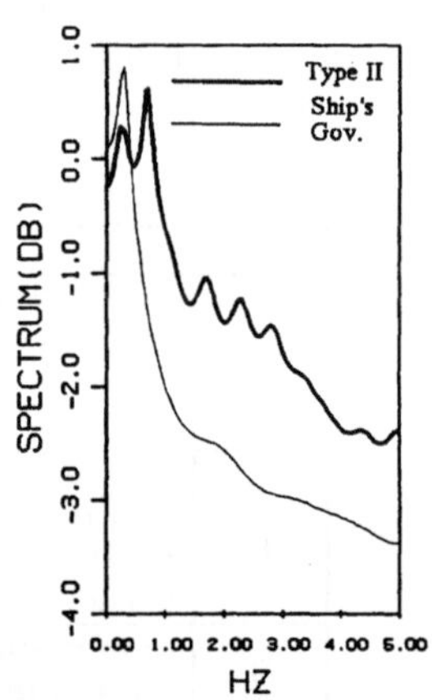

Fig.24 Power spectra of rack position

Fig.21 demonstrates typical examples of the time histories of the rate of propeller revolution, propeller torque and the fuel oil rack position in controlling with the optimal governor and those with the ship's governor. According to the table and the figures, it is shown that:

1. The variance of the rate of propeller revolution in the optimal governor is about 20% smaller than that of the ship's one, notwithstanding that the system must newly regulate the propeller torque in addition to the propeller rpm,

2. The variance of propeller torque in the optimal governor also reduces to about 20% smaller than that of the ship's one ,

3. The variance of necessary movement of the fuel oil pump rack position to realize the above performance slightly increases in the optimal governor than that in the ship's one.

However, we must notice that the motions of the propeller rpm and the rack position associate with harmful high frequency wave elements in the Type II governor. It might be necessary for the system to supplement some filtering facility to improve this deficiency or to introduce the new criterion like (18).

Fig.22 shows the comparisons between the power spectra of the rate of propeller revolution in controlling with the type II governor and that with the ship's governor. Fig.23 and Fig.24 show the comparisons between the power spectra of propeller torque and the fuel oil rack position with both the system, respectively. We can see that the power spectrum of propeller torque with the Type II governor reduces in low frequency domain, comparing with the ship's one.

7. A Unified Main Engine Control System

Fig.25 shows the relative noise contribution function (Akaike (1968)) to the rate of propeller revolution from other ship's motions (Nakatani,Ohtsu and Horigome(1988)). In this figure, we might find the following three causes of the fluctuations of propeller revolution.

The first one is the disturbances induced in the main engine itself. This disturbances bring out generally fluctuations with high frequency mode, as seen in higher frequency domain than 0.7Hz (RPM area in the figure). But it can not be observed.

The second one is the relative vertical displacement of propeller's position with the sea surface, which is mainly induced by ship's pitching and/or heaving motions. This disturbance brings out fluctuations with the lower frequncy mode as seen in 0.3 to 0.6 Hz in the figure and is more dominant than the first one. These ship's motions are measurable and predictable if some measurement instrument is installed and the appropriate model is obtained. Especially in rough sea, strong ship's vertical motion induces the excess of propeller torque shaft (torque rich), propeller racing and turbo charger's surging. The conventional governor does not always provide the ability to cope with such a strong disturbance.

The last one is the rudder motion as seen around 0.05 Hz. The stream field around a rudder always changes, associating with the rudder motion. This motion may be fundamentally observable and predictable, if we can design a unified computer controlled system connecting the governor system with an autopilot one.

The governor system which was developed in the last chapter, is rather designed so as to regulate the fluctuations of propeller revolution and/or propeller torque in the high frequency domain. However, our new governor system must also aim at regulating the fluctuations of them in the lower frequency domain induced by ship's motions and rudder one.

The possibility to realize such a system is shown in Fig.26. This system is composed of double feedback loops. The below loop sends the optimal control signal to an actuator of the governor in order to regulate directly fluctuations of the rate of propeller revolution and/or propeller torque. On the other hand, the above loop generates the optimal reference signal for the rate of propeller revolution and/or propeller torque to cope with the fluctuations of them in the low frequency domain, induced by ship's motions and rudder motion.

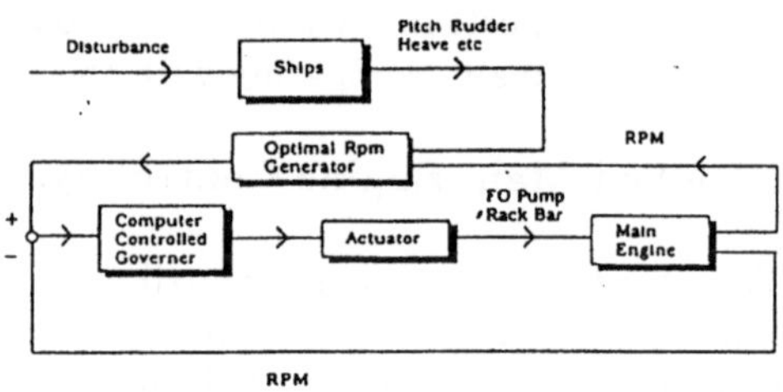

Fig.26 Unified control system of marine system

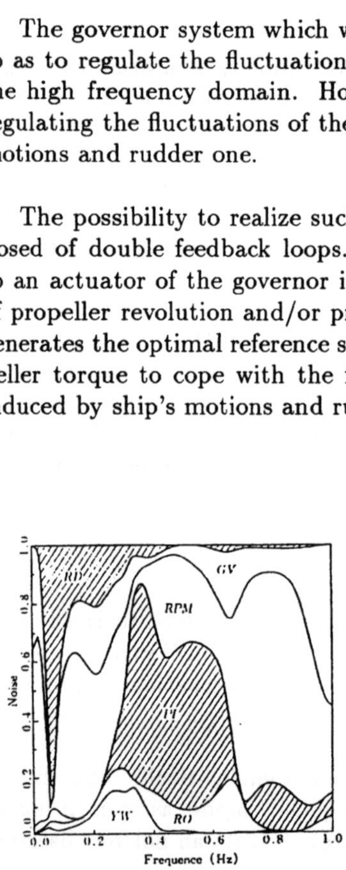

Fig.25 Noise contribution function of ship's motions to propeller rpm (RPM:propeller rpm,YW:yaw,RO:roll,PI:pitch,GV:governor and RD:rudder)

Thus, we reach a unified governor control system taking account of the relevant ship's motions which are observed through appropriate sensors and the rudder motion obtained through an autopilot system, for example, the AR autopilot stated here.

ACKNOWLEDGEMENTS

The authors express their hearty thanks to Dr.Akaike,H. of the Institute of Mathematical Statistics and Dr.Yamanouchi,Y. of the Mitsui Zosen Akishima Lab. for their warm encouragement. Their sincere thanks are also due to the colleagues of the Tokyo University of Mercantile Marine,Dr.Horigome,M., Mr.Hara,M. and the successive captains and the crew of training ship Shioji Maru II and III, especially Chief Eng.Hotta,T.. They also devote their thanks to Mr.Ishizuka,M. of the NYK Line Co.Ltd and Dr.Oda,H. in the Mitsui Zosen Akishima Lab.. Thanks are also due to their colleagues of the Institute of Mathematical Statistics.

REFERENCES

Akaike,H.(1968). On the Use of a Linear Model for the Identification of Feedback Systems. *Ann.Inst.Statis.Math.,* Vol.20,pp.425-439.

Akaike,H.(1971). Autoregressive Model Fitting for Control. *Ann. Inst. Statis Math.* , Vol.23, pp.163-180.

Akaike,H.(1974). A New Look at the Statistical Model Identification. *IEEE, Trans., Automa. Contr.* AC-19, pp.716-723.

Akaike,H. and Nakagawa,H.(1988). *Statistical Analysis and Control of Dynamic Systems.* Kluwer Academic Publishers.

Åström,K.J. and Wittenmark,B.(1984). *Computer Controlled Systems: Theory and Design.* Prentice-Hall,Englewood Cliffs,NJ.

Cowley,W.E.(1974). Development of an Autopilot to Control Yaw and Roll. *The Naval Atchitect.*

Ishizuka,M.,Ohtsu,K.,Hotta,T. and Horigome,M.(1991).Statistical Identification and Optimal Control of Marine Engine System (Part 1), *Journal of The Society of Naval Architects of Japan,* Vol.170, pp.211-220.

Ishizuka,M.,Ohtsu,K.,Hotta,T. and Horigome,M.(1992). Statistical Identification and Optimal Control of Marine Engine System (Part 2),*Journal of The Society of Naval Architects of Japan,* Vol.171, pp.241-255.

Kroukowsky,K.(1961). *Theory of Seakeeping,* Society of Naval Architects and Marine Engineers, New York.

Minorsky,N.(1922). A Directional Stability of Automatically Steered Bodies, *Trans. Society of Naval Architects and Marine Engineers.*

Nakatani,T.,Ohtsu,K. and Horigome,M.(1988).A Statistical Analysis of Ship's Main Engine's Propeller Revolution and its Optimal Control, *Journal of Japan Institute of Navigation,* Vol.79.

Oda,H.,Ohtsu,K.(1992). A Rudder Roll Stabilization Control System through Multivariate Autoregressive Model, *IFAC Workshop, CAMS'92,* Genova.

Ohtsu,K.,Horigome,H. and Kitagawa,H.(1979). A New Ship's Auto Pilot Design through a Stochastic Model, *Automatica,* Vol.15, No.3, pp.255-268.

Ohtsu,K.,Horigome,H. and Kitagawa,H.(1979). A Robust Autopilot System against the Various Sea Conditions - Noise Adaptive Autopilot - , *Proceed. of 3rd ISSOA Sympo.,* Tokyo.

Ohtsu,K.,Ishizuka,M.,Kitagawa G.,Hotta,T. and Horigome,M.(1989). An Optimal Marine Engine Control Using Multi-Variate Auto-Regressive Model, *IFAC Workshop, CAMS '89,* Denmark.

Ohtsu,K. and Ishizuka,M.(1992). Statistical Identification and Optimal Control of Marine Engine, *IFAC Workshop, CAMS'92,* Genova.

Ootomo,T.Nakagawa,and Akaike,H.(1972). Statistical Approach to Computer Control of Cement Rotary Kiln, *Automatica,* Vol.8, No.5, pp.35-48.

Ozaki,T. and Tong,H.(1975). On the Fitting of Non-Stationary Autoregressive Models in Time Series Analysis, *Proc. 8th Howaii Conference on System Science.*

STATISTICAL MODELS FOR EARTHQUAKE OCCURRENCE: CLUSTERS, CYCLES AND CHARACTERISTIC EARTHQUAKES

D. VERE-JONES
Institute of Statistics and Operations Research
Victoria University of Wellington
PO Box 600, Wellington, New Zealand

1. Introduction

This paper reviews different approaches to modelling two key but opposing features of earthquake data, their tendency towards clustering and their tendency towards regular or periodic behaviour.

In recent years great efforts have been made to produce better information about the physical processes underlying earthquake occurrence. Extensive and sophisticated instrumentation; new lines of geological and historical research; global coverage of major and even moderate-sized earthquakes; all these and other efforts are providing great quantities of data for analysis and interpretation. At the same time public and political interest in the problems of earthquake prediction have fostered a new concern to express the results of this research in the language of probabilities and risk.

Despite this great activity, rigorously formulated and widely accepted stochastic models are few and far between. The subject remains obdurately resistant to simple explanations.

One reason for this situation is the extraordinary difficulty of directly observing the process of earthquake genesis. Models can be tested at best indirectly, in terms of their ability to explain and predict the effects which can be observed on the earth's surface.

H. Bozdogan (ed.), Proceedings of the First US/Japan Conference on the Frontiers of Statistical Modeling: An Informational Approach, 105–136.

A second reason, the importance of which is only slowly emerging, is the tightly interacting, near critical character of the whole family of processes taking place in the earth's crust. Analogies between the earth and a living organism, between earthquake prediction and medical prognosis, are helpful not only to the scientific journalist but also to the statistician looking for methodologies developed elsewhere that may be applicable to earthquake risk. The traditional models of stationary processes, with limited, short range dependence, are challenged by attempts to bring into the statistical camp such concepts as fractals, chaos and self-organizing criticality, where long-term dependence and self-similarity are characteristic features. It has never been a more exciting field for a statistician to work in, but it is still not one where progress can be made easily.

The present paper attempts to compare and review some of these contrasting viewpoints. It contains four further sections. The first two provide a brief synopsis of background information concerning earthquake occurrence (Section 2) and point process modelling (Section 3). The second two review some specific models for aspects of clustering (Section 4) and periodic or quasi-periodic behaviour (Section 5).

If I may conclude this introduction with some remarks of a more personal character, I would like to say first how privileged I feel to be able to contribute to this conference celebrating Professor Akaike's sixty-fifth birthday. This is not only because of his many outstandingly perceptive contributions to statistics and stochastic modelling, but also because his invitation to visit the Institute of Statistical Mathematics in 1976, and the continuing links and contacts that have come from that visit, have been of major importance in my own work and personal life. Indeed, the opportunity of working with some of his able young colleagues was not only a major personal stimulus, but has led, in their hands and with his guidance, to a whole train of new research, ably expounded in this meeting by Yosi Ogata, Tohru Ozaki and others. For this opportunity I remain extraordinarily grateful.

2. Earthquake Data

We shall be concerned in this paper almost exclusively with the modelling of earthquake catalogues, the lists of occurrence times, epicentres (latitude and longitude), magnitudes, depths, and various further characteristics routinely compiled and published by national agencies, and also on a world-wide basis by organizations such as the International Seismological Data Centre in the UK, and the US Geological Survey.

It may be worthwhile emphasizing that these are by no means the seismologists' primary data, but are reconstructed through a difficult and often uncertain process of inversion from primary measurements on earthquake waves. However, we shall leave aside, for reasons of space, any detailed discussion of how the earthquake parameters are obtained. Likewise, ancillary measurements of fault movements, stress and strain, perturbations of the earth's magnetic and electric fields etc., while undoubtedly relevant to the process of earthquake occurrence, will be left aside, partly for lack of space, but even more so because the presently existing data is too piecemeal for them to be used easily in the statistical analysis of a catalogue.

Even within the realm of catalogue data there are major limitations. The value of a catalogue for meaningful statistical analysis is crucially linked to its completeness and homogeneity in time. It must refer to a well-defined and constant geographical region, to a specified time interval, and to a defined lower magnitude threshold. Within these boundaries, the catalogue must be complete, and recorded in a consistent manner, otherwise physical features become confounded with artefacts of the observation process.

These are stringent requirements, and reality sometimes falls short of the ideal. To the extent that it does so, the conclusions and interpretation of any statistical analysis must be treated with considerable caution.

Of the parameters with which we shall be concerned in detail, the origin time (conventionally reckoned as the time of first motion of the earthquake) is determined with the greatest relative precision, usually at least to 1 part in 10^6 of the observation period. Epicentral latitude and longitude (location of first motion, projected upwards onto the earth's surface) are typically determined to a precision of 1 part in 10^3 or somewhat less; uncertainties in the travel times, linked to incomplete knowledge of the velocity structure of the rocks through which the earthquake waves travel, form the main limiting factor. Depth is the least well-determined of the three spatial coordinates, and is particularly sensitive to errors in travel time estimates. Since most damaging earthquakes are relatively shallow (typically 5-15 km in depth) it plays a lesser role in most discussions of risk.

Magnitude is a major source of difficulty. Several magnitude scales are in use, none of which is entirely satisfactory. All scales are crudely related to the logarithm of the energy release, a typical relation (for Richter's original magnitude scale) being

$$\log_{10}E = 1.5M + \text{const.} \tag{2.1}$$

Most instrumental catalogues list events with $M \geq 5$ reasonably consistently over the last 30-50 years. Beyond that, and particularly if historical (non-instrumental) data is used, estimation of magnitudes and evaluation of catalogue completeness is a specialised art, requiring historical and engineering as well as seismological knowledge. Similar comments apply to the use of geological data. Most historical catalogues aim at completeness for $M \geq 6$ or $M \geq 6.5$ in the region under consideration. Accuracies of instrumental magnitudes are at best in the range of ± 0.1, with discrepancies of ± 0.2 or more not uncommon in historical data.

Despite its manifest shortcomings the magnitude remains an essential control variable for statistical purposes. Its replacement by seismic moment, having a somewhat sharper definition in terms of the physical model of fault mechanism, has long been advocated by the seismologists. From a statistical point of view, the overriding consideration is to have consistently available an agreed measure of the size of an earthquake; the shortcomings of the measure itself are of somewhat secondary importance, granted that in any case its measurement will be subject to substantial errors, and that any notion of "size", on a 1-dimensional scale, must be an inadequate description of a complex phenomenon.

Additional variables relate to the orientation of the fault and direction of slipping; the lengths of fault breaks at the surface (if any) and the extent of their upward or lateral movement; the duration and character of shaking; long-term deformations of the ground following the earthquake; associated phenomena including damage and reports of unusual incidents. All are potential grist to the statistician's mill, although many fall foul of lack of completeness or consistent measurement. Again there will be space here to refer to such variables only in passing.

There is, of course, a vast literature on different aspects of seismology. Gutenberg and Richter (1949) is the classical introduction to the study of seismicity. Bolt (1988) and Mogi (1985) may be recommended as non-technical introductions. Statistical problems are reviewed in Vere-Jones and Smith (1981).

3. Modelling Earthquake Occurrence

In this section we describe some general features of models for catalogue data.

3.1 POINT PROCESS MODELS

Insofar as a catalogue comprises a list of events in time sequence, it is naturally regarded as a realization of a point process, a subject on which there now exists a considerable literature: see, for example, Cox and Lewis (1966) and Karr (1986) for statistical aspects, Cox and Isham (1980) or Daley and Vere-Jones (1988) for modelling and theoretical aspects respectively; all give extensive further references.

Early applications to earthquake data (e.g. Vere-Jones and Davies (1966)) concentrated on matching descriptive aspects of model and data in a time series context, assuming stationarity. Interest was focused on second order aspects such as the second-order intensity

$$m(\tau)\,d\tau \;=\; E[dN(t+\tau) \mid dN(t)=1], \tag{3.1}$$

the covariance density

$$c(\tau)\,d\tau\,dt \;=\; \mathrm{Cov}[dN(t+\tau)\,dN(t)], \tag{3.2}$$

and the point process spectrum (Bartlett spectrum in the terminology of Daley and Vere-Jones (1988))

$$f(\lambda) \;=\; \frac{1}{2\pi}\left\{m + \int_{-\infty}^{\infty} e^{i\lambda\tau}\,c(\tau)\,d\tau\right\} \tag{3.3}$$

(see the references quoted for further discussion and explanation).

A major development of the early 1970s, which opened the way to a more satisfactory and systematic development of estimation and model-testing procedures for point processes, was the recognition of the crucial role played by the "conditional intensity function". It holds the key to likelihood analysis, simulation and forecasting. For a point process in time this function takes the form

$$\lambda(t)\,dt \;=\; E[dN(t) \mid \mathcal{H}_t] \tag{3.4}$$

where $\mathcal{H}_t$ is short for the history of the process up to time t, meaning mathematically the σ-algebra of events defined by observations over the period (0,t). The likelihood for a set of observations $\{t_i\}$ in (0,T) can then be written (for brevity suppressing the explicit dependence of $\lambda(t)$ on $\mathcal{H}_t$) as

$$L(t_1,\ldots,t_{N(T)}) = \Big[\prod_1^{N(T)} \lambda(t_i)\Big] \exp\Big\{-\int_0^T \lambda(u)\,du\Big\}. \tag{3.5}$$

This is the same formal expression as the likelihood for a non-stationary Poisson process with time-varying intensity $\lambda(t)$. In effect, the multiplication rule for independent events is replaced by the corresponding rule for a chain of conditional probabilities.

Focussing attention on the conditional intensity function $\lambda(t)$ forces the modeller to concentrate on how the current risk $\lambda(t)$ is affected by past information, which is also the key to prediction and simulation methods. To be more precise, $\lambda(t)$ plays the role of a conditional hazard function for the time to the next event. Denoting this by T^* we can write

$$P\{T^* > \tau\} = \exp\Big\{-\int_0^\tau \lambda^*(u)\,du\Big\} \tag{3.6}$$

where, denoting the present by t, $\lambda^*(u)$ is obtained by extrapolating $\lambda(t)$ forward to $\lambda(t+u)$ on the assumption that no additional event occurs in (t, t+u). The same idea can be used to iteratively simulate a sequence of events with a prescribed form of conditional hazard (see Lewis and Shedler (1976), Ogata (1981), and, for an earthquake example worked out in detail, Wang et al. (1991)). Evidently, there will be major benefits in both analysis and interpretation when the conditional intensity can be written down explicitly, so that recent research has concentrated on classes of processes for which this is possible.

Once the likelihood (3.5) can be written down explicitly, routine though sometimes technically demanding optimization procedures can be used to obtain estimates of both parameters and (through evaluation of the Hessian) their standard deviations. Within this likelihood framework, and despite some limitations, the A.I.C. provides an extraordinarily useful tool for comparing models of varying degrees of complexity. Vere-Jones (1978) sets a scenario for this type of approach in the earthquake context. Many examples worked through in detail can be found in papers by Ogata, Ozaki, Zheng, and others (see Sections 4 and 5 below and in the present volume the tutorial paper by Ogata). General discussions of inference in this context are given by Karr (1986) and Kutoyants (1984).

It is important that this formalism extends to marked point processes in time. This is ultimately the most appropriate context for earthquake studies, the mark comprising at least the magnitude, probably the location (epicentral latitude and longitude), and possibly additional variables. Denoting the mark by a vector $\underset{\sim}{x}$, the conditional intensity becomes

$$\lambda(t, \underset{\sim}{x})\,dt\,|d\underset{\sim}{x}| = E[N(dt \times d\underset{\sim}{x}) \mid \mathcal{H}_t] \tag{3.7}$$

where $dt \times d\underset{\sim}{x}$ is shorthand for an infinitesimal region in time-mark space, dt and $|d\underset{\sim}{x}|$ are used for the Lebesgue measure of such regions (assuming densities are taken with respect to this measure), and the past $\mathcal{H}_t$ now includes information about the marks as well as the occurrence times. The likelihood (3.5) extends to

$$L((t_1, \underset{\sim}{x}_1), (t_2, \underset{\sim}{x}_2), \ldots, t_{N(T)}, \underset{\sim}{x}_{N(T)}) \tag{3.8}$$

$$= \left[\prod_1^{N(T)} \lambda(t_i, \underset{\sim}{x}_i)\right] \exp\left\{-\int_0^T \int_{\mathcal{X}} \lambda(u, \underset{\sim}{y})\, du\, d\underset{\sim}{y}\right\}$$

where $\mathcal{X}$ denotes the mark space. Multivariate point processes can be treated as the special case where one component of the mark is discrete. The formal simplicity of these extensions does not mean that numerous technical problems may not arise, with convergence of the likelihood optimization routines, for example, or with boundary effects, but they do provide a very general and powerful formalism for handling a wide range of models.

One context to which the formalism does not extend so simply is that of pure spatial processes, where there is no time-like variable to order the conditional expectations. Only for the special case of a Poisson process (non-homogeneous in general) can the likelihood be written in the form (3.5). Despite these difficulties, there have been considerable advances in the treatment of pure spatial processes also, as witness recent books by Ripley (1981), Diggle (1983) and others. Several papers by Ogata, Tanemura and others (see in particular Ogata and Tanemura (1984), as well as Ogata's paper in the current collection) illustrate what can be done in the earthquake context.

Another extension is to problems of smoothing and interpolation, whether in space, time, or space and time jointly. Here there is a close connection with work on density estimation. Vere-Jones (1992; see also §4.3 below) provides a recent review of this direction in the earthquake context.

3.2 FRACTALS AND SELF-SIMILAR PROCESSES

The point process framework described in the previous section was developed for the traditional paradigm of stationary processes with short-term dependence. In recent years a quite different paradigm has come into prominence, particularly following the publication of Mandelbrot's books on fractals (Mandelbrot (1973, 1982)). In these Mandelbrot draws attention to the wide range of natural processes exhibiting different types of scale invariance properties, in particular processes whose trajectories form "fractal sets".

Such a set is defined by Mandelbrot in terms of its Hausdorff dimension. Suppose the set in question is covered with a (minimum) number $N(\Delta)$ of spheres (or other sets of standard shape) with linear dimension Δ. If the limit as $\Delta \to 0$ of the ratio

$$d_\Delta = \log N(\Delta)/|\log \Delta|. \tag{3.9}$$

exists, it is called the "capacity dimension" of the set, and then coincides with the Hausdorff dimension. The Hausdorff dimension itself can be defined more generally, but at the expense of greater technicality.

Falconer (1990) gives a clear and helpful account of these and other definitions of dimension, as of fractal concepts generally.

A set is said to be a fractal if its Hausdorff dimension differs from its ordinary (topological) dimension. A well known example is the Cantor set, which has topological dimension 0 and Hausdorff dimension approximately 0.64.

Many fractal sets are characterized by a "scale invariance" property: their main structural features are preserved on whatever scale the set is examined. From a stochastic modelling viewpoint, the task is to describe, and find methods of fitting and testing, those stochastic processes whose realisations have features of this kind.

This is a task which is still very far from complete. The underlying questions have been blurred in applications by an over-hasty desire to label everything in sight a fractal. For example it is characteristic of many fractal and self-similar processes that they exhibit either power-law correlations or power-law tails in their distributions. There is, however, a vast array of processes with realisations exhibiting either one or both of these features, but not forming a fractal in the strict sense above.

Part of the difficulty is that the definitions, even of self-similarity, are more restrictive than might appear at first sight. For example, it is not possible for a point process (in view of the integer character of the counting process) to be self-similar in the precise sense defined below. Some of Mandelbrot's most interesting examples, such as the "Levy dust", though their realisations are sets of discrete points, are not point processes because they fail to satisfy the basic requirement that the number of points in a bounded set is finite with probability one. A more general theory of random sets may be needed to describe their statistical properties adequately.

If the points are marked, an easier approach is to treat such examples as random measures, defining the measure ξ by

$$\xi(A) = \sum_{i:\, \underset{\sim}{x}_i \in A} M_i \tag{3.10}$$

where M_i denotes the mark and $\underset{\sim}{x}_i$ its location in space, time, or both. Such a random measure can be said to be self-similar if a change of scale in $\underset{\sim}{x}$ can be compensated for by a change of scale for the marks; or more precisely if there exists a constant H such that

$$\xi_r(A) = r^{-H}\xi(rA) \tag{3.11}$$

has the same probability structure as the original random measure ξ.

There are two well-known classes of processes which are self-similar in this sense, or in the closely related sense that the cumulative process has stationary and self-similar increments. The first is the class of *stable processes*, with the complete independence (independent increments in the 1-D case) property that the measures allocated to disjoint sets are independent random variables, and with distributions belonging to the class of stable distributions. The other class comprises the *fractional Brownian motions* and processes simply derived from these. A fractional Brownian motion has Gaussian (short-tailed) increments but long-tailed correlations. The two classes are therefore diametrically opposed: one has independence (zero correlations) and long tails; the other has long-term dependence but short tails. Clearly one may anticipate many processes which have both long-term correlations and long-tailed distributions.

Despite the great interest in the mathematical structure of such processes, there is at present no very well-established approach to the statistical analysis of fractal or self-similar processes, particularly to questions of estimation and testing. Many discussions reduce to a verification of one or other of the power law features already mentioned, from which, with further assumptions, a value for the fractal dimension or scaling parameter H may be deduced. An alternative is to go back to the defining equation (3.9) and attempt to estimate d_Δ, and its limit d as $\Delta \to 0$, from empirical covering studies. Here, among others, recent Russian work on "multifractals" may be mentioned. Multifractals (see Mandelbrot (1988)) are to fractals roughly what a marked point process is to an ordinary point process. In place of counting just the number $N(\Delta)$ of standard sets required to cover the underlying set, each covering set is given a weight related (in the earthquake context) to the number or possibly the energies of the events it contains. Typically, this weight will grow locally as a fractional power of the linear dimension Δ, and the value of this power, or the distribution of such powers if they are not identical in all parts of the set, gives additional information about the structure of the set under study. Geilikhman et al. (1990) give further details; see also §4.3 below.

The interest of such models lies not only in the probabilistic and statistical questions to which they give rise, but also in their implications for the underlying physical processes. Many of the dynamic processes taking place in the earth's crust bear striking resemblances to those taking place in the atmosphere and oceans, notwithstanding the obvious differences in material properties. All involve a transfer of energy from larger to smaller scales. All exhibit the long-term dependence structure first studied for turbulence, as well as long-tailed distributions for energies and related physical variables. They are all part of a dynamic system in delicate equilibrium, close to critical values or "phase transitions" marked by sudden redistribution of energies, such changes forming part of the mechanism of energy transfer between scales. Nearly all exhibit some form of scale invariance or self-similarity.

Such features lie behind what are still largely qualitative attempts to develop a broader framework for the study and modelling of geophysical processes. As an example, Sadovsky, Pisarenko and others (see, for example, Sadovsky et al. (1984)) have developed the concept of a "hierarchical block structure" as a framework for discussing processes in the crust. No matter what scale is chosen, the observer sees a picture consisting of a few large blocks, dominating his field of view, separated by relatively narrow cracks or spaces. On closer examination these spaces themselves are seen to contain a few larger blocks separated by yet narrower gaps, etc. Interest focuses on the reaction of such systems to large-scale external forces. In somewhat similar vein, Knopoff and coauthors (e.g. Burridge and Knopoff (1967)) have examined the behaviour of analogues to an earthquake faulting system in the form of blocks connected by elastic springs and sliding over a rough surface. Both models show features of the phenomenon of "self-organizing criticality" studied by Bak et al. (1990) and others, and exhibit power-law distributions for energies etc, reminiscent of those observed for earthquakes.

Important though such ideas may be for the qualitative understanding of certain aspects of earthquake occurrence, particularly perhaps the fracture process itself, they are not necessarily very helpful to the seismologist or engineer faced with immediate practical demands for risk forecasts or estimates of the expected level of ground shaking. For such purposes more traditional models may be useful for some time to come. Indeed, from a practical point of view there may be little difference. Davies and Harte (1987), for example, have pointed to the virtual impossibility of distinguishing between fractional Brownian motion and a moderate order (5-10 parameters) autoregression on the basis of data sets of several hundred points only. Similarly the more traditional point process models may be quite adequate for developing the smoothing procedures used in studies of regional seismicity and earthquake risk. Finally it must be borne in mind that the fractal models studied to date tend to be extremely simple, idealized examples.

With these general remarks in mind, we turn in the last two sections to some more specific earthquake phenomena and the models that have been used to describe them.

4 . Models for Clustering

At all scales, clustering appears as a striking feature of earthquake occurrence, to the extent that any account of clustering runs the danger of embracing the whole subject of earthquake occurrence. To limit the scope of the discussion in this section we shall consider three aspects only. First we look at cluster *structure*, with some sidewards glancing at the question of whether the clusters have real physical meaning, or are just an artefact of highly irregular but purely random mechanisms. Then we consider two broad types of cluster structures: Poisson cluster models, and fractal models.

4.1 MODELS FOR CLUSTERS : AFTERSHOCKS, FORESHOCKS AND SWARMS

The most noticeable and widely studied examples of clusters are aftershock sequences. From the earliest times it was recognized that major earthquakes were likely to be followed by many further episodes of shaking before the earth again lapsed into quiescence. Such observations have been fully confirmed by instrumental studies. The great majority of large shallow earthquakes are followed by aftershock sequences, which are commonly the subject of special studies in their own right, and listed as such in the catalogues. From the comparative study of many such sequences (see, for example, Utsu (1961) and many subsequent articles) a number of characteristic statistical relationships for aftershocks have emerged, of which the two most important are the following.

(i) Omori's Law: the frequency $\nu(t)$ of events above a specified magnitude threshold decays hyperbolically with time t after the main shock:

$$\nu(t) = b(c+t)^{-\alpha} \qquad (4.1)$$

where b, c are constants and α is commonly close to 1 and almost always with the range $1 \leq \alpha \leq 2$. Heightened activity of small events can persist for many years after the main event.

(ii) Utsu-Seki relationship: the aftershocks fill out a well-defined area (often rather elliptical in shape, with the main event eccentrically placed) which is related to the magnitude of the main shock through an equation of the form

$$\log_{10}A \approx M - 4 \approx \text{const.} + \tfrac{2}{3}\log_{10}E \qquad (4.2)$$

where E is the energy release from the main shock (Utsu and Seki (1954/5)).

In terms of stochastic models, a simple interpretation of equation (4.1) is to regard $\nu(t)$ as a parametric expression for the intensity of a time-varying Poisson process. This follows the original suggestion of Jeffreys (1938), and has been the study of recent detailed studies by Ogata and coauthors (Ogata (1983), Ogata and Shimazaki (1984), and, in the present volume, Ogata (1992)). A feature of these studies is their use of "residual analysis", meaning here the use of a time transformation to study details of the fit to the proposed parametric form. This enables the authors to study features such as secondary aftershock sequences, periods of relative quiescence, etc.

Such well-established results suggest that aftershocks have a clear claim to physical existence as phenomena in their own right. The counter argument is that in the final analysis aftershocks are just earthquakes; as individual events they do not differ in any significant way from other earthquakes. Thus the distinction between "aftershocks" and "background activity" in the later stages of an aftershock sequence is a purely arbitrary one, although some guide may be available in the form of the radiation patterns and seismograms which, for aftershocks, often follow a consistent pattern. A similar situation arises in identifying the "aftershocks" of smaller events.

This counter-viewpoint is reinforced by noting that the relationship (4.2) is not dependent on the absolute value of the magnitude M, but has the nature of a scaling relationship between the variables E and A, which holds for all earthquakes at all scales. Recent studies by Yamashita and Knopoff (1987) suggest that Omori's Law also has a scale invariant character, and can be related to the time intervals required to wear down (perhaps by stress corrosion) residual barriers either on the main fracture surface or between the main fracture and smaller fracture surfaces outside it. If the dimensions of these barriers follow a power law relationship, as would be anticipated on general grounds from distributions of energies, fault lengths, etc., then the distribution of time delays will also follow a power law. A somewhat similar explanation, based on the observation that first passage times in a 1-dimensional Brownian motion follow a stable law of parameter 1/2 (Holtzmark distribution), is offered by Kagan and Knopoff (1984).

The extreme form of this viewpoint is that every earthquake is at one and the same time the product of events which have preceded it, and the producer of events that follow, so that the distinction between main shock and aftershock becomes blurred. A complication is that aftershocks are not the only kinds of clusters which are observed. Foreshocks, usually smaller in number and less structured in their occurrence patterns than the aftershocks, are sometimes associated with major shallow events, while earthquake swarms, which do not show any clear initiating event, are common in volcanic regions and can occur elsewhere. No doubt these phenomena are all in some sense the response to local fluctuations in stress levels, just as the aftershocks themselves can be regarded as a response to the particularly sharp stress concentrations produced in the vicinity of a major event. A comprehensive physical model for clustering is still missing, however.

4.2 POISSON CLUSTER MODELS

We turn next to models which incorporate clusters as part of an overall stochastic model for a regional earthquake catalogue. The earliest such models were simple variants on a stationary Poisson process, for example a "trigger model" (Vere-Jones and Davies (1966)), in which "trigger events", or "cluster centres", occur according to a simple Poisson process and the cluster members are closely distributed about them, perhaps following a local non-stationary Poisson process with intensity of the type (4.1). Attention at that time was focussed on the second order properties, such as the second order intensity (point pair correlation function), the variance time curve, and the point process spectrum, all of which gave useful visual characteristics of the process.

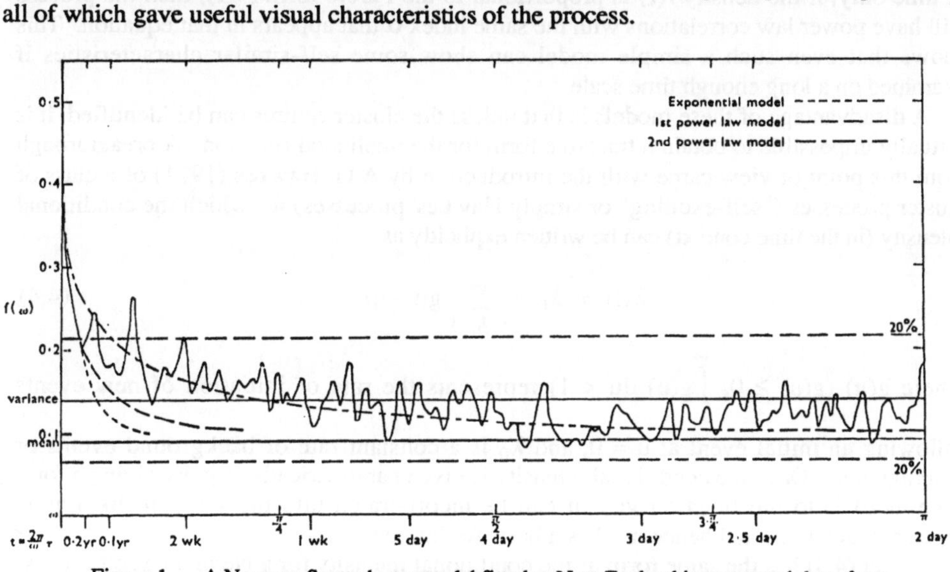

Figure 1. A Neyman-Scott cluster model fitted to New Zealand instrumental data, from Davies and Vere-Jones (1966). The periodogram shown here is better fitted by clusters with power-law frequency decay. No evidence here of periodic effects.

Typically, instrumental catalogues studied in this way show few significant features beyond clustering, although it is important that this is prominent at all ranges.. Initial studies were limited to time characteristics, but were later extended to include studies of space and space-time correlations as well, for example in the pilot study by Vere-Jones (1978b), and Kagan and Knopoff (1980); Kagan (1981a,b) also considered third and fourth order correlations.

Model details for Poisson cluster processes were extensively described in the earlier papers, for example Vere-Jones (1971). A typical result for the Neyman-Scott process (stationary Poisson cluster centres; cluster members i.i.d. about the cluster centre with common d.f. $f(\underset{\sim}{x})$) is for the ratio $r_2(\underset{\sim}{x})$ of the 2nd order intensity to the intensity of a simple Poisson process with the same average rate. This takes the form (see, for example, Daley and Vere-Jones (1989), p.364 et.seq.)

$$r_2(\underset{\sim}{x}) = 1 + \frac{E[N(N-1)]}{E(N)} \int f(\underset{\sim}{x} + \underset{\sim}{u})\, f(\underset{\sim}{u})\, d\underset{\sim}{u} \tag{4.3}$$

where N is the number (in general random) of points in a cluster. The fact that this ratio is greater than unity for all positive $\underset{\sim}{x}$ is an indication of the *overdispersion* of this process relative to the Poisson, i.e. its greater variability.

From this representation it can be seen that for large $\underset{\sim}{x}$ the asymptotic form of the correlations are determined by the tail of the distribution $f(\underset{\sim}{x})$. For example, for a process in time only, if the density f(t) is proportional to the Pareto form (4.1), then the process will have power law correlations with the same index α that appears in that equation. This shows that even such a simple model can show some self-similar characteristics if examined on a long enough time scale.

A disadvantage of these models is that unless the cluster centres can be identified, it is virtually impossible to obtain a tractable form for the likelihood function. A breakthrough from this point of view came with the introduction by A.G. Hawkes (1971) of a class of cluster processes ("self-exciting" or simply Hawkes' processes) for which the conditional intensity (in the time context) can be written explicitly as

$$\lambda(t) = \lambda_0 + \sum_{i:t_i<t} g(t - t_i) \tag{4.4}$$

where $g(u)$ ($g(u) \geq 0$, $\int_0^\infty g(u)\, du < 1$) represents the rate of initiation of new events following an initial event at $u = 0$, and λ_0 is a constant rate of background events or "immigrants". Once the conditional intensity is given parametrically (e.g. by giving a form such as (4.1) to the function $g(\cdot)$) it can be incorporated into the standard likelihood representation (3.5) and hence used as a basis for inference.

In fact (4.4) has the same form as the conditional intensity for a Neyman-Scott process given that the t_i are restricted to points known a priori to be cluster centres. The feature of the Hawkes process is that the sum in (4.4) extends over *all* prior events $\{t_i\}$ That it is still a Poisson cluster process derives from its interpretation (Hawkes and Oakes (1974)) as a branching process with immigration: each "immigrant" sets up its own "family tree" of events initiated by the immigrant and its descendants; the overall process is then the superposition of such family trees, which form the clusters.

The self-exciting feature embodied in (4.4) makes the process an attractive basis for earthquake modelling quite apart from its statistical advantages. It was first applied in the earthquake context by Hawkes and Adamopoulos (1973), but the conditional intensity approach, with many extensions and refinements, was mainly developed by Ozaki, Ogata and other Japanese writers. See Figure 2 for an example. Ogata (1988) gives an interesting comparison between the Hawkes model and the trigger process with known trigger events, and concludes that the Hawkes model is marginally superior in its ability to fit standard catalogues. Further aspects are discussed in Ogata et al. (1982), Ogata and Katsura (1986), Ogata (1989), and Ogata's paper in the present volume, which contains many further references.

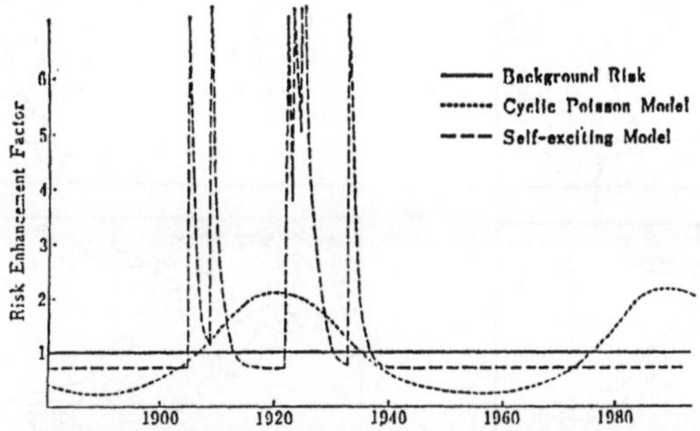

Figure 2. Cyclic Poisson and self-exciting cluster models fitted to Kawasumi's Tokyo data, from Vere-Jones and Ozaki (1982). Both represent considerable improvements over the simple Poisson model. If the cyclic model is tested against a cluster alternative, the significance largely disappears. The graph shows the normalized conditional intensity functions.

Perhaps the most challenging extension of Hawkes' original model is to processes involving spatial as well as time dimensions. A preliminary study of this kind was made by Musmeci and Vere-Jones (1992), using a space-time version of the Hawkes' model with conditional intensity of the form

$$\lambda(\underset{\sim}{x},t) = \lambda_0(\underset{\sim}{x}) + \sum_{t_i<t} g(t-t_i, \underset{\sim}{x}-\underset{\sim}{x}_i \mid M_i) \tag{4.5}$$

where $(t_i, \underset{\sim}{x}_i, M_i)$ are the coordinates of the i-th event in time,

$$g(u, \underset{\sim}{x} \mid M) = Ae^{\beta M} D(u,\underset{\sim}{x})e^{-\lambda u}$$

and $D(u,\underset{\sim}{x})$ is a generalized diffusion kernel

$$D(u, \underset{\sim}{x}) = |u|^{-d}\, k(\underset{\sim}{x}/u).$$

A typical choice for the probability kernel k is the Gaussian form

$$k(\underset{\sim}{y}) = \frac{1}{2\pi\sigma_1\sigma_2} \exp\left\{-\frac{1}{2}\left(\frac{y_1^2}{\sigma_1^2} + \frac{y_2^2}{\sigma_2^2}\right)\right\}.$$

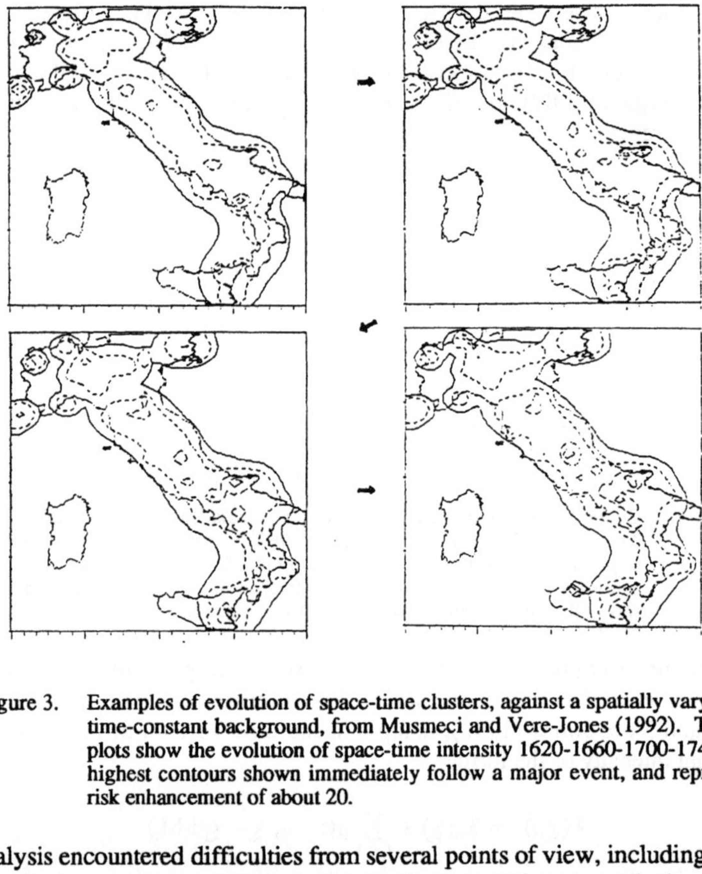

Figure 3. Examples of evolution of space-time clusters, against a spatially varying but time-constant background, from Musmeci and Vere-Jones (1992). The four plots show the evolution of space-time intensity 1620-1660-1700-1740. The highest contours shown immediately follow a major event, and represent a risk enhancement of about 20.

The analysis encountered difficulties from several points of view, including the stability of the numerical procedures, defects in the particular catalogue used, the treatment of boundary effects, and the interpretation of the fitted model, which suggested a supercritical rather than a subcritical branching process. There is considerable scope for further work with such models, including the extent to which they exhibit approximate self-similar behaviour.

4.3 SELF-SIMILAR AND FRACTAL MODELS

The primary evidence for self-similarity in the earthquake process is visual: seismicity maps on many different scales look broadly similar; as do patterns of faults or cracks. In both contexts power laws, which are generally associated with self-similarity, abound, both for correlational and distributional aspects. The outstanding example is the Gutenberg-Richter frequency-magnitude law, which in probabilistic language can be stated in the form

$$P(\text{Mag} > M \mid \text{Mag} > M_0) = e^{-\beta(M-M_0)} = 10^{-b(M-M_0)} \tag{4.6}$$

where b, although varying from region to region and possibly over a given region within time, is often close to unity. Bearing in mind (2.1) this is approximately equivalent to

$$P(\text{Energy} > E \mid E > E_0) = (E/E_0)^{-\alpha} \tag{4.7}$$

where

$$\alpha \approx \frac{2}{3} b \approx \frac{2}{3}.$$

Distributions of the same general form hold for fault lengths, seismic moments, rupture areas, etc. Taken together, the family of such distributions provide rather strong support for the idea that earthquake processes are self-similar in character, for they strongly suggest that the relationships between fault length, energy, moment, etc. are independent of scale.

There is a considerable literature which attempts to explain models (4.6) and (4.7). One of the simplest, first suggested by Saito et al. (1973), and in a more general formulation by Vere-Jones (1976), treats the developing fracture as a branching process in critical or near critical condition, supposing the total energy release to be proportional to the total "progeny". In the critical case, and independently of the form of the offspring distribution, this leads to a version of (4.7) with $\alpha = 1/2$. For a subcritical but close to critical process, this is replaced by the more general form (Vere-Jones (1977))

$$P(\text{Energy} > E) \sim C\, E^{-1/2}\, e^{-\beta E}. \tag{4.8}$$

When β is small, the power law form may dominate over quite an appreciable range of values of E.

In fact distributions of the general form (4.8) have been suggested by Kagan (1991 b, c) as an appropriate 2-parameter alternative for (4.6) if it is desired to accommodate an apparent turn-down in the extreme far tail of the magnitude law. Moreover, his recent studies of seismic moments suggest that the value $\alpha = \frac{1}{2}$ (corresponding roughly to $b = \frac{3}{4}$) may be closer to the empirical values than was at first appreciated (Kagan 1991 b).

Numerous writers have suggested percolation as an alternative model, but the interpretation is still ambiguous – see Molchanov et al. (1986) and Bebbington et al. (1990) for discussions. Nearly all such models illustrate the general principle that processes close to criticality generate the kinds of power law distributions that are frequently observed in geophysics. The difficulty comes in relating the highly idealized concepts which define the mathematical model to any meaningful (and measurable) physical quantities.

In the last few years many studies have sought to establish the fractal character of earthquakes and earthquake faults. One approach is to attempt to verify the definition (3.9) empirically, from measurements on maps. For example Aviles et al. (1987) and Ukubo and Aki (1987) have examined in this way parts of the San Andreas fault system in California, and Hirata (1989) has looked at fault traces in Japan. Alternatively, the definition can be applied directly to maps of epicentres, as in Sadovsky et al. (1984).

It is in this context that the use of "multifractals" has been suggested by Geilikman et al. (1990). Strictly speaking, the multifractal concept does not apply to sets as such, but to measures, and is effectively equivalent to studying the family of "entropy dimensions" defined for a probability measure $p(\cdot)$ over a k-dimensional cube as the limits as $\Delta \to 0$ of the ratios

$$d_1(\Delta) = \Big[\sum_i p(\Delta_i) \log p(\Delta_i)\Big]/\log \Delta \tag{4.9a}$$

$$d_q(\Delta) = \frac{1}{q-1} \log\Big[\sum_i p(\Delta_i)^q\Big]/\log \Delta \quad (q \neq 1), \tag{4.9b}$$

where $\{\Delta_i\,;\, i = 1,2,\dots,n^3\}$ is a subdivision of the original cube into equal smaller cubes of side Δ equal to $(1/n)$ of the side of the original cube, and the sum is restricted to subsets with $p(\Delta_i) > 0$.

If $p(\cdot)$ has a density with respect to Lebesgue measure in k-D space, the ratios in (4.9) tend to a common limit k, and similarly they tend to a common limit $k' < k$ if the measure is singular but has a density on a smooth lower-dimensional manifold. In the case of highly irregular measures, however, the limits may exist but take different values for different q. In this case the family of "entropy dimensions" $d_q = \lim_{\Delta\to 0} d_q(\Delta)$ provide a kind of functional description of the extent and nature of the irregularities. The papers cited gives further developments and interpretations in terms of local "critical indices". Note that for $q = 0$, (4.9b) reduces to the definition of the fractal dimension of the support of $p(\cdot)$.

As examples of such "multifractal measures" Pisarenko and Pisarenko (1990, 1991) and Mandelbrot (1989) examine the properties of the family of probability measures on [0,1] associated with random variables of the form

$$U = \sum_1^\infty \xi_k/m^k \tag{4.10}$$

where the ξ_k are i.i.d. with common multinomial distribution defined on the integers 0,1,...,m–1. The random variable whose distribution is the "Cantor staircase" is a special case.

While it is not claimed that the measures associated with (4.10) are directly applicable as models for earthquake occurrence, their qualitative properties may provide a guide to likely properties if the earthquakes follow a fractal or self-similar pattern. To verify this Geilikhman et al. take the measure $p(\cdot)$ in (4.9) to be the proportion of events from a given catalogue falling in a given subset of the region, and plot the numerator and denominator of (4.9) for different values of Δ with fixed q, showing linear behaviour in the log-log plot for a limited range of values of Δ, with slopes depending on q.

A different class of fractal models is suggested by Mandelbrot (1983) in the context of clustering of galaxies. Let T(t) be a (cumulative) process with positive stable increments, $\underset{\sim}{X}(t)$ a continuous self-similar process such as a fractional Brownian motion in 1, 2, or 3 dimensional space, and consider the trajectories of the "subordinated" process

$$\underset{\sim}{Y}(t) = \underset{\sim}{X}[T(t)].$$

Since T(t) is necessarily a pure jump process – any non-zero interval contains a countable set of jumps with convergent sum – $\underset{\sim}{Y}(t)$ moves in a series of jumps along the path of $\underset{\sim}{X}(\cdot)$. The points at which the process pauses between jumps form a highly clustered but often very sparse set; some dramatic illustrations are given in Mandelbrot's book. They do not form a point process, since they fail the requirement that there should be only a finite number of points, with probability one, in any bounded set. They do, however, form self-similar fractal sets with properties reminiscent of those of the Cantor set. I am not aware of their having been used directly as models for earthquake occurrence.

Finally, one should note a series of studies, notably the papers by Kagan and Knopoff (1980), Kagan (1981a,b) and more recently Kagan and Jackson (1991), Kagan (1991 a), in which fractal properties of the earthquake process are adduced from correlation or moment functions. Any random measure which is stationary and self-similar in the sense of (3.11) will have its covariance functions

$$C_{AB}(\underset{\sim}{x}) = \mathrm{Cov}(\xi(A), \xi(\underset{\sim}{x} + B))$$

forced into power law form by the self-similarity requirement, say $C_{A,B}(\underset{\sim}{x}) \sim |\underset{\sim}{x}|^{-\alpha}$, where α is related to the self-similarity index H by

$$\alpha = 2 - 2H.$$

Fractal sets such as the Levy dust just referred to play a somewhat maverick role in this discussion, since they are not strictly stationary and such covariance functions are not well-defined. However, as Mandelbrot points out, they do enjoy the weaker property that distributions relative to a given point in the process are independent of the choice of that particular point ("Palm-stationarity" in the terminology of Daley and Vere-Jones (1989)).

Thus moment distributions based on interpoint distances may be expected to provide useful information, and Kagan has studied 2-point, 3-point and 4-point correlation functions of this kind in the three papers cited. In Kagan (1987) he develops an extension of the branching model which incorporates fault development in time and space as well as energies, and leads to appropriate power law behaviour in all key variables.

4.4 MODELS FOR SMOOTHING AND INTERPOLATION

It is a common requirement to produce from an earthquake catalogue not only seismicity maps showing the location of epicentres, but also smoothed contour maps which may be used for display purposes in the course of developing zoning studies, risk maps, etc. Such maps may be required in space, time, or both together. The possibility of producing time dependent maps for earthquake risk, using current information on precursory variables, is also of potential interest.

All such tasks may be considered variants on the classical problem of estimating a probability density function. In the simplest case of estimating contours for the intensity of a non-homogeneous Poisson process, they are formally identical, at least up to a scale factor. The special features which arise with earthquakes are chiefly due to the clustering, which is equivalent to a kind of dependence between observations in the density estimation context. The high dimensionality, specifically the space-time aspects, also introduces aspects not normally taken into account in density estimation. Insofar as Vere-Jones (1992) contains a rather detailed account of these questions, we shall make only a few broad comments here.

The main techniques which have been used for density estimation are kernel estimation, fitting by a series of orthogonal functions, and spline fitting. All of these can and have been used for earthquakes. In general, the contents of any earthquake catalogue can be fitted to an arbitrary degree of closeness by a non-homogeneous Poisson process, up to the degenerate limit where each point in the catalogue is fitted by a delta function. It is only prior assumptions of smoothness, including some assumptions about the degree of clustering, which counteract this progression and allow some rational choice of optimal smoothing. In all cases the crucial decision relates to the degree of smoothing, which is controlled by the choice of bandwidth in kernel estimation, the number of terms in the orthogonal expansion, and the spacing between knots in spline representations.

The main techniques used to resolve this problem may be listed as follows:

- (i) Leave-one-out cross validation.
- (ii) Training and evaluation (i.e. one part of the data is used to fit a set of contours and the remaining part is used to evaluate their effectiveness).
- (iii) Simple AIC used in conjunction with a simplified likelihood such as the non-homogeneous Poisson to determine the optimal number of coefficients to include in the model.
- (iv) Bayesian versions of AIC, which are tantamount to the use of penalized likelihoods with a suitable smoothness penalty.

Numerous references are given in the review paper cited, but it would be improper in the present meeting to omit mention of Akaike's pioneering development of an automatic density estimation package in Akaike and Arahata (1978), and the substantial series of papers utilizing the "objective Bayesian approach" (Akaike (1980)) in conjunction with spline fitting methods. This has been applied to earthquake catalogues by Ogata and Katsura (1988), Ogata, Imoto and Katsura (1991), among other papers.

Returning to the special difficulties associated with clustering, its essential effect is to increase the variability of the parameter estimates and hence to alter the choice of optimal smoothing parameter. A difficult problem is the proper attribution of observed clustering between transient clusters and permanent inhomogeneities. The two have very different consequences for prediction or risk forecasting but not necessarily for displaying the contents of a given catalogue. One's approach to smoothing may therefore depend on the purpose in view.

A slightly different approach is taken by Diggle and Marron (1988), under the assumption that the observed process is a Cox process (doubly stochastic Poisson process) and that the aim is to optimally depict the clusters rather than to smooth them over. In effect the large-scale variation is used to estimate the extent of variability to be expected within the doubly stochastic framework, and this information is used to control the choice of smoothing parameter.

Once again, there is considerable scope for further elucidation of these questions.

5. Cycles and Characteristic Earthquakes

The next group of models incorporate what might appear to be the opposite tendency to clustering, namely a tendency towards regular or periodic behaviour. In fact the two tendencies may be less contradictory than they sound. For example, a regular succession of periods of heightened activity could also be viewed as a sequence of clusters. A renewal process, with independent intervals between events, produces a sequence that is more regular than a Poisson process if the coefficient of variation (ratio of standard deviation to mean) is less than unity, but is more irregular (clustered) than a Poisson if this coefficient exceeds unity. Moreover most catalogues contain a superposition of events from an indeterminate number of sources which may or may not be acting independently; in such circumstances periodic effects in individual components may easily be obscured.

Most of the evidence for periodic effects comes either from very large events preserved in the historical or geological records (in which evidence for aftershocks etc. may be lost) or from sequences in rather special geological settings. It is possible that periodicity only appears in situations where the observed events largely exhaust the energy-bearing capacity of a relatively isolated system.

In the next few sections we examine some specific models and their possible applications.

5.1 PERIODIC POINT PROCESSES

Under this heading we shall understand a regularity in the occurrence of events which is tied to a fixed lattice in absolute time, as distinct from examples, such as relaxation or renewal processes, which retain some regularities in the relative occurrence times of events, but drift away from regularity on an absolute time scale. The simplest periodic models can be brought within the framework of stationary point processes by supposing the beginning of the observation period to be randomized with respect to the regular lattice of time points. The model is then characterized by a Bartlett spectrum with mass concentrations (line spectrum) at the lattice frequency, and commonly at its harmonics as well. However the second order features are not such an adequate descriptor of point process structure as they are of the structure of continuous Gaussian processes, so that, for example, several different point processes may have the same Bartlett spectrum. (See Chapter 11 of Daley and Vere-Jones (1988) for an account of the spectral theory of point processes.)

The two simplest 1-dimensional models are:

(i) the jitter process, in which events at a regular lattice of time points are independently perturbed so that the actual sequence of events has the form

$$t_i = b + ai + \varepsilon_i \quad (i = 0, \pm 1, \ldots;\ \varepsilon_i \text{ i.i.d.}); \tag{5.1}$$

(ii) the Poisson process with periodic rate function of the form

$$\lambda(t) = a + b\cos(\alpha t + \phi) \quad (a > 0,\ 0 \le b < a) \tag{5.2}$$

or

$$\lambda(t) = \exp\{a + b\cos(\alpha t + \phi)\}. \tag{5.3}$$

Both types of model give rise to Bartlett spectra of broadly similar form. Thus the model (5.1) is associated with a spectrum (Daley and Vere-Jones, Ex. 11.2.4)

$$\gamma(\omega) = [|\tilde{f}_\varepsilon(\omega)|^2/a^2] \sum_{k \ne 0} \delta(\omega - \frac{2\pi k}{a}) + \frac{1}{2\pi a}[1 - |f_\varepsilon(\omega)|^2]$$

where $\delta(x)$ is the Dirac delta function and $\tilde{f}_\varepsilon(\cdot)$ is the characteristic function of distribution of the perturbations ε_i. (5.2) on the other hand has Bartlett spectrum

$$\gamma(\omega) = \frac{1}{2\pi}\{a + \tfrac{1}{2} b^2\,\delta(\omega - \alpha)\},$$

with a single spike at α, while (5.3) gives rise to spikes of height proportional to the Fourier coefficients of $I_0(2b\cos\frac{1}{2}\alpha x)$ at the various multiples of α.

Both types of models show a series of spectral lines superimposed on a slowly varying continuous spectral background.

A more telling distinction between the models relates to the variance-time curve

$$V(\tau) = \mathrm{Var}[N(\tau)] \tag{5.4}$$

where $N(\tau)$ is the number of points in an interval of length τ. It is not difficult to see that if $E(|\varepsilon|) < \infty$, the jitter model (5.1) exhibits what Lewis and Govier (1964) called "*controlled variability*": $V(\tau) \le C < \infty$. This is in sharp contrast to any of the Poisson or renewal process models in which on average $V(\tau)$ grows linearly with τ.

Likelihoods for the processes (5.2) and (5.3) can be written down immediately using the general form (3.5); see Vere-Jones and Ozaki (1982) and Vere-Jones (1982) for further discussion of (5.3) and Kutoyants (1984) for (5.2). In particular, the papers cited show that the locally optimal test for $b \neq 0$ in (5.3) is approximately equivalent to testing the maximum of the periodogram, and that the location of the maximum provides an efficient estimate of α.

Applications of the cyclic Poisson model to earthquake data are given in Vere-Jones and Ozaki (1982) and Vere-Jones and Deng (1989), which treat historical events from the Tokyo region of Japan and North China respectively. One of the problems in testing for periodicity with such data is the nuisance effect of clustering; this means that the appropriate null hypothesis is not really a stationary Poisson process but rather a cluster process. This circumstance dramatically reduces the power of the tests. Another problem is that tests for the *maximum* of the periodogram (5.3) present special difficulties in a point process context. Vere-Jones and Ozaki (1982) present a heuristic argument but a rigorous approach has still to be developed. Both of the data sets cited show large spectral peaks, but both data sets are small (30-60 events), and by the time the various effects above have been taken into account the significance of the peaks is marginal (Fig 4).

My personal inclination is to feel that the periodic effects are real, but better explained by a relaxation model of the type discussed in the following section.

5.2 STRESS RELEASE MODELS

The physical basis for stress release models goes back to the classical description of the elastic rebound model for earthquakes formulated by Reid (1910) following the 1906 San Francisco earthquake. Stress along a fault accumulates, as a result of large-scale tectonic movements, until the breaking strength of the crust is released, at which point an earthquake occurs, most of the stored energy is released in the form of elastic waves, and the cycle begins again.

This idea can be brought within the orbit of the point process models by representing the conditional intensity in the form

$$\lambda(t) = \psi[X(t)] \tag{5.5}$$

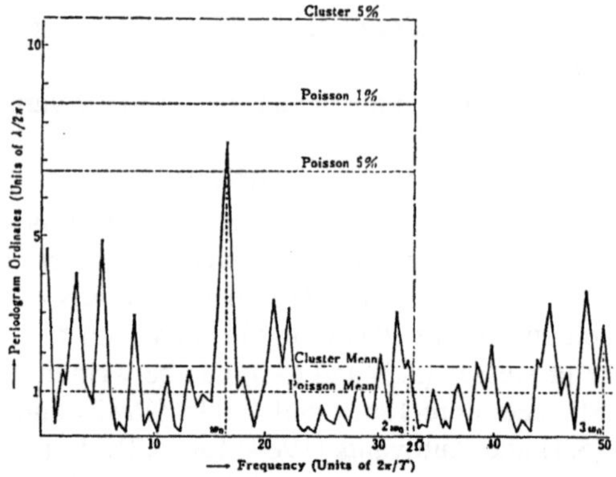

Periodogram for Kawasumi's Data (full set).

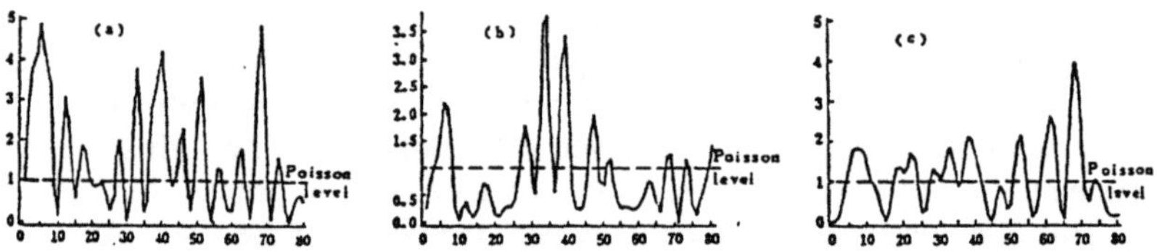

Mean corrected periodogram. a. full data set; b. data 1480 – 1930; c. events with M ≥ 6.5

Figure 4. Periodicity for Japanese and Chinese historical data, from Vere-Jones and Ozaki (1982) and Vere-Jones and Deng (1989). The Japanese data shows the effect on significance levels of allowing for clusters. Obvious clusters have been removed from the Chinese data, but it is still sensitive to variations in the data set. Japanese data above, Chinese data below.

where X(t) models a stress release level (in general unobservable) and $\psi(x)$ is a monotonic increasing risk function. Typically X(t) is taken to be a process which increases linearly between events and instantaneously decreases by an amount equal to a presumed "stress release" S_i at the time of an event. It then has the structural form

$$X(t) = X(0) + bt - \gamma S(t) \tag{5.6}$$

with

$$S(t) = \sum_{i\,:\,0 \le t_i < t} S_i . \tag{5.7}$$

The function $\psi(x)$ should be low for low stress levels and increase rapidly as the breaking stress of the crust is approached; however, simple choices of convenience such as

$$\psi(x) = e^x \tag{5.8}$$

seem to give reasonable results in practice.

Treating X(0) as an additional unknown parameter then gives rise to a 3-parameter model for which the likelihood can be written down in the standard way following (3.5), taking the stress drops S_i as given.

In fact the stress drops should also be considered part of the overall model. It seems reasonable to suppose that their distribution is related to the current stress level X(t), through a family of joint distributions

$$F(s \mid x) = \text{Prob}(S \leq s \mid X(t) = x \text{ when jump occurs}). \tag{5.9}$$

With these assumptions the process X(t) is Markovian and its ergodic properties can be studied (Knopoff (1971); Zheng (1990) gives a thorough treatment with references to earlier work). This said, there are still many features of the model which remain unexplored. No detailed results for its second order properties are known, although Vere-Jones (1988) shows that it has the controlled variability property if and only if the stress drops are constant, and obtains some partial results for a simple special case. Wang et al. (1991) give simulation results for different forms of the stress drop distribution and examine numerically the anomalous behaviour of the likelihood ratio test against a Poisson null hypothesis (see also Ogata and Vere-Jones (1984), Vere-Jones and Ogata (1984)). When the stress drops follow a long-tailed distribution the trajectories visually suggest that the process has self-similar properties, at least within certain ranges; it would be of interest to examine this tendency in greater depth (Fig. 5).

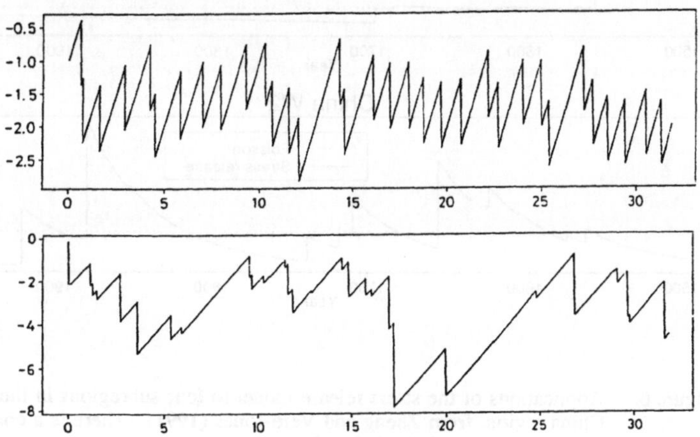

Figure 5. Simulated stress release models, from Wang et al. (1991). The graphs show simulations of X(t) (equation (5.6)), the first for constant stress drops and the second for power-law stress drops, both with exponentially increasing risk function (5.8).

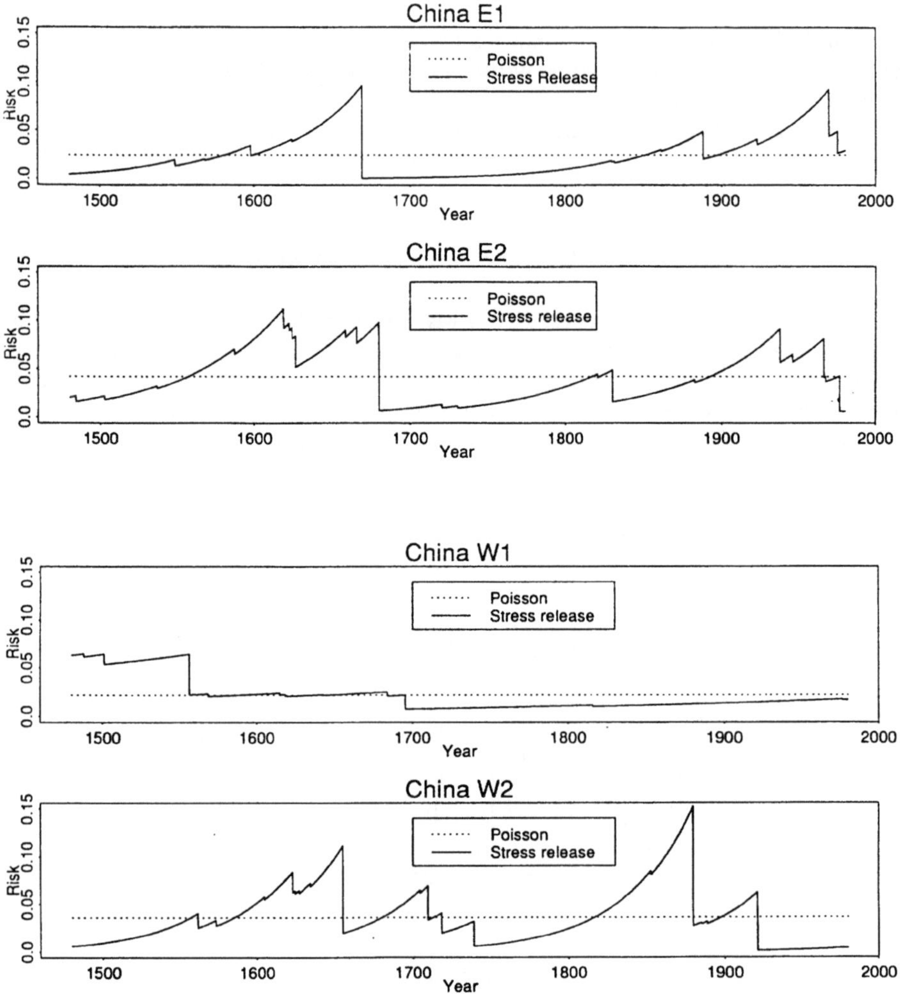

Figure 6. Applications of the stress release model to four subregions in the Northern China region, from Zheng and Vere-Jones (1991). There is a considerable variety in the quality of fit. Fitting the four subregions separately gives much better results than fitting the region as a whole.

No really plausible extension of the model to a space-time context has yet been developed, although multivariate extensions are developed in Vere-Jones and Zheng (1990) and Zheng and Vere-Jones (1991), and applied to a range of historical catalogues. Here the model is applied simultaneously to data from several different subregions or magnitude ranges, and the AIC criterion is used to judge the effectiveness of the subdivisions; details of the multivariate models and the inference process are given in the papers cited.

The geophysical conclusions which seem to emerge from these studies are themselves quite interesting. The stress release model represents a very substantial improvement over the Poisson model with constant intensity or constant with simple trend for the Chinese data, for the Tokyo region of Japan, and possibly for some parts of the Persian complex (see Fig. 6). In all cases the multivariate versions of the model, which allow different subregions to develop their own individual regimes, perform better than a single model fitted to the whole region. In other subregions, however, the model does not fit at all well. It appears, in fact, as if the model can be quite a useful discriminatory tool.

The difficulty of extending the model to the space-time context does not seem to hinge on technical problems, but suggests rather the need for a substantial new insight into the physical mechanism of spatial dependencies. As we shall see in the following section, the "time predictable" and "slip predictable" models used in describing the behaviour of individual fault segments are variations on the same theme, with the problem of linkage between segments still a key problem.

A crucial feature of the stress release models, not always shared by the models of the succeeding section, is that the accumulated stress is a process of controlled variability, for we have

$$S(t) = \sum_{i:\,0\le t_i<t} S_i - \gamma^{-1}[bt + X(0) - X(t)] \tag{5.10}$$

and if X(t) is ergodic all terms on the right side have bounded variability as $t \to \infty$.

5.3 CHARACTERISTIC EARTHQUAKES

In its crudest form, the characteristic earthquake model reduces to the hypothesis that earthquakes along a given fault or fault segment repeat themselves in virtually identical form at virtually identical time intervals. Although this sounds a clear-cut and readily verifiable or refutable hypothesis, the evidence is still the subject of controversy. There are two basic reasons for this apparently paradoxical situation. The first is that the evidence comes mainly from short sequences of events along a variety of faults associated with widely different sizes and time scales. Some of the evidence is derived from geological data, where timing by radiocarbon methods (see Sieh (1981), and Sieh et al. (1988)) is subject to statistical error. The second reason is that the proposed models often do not include a specification of the variability that is implicit in the qualifier "virtually". In this sense they are not yet properly specified statistical models and cannot be subjected to a rigorous statistical analysis. This is particularly true of the geological studies, which in other respects offer perhaps the most convincing evidence for an effect of this type. We shall return to these studies at the end of the section, but in the meanwhile consider some simple models for the time series of events which have been proposed for use in earthquake prediction and risk estimation studies.

A convenient starting point is the paper by Shimazaki and Nakata (1980) who suggest two variants – the "time predictable" and "slip predictable" models. Both assume a linear input of stress (or possibly seismic moment – the exact interpretation of the variable does not play a role in the structural description of the model, only in the process of fitting it to data), and a sudden stress drop at the time of the earthquake, as in the stress release model. In the time-predictable model the event is supposed to occur as soon as the stress level reaches a specific value construed as the breaking strength of the crust. The stress drop itself is allowed to be variable but since the stress drop can be observed, the time taken for the process to recover to the critical stress value can in principle be calculated. In the slip predictable model the time at which the event occurs is allowed to be variable, but the event is supposed to always release the accumulated stress, so that after each event the process is returned to the same base level.

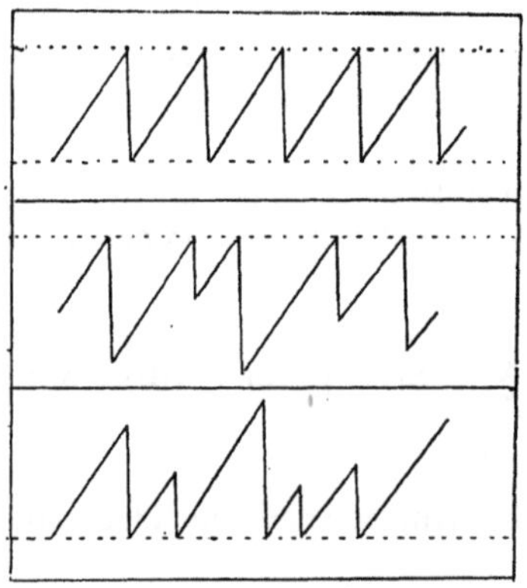

Figure 7. Stochastic representation of characteristic earthquakes (top line) time-predictable (middle line) and slip-predictable (bottom line) model, after Shimazaki and Nakata (1980).

Numerous suggestions have been made as to how these basic ideas could be incorporated into more realistic stochastic models. Nishenko and Buland (1987) examined a number of earthquake sequences and suggested that a renewal model with log normal interval distribution having coefficient of variation around 25% could be used quite generally. Their procedure was to first normalize each sequence by dividing by the mean time interval (equivalently the range between first and last events), a procedure which produces a strong downwards bias in the coefficient of variation, particularly where only three or four intervals are involved. While the distribution indeed appears to be skewed to the right there is barely sufficient information to distinguish the log normal from other long-tailed distributions.

A renewal process is exactly what one would expect from the time predictable model if the jump sizes were independently and randomly distributed. However in this case the time intervals would be proportional to the stress drops and should therefore follow a power-law distribution rather than a log normal distribution (see the discussion around (4.6)). On other grounds, such long-tailed distributions seem quite inappropriate (Davis et al. (1989)).

In the papers by the Working Group on California Earthquake Probabilities (Working Group (1988, 1990)) a slightly different variant was proposed. Here it was suggested that the times T_i and slips S_i are related by

$$T_i = S_i \varepsilon_i / V \tag{5.11}$$

where V is a constant input rate, and the ε_i are i.i.d. log normal with unit median, or equivalently $U_i = \log \varepsilon_i$ are i.i.d. normal with zero mean. Thus the conditional distributions of the time intervals, given the values S_i, are log normal. If in addition the S_i are independent and log normal then so is the product (5.11), and the overall process is again of renewal type with log normal intervals.

A difficulty with this variant as a model for the long run behaviour of the process is that it leads to arbitrarily large fluctuations of the accumulated stress. Thus the accumulated stress after N cycles is

$$V\sum_1^N T_i - \sum_1^N S_i = V\sum_1^N S_i(\varepsilon_i - 1) \tag{5.12}$$

where the variance of the right side grows indefinitely with N.

If the time-predictable model has been suggested in connection with the Californian data, the slip-predictable model has been suggested with data from the middle American Trench (Kiremidjian and Anagnos (1984)). In the simplest version successive time intervals are assumed independent and the following stress drops assumed proportional to them. Again this gives rise to a renewal process with the interval length distribution having the same form as the stress drop distribution. Insofar as exact proportionality is not observed the authors cited suggest a type of semi-Markov dependence in which the distribution of the time interval is allowed to depend on the stress drop of the *succeeding* event, the sizes of these events following a simple Markov chain. The time intervals between a pair of events of prescribed magnitudes is supposed to follow a Weibull distribution with parameters depending on the magnitudes.

Although the model here is more complex, a similar problem arises with the accumulated stress release, which is again representable by a sum analogous to (5.12) and has unbounded variability as the number of cycles N increases.

The thinking underlying the characteristic earthquake model is more fully revealed in geological studies of the movements of underlying faults (see, for example, Sieh (1981), Schwarz and Copperfield (1984)). Here we enter a different modelling world, where the stochastic elements are not so much points as segments along a given fault or fault system. If ruptures are generally terminated by an "asperity" (bend or other barrier) it would not be surprising if particular segments of a fault ruptured repeatedly between the same end-points, producing a series of earthquakes of similar characteristics. On occasion, however, the barrier may be surpassed, and a larger-than-normal earthquake can result.

While various suggestions have been made, a satisfying stochastic model for these processes seems to me still lacking. As an example, a recent paper by Ward (1991) suggests a simulation model along the following lines. Initial values of stress and strength are assigned at random to different segments along a fault. At each time step, the stress levels at all segments are increased by a fixed small amount. If this causes the stress level to exceed the strength at some point on a segment, a rupture occurs which spreads until all stress levels are reduced below the relevant strengths; then a further time increment is taken and the cycle repeated. The main technical difficulties relate to the procedure for estimating the extent of spread of the rupture and associated modifications to the stress levels.

Strictly speaking, the model is only random insofar as the choice of initial values is concerned, a feature it shares with the block and spring models of Burridge and Knopoff (1967), and indeed with the much wider class of so-called "chaos" models. However, the output mimics many observed features of earthquake occurrence. It seems to me over-optimistic to suppose that the initial conditions are the only place where uncertainty concerning the evolution of the process needs to be inserted, but it is clearly of importance that many qualitative features of the process can be derived with very little explicit statistical modelling. Just how such systems should best be handled, particularly for purposes of prediction and forecasting, is a major theme in many areas of geophysics at the present time. The recent series of papers on Chaos in the *Journal of the Royal Statistical Society* (1992, Series B) give some idea of the ramifications which might be involved, but it lies beyond the scope of this paper to treat this question in detail.

Acknowledgments

In addition to the Conference organizers, I am indebted to my colleagues Zheng Xiaogu and Fabio Musmeci for their help in recent work; to the International Institute of Earthquake Prediction and Mathematical Physics in Moscow for an invitation which opened up for me the wide scope of their recent work on self-similarity and related ideas; and to many long-suffering friends and colleagues who keep me informed of their recent work with a minimum of response or apparent appreciation on my part.

References

Akaike, H. (1977). On Entropy Maximization Principle, *Applications of Statistics*, (ed. P. R. Krishnaiah), 27-41, North Holland, Amsterdam.

Akaike, H. (1980). Likelihood and Bayes Procedure, *Bayesian Statistics*, (eds. J. M. Bernard, M. H. de Grott, D. V. Lindley and A. F. M. Smith), 141-204, Valencia U.P.

Akaike, H. and Arahata, E. (1978). GALTHY : A Probability Density Estimation, *Computer Science Monographs*, 9, Institute of Statistical Mathematics, Tokyo.

Aviles, C. A., Scholz, C. H. and Boatwright, J. (1987). Fractal Analysis Applied to Characteristic Segments of the San Andreas Fault, *J. Geophys. Res.*, 92, 331-344.

Bak, P., Tang, C. and Wiesefeld, K. (1988). Self-Organized Criticality, *Phys. Rev.*, A. 38, 364-374.

Bebbington, M., Zheng, X. and Vere-Jones, D. (1990). Percolation Theory : A Model for Rock Fracture? *Geophys. J. Int.*, 100, 215-220.

Bolt, B. A. (1988). *Earthquakes* (revised edition), W. H. Freeman, New York.

Burridge, J. and Knopoff, L. (1967). Model and Theoretical Seismicity, *Bull. Seism. Soc. Am.*, 57, 341-371.

Cox, D. R. and Isham, V. (1980). *Point Processes*, Chapman and Hall, London.

Cox, D. R. and Lewis, P. A. W. (1966). *The Statistical Analysis of Series of Events*, Methuen, London.

Daley, D. J. and Vere-Jones, D. (1988). *An Introduction to the Theory of Point Processes*, Springer, New York.

Davies, R. B. and Harte, D. (1987). Tests for Hurst Effect, *Biometrika*, 74, 95-101.

Davis, P. M., Jackson, D. D. and Kagan, Y. Y. (1989). The Longer it Has Been Since the Last Earthquake, the Longer the Expected Time Till the Next? *Bull. Seism. Soc. Am.*, 79, 1439-1496.

Diggle, P. J. (1983). *Statistical Analysis of Spatial Point Patterns*, Academic Press, London.

Diggle, P. J. and Marron, J. S. (1988). Equivalence of Smoothing Parameter Selectors in Density and Intensity Estimation, *J. Am. Stat. Soc.*, 83, 793-800.

Falconer, K. (1990). Fractal Geometry : Mathematical Foundations and Applications. Wiley, Chichester.

Geilikhman, M. B., Golubeva, T. V. and Pisarenko, V. F. (1990a). Multifractal Patterns of Seismicity, *Earth and Planetary Sciences Letters*, 99, 127-132.

Geilikhman, M. B., Golubeva, T. V. and Pisarenko, V. F. (1990b). Self-Similar Hierarchical Structure of the Distribution of Earthquake Epicentres. In: Computer Analysis of Geophysical Fields, *Computational Seismology*, 23, 123-139, Nauka, Moscow.

Geilikhman, M. B. and Pisarenko, V. F. (1989). On Self-Similarity in Geophysical Phenomena, *Discrete Properties of the Geophysical Medium* (ed. M. A. Sadovskiy), 109-131, Nauka, Moscow.

Gutenberg, B. and Richter, C. F. (1949). *Seismicity of the Earth*, Princeton University Press, Princeton.

Hawkes, A. G. (1971). Spectra of Some Self-Exciting and Mutually Exciting Point Processes, *Biometrika*, 58, 83-90.

Hawkes, A. G. and Adamopoulos, L. (1973). Cluster Models for Earthquakes : Regional Comparisons, *Bull. Int. Stat. Inst.*, 45, 454-461.

Hawkes, A. G. and Oakes, D. (1974). A Cluster Representation of a Self-Exciting Process, *J. Applied Prob.*, 17, 493-503.

Hirata, T. (1989). Fractal Dimensions of Fault Systems in Japan : Fractal Structure in Rock Fracture Geometry at Various Scales, *Pageoph.*, 131, 157-170.

Ito, K. and Matsuzaki, M. (1990). Earthquakes as Self-Organized Critical Phenomena, *J. Geophys. Res.*, 95, 6853-6860.

Jeffreys, H. (1938). Aftershocks and periodicity in earthquakes, *Gerlands Beiträge fur Geofysik*, 53, 111-139.

Kagan, Y. Y. (1981a). Spatial Distribution of Earthquakes : The Three-Point Correlation Function, *Geophys. J. Roy. Astron. Soc.*, 67, 697-718.

Kagan, Y. Y. (1981b). Spatial Distribution of Earthquakes : The Four-Point Correlation Function, *Geophys. J. Roy. Astron. Soc.*, 67, 719-733.

Kagan, Y. Y. (1987). Random stress nd earthquake statistics: time dependence. *Geophys. J. Roy. Astron. Soc.*, 88, 723-731.

Kagan, Y. Y. (1991). Fractal Dimensions of Brittle Fracture, *J. Non Linear Science*, 1, 1-16.

Kagan, Y. Y. and Jackson, D. (1991). Long-Term Earthquake Clustering, *Geophys. J. Int.*, 104, 117-133.

Kagan, Y. Y. (1991b). Seismic moment distribution. *Geophys J. Int.* 106, 123-134.

Kagan, Y. Y. (1991c). Likelihood analysis of earthquake catalogues. *Geophys J. Int.* 106, 709-716.

Kagan, Y. Y. and Knopoff, L. (1980). Spatial Distribution of Earthquakes : The Two-Point Correlation Function, *Geophys. J. Roy. Astron. Soc.*, 67, 303-320.

Kagan, Y. Y. and Knopoff, L. (1984). A stochastic model of earthquake occurrence. *Proc. 8th Int. Conf. Earthquake Engineers*, 1, 275-302.

Karr, A. F. (1986). *Point Processes and Their Statistical Inference*, Marcel Dekker, New York.

Kiremidjian, A. S. and Anagnos, T. (1984). Stochastic Slip Predictable Model for Earthquake Occurrences, *Bull. Seism. Soc. Am.*, 74, 739-755.

Knopoff, L. (1971). A Stochastic Model for the Occurrence of Main Sequence Earthquakes, *Rev. Geophys. and Space Phys.*, 9, 175-188

Kutoyants, Yu. A. (1984). *Parameter Estimation for Stochastic Processes* (translated from revised Russian edition by B. Prakasa Rao), Helderman, Berlin.

Lewis, P. A. W. and Shedler, G. S. (1976). Simulation of Non-Homogeneous Poisson Processes with Log-Linear Rate Function, *Biometrika*, 63, 501-506.

Lewis, T. and Govier, L. J. (1964). Some Properties of Counts of Events for Certain Types of Point Processes, *J. Roy. Stat. Soc. B*, 26, 325-337.

Mandelbrot, B. (1977). *Fractals : Form, Chance and Dimension*, W. H. Freeman, San Francisco.

Mandelbrot, B. (1983). *The Fractal Geometry of Nature*, W. H. Freeman, San Francisco.

Mandelbrot, B. (1988). An Introduction to Multifractal Distribution Functions, *Fluctuations and Pattern Formation*, (eds. H. E. Stanley and N. Ostrowsky), Kluwer Academic, Amsterdam.

Mandelbrot, B. (1989). Multifractal Measures, Especially for the Geophysicist. PAGEOPH, Vol. 131, Nos Y2 (1989) pp 5-42.

Mogi, K. (1985). *Earthquake Prediction*, Academic Press, Tokyo.

Molchan, G. M. (1991). Identification of Aftershocks : Review and New Approaches. In: Contemporary Methods of Interpretation of Seismological Data, *Computational Seismology*, 24, 19-49.

Molchanov, S. A., Pisarenko, V. F. and Reznikova, A. Ya. (1986). On the Percolation Approach to Fracture Theory. In: Mathematical Methods in Seismology and Geodynamics, *Computational Seismology*, 19, 3-7.

Musmeci, F. and Vere-Jones, D. (1992). A Space-Time Clustering Model for Historical Earthquakes, *Ann. Inst. Stat. Math.*, 44, 1-11.

Nishenko, S. P. and Buland, R. (1987). A Generic Recurrence Interval Distribution for Earthquake Forecasting, *Bull. Seism. Soc. Amer.*, 77, 1382-1399.

Ogata, Y. (1983). Estimation of the Parameters in the Modified Omori Law for Frequencies by the Maximum Likelihood Procedure, *J. Phys. Earth*, 31, 115-124.

Ogata, Y. (1988). Statistical Models for Earthquake Occurrence and Residual Analysis for Point Processes, *J. Amer. Stat. Ass.*, 83, 9-27.

Ogata, Y., Akaike, H. and Katsura, K. (1982). The Application of Linear Intensity Models to the Investigation of Causal Relations Between a Point Process and Another Stochastic Process, *Am. Inst. Stat. Math.*, 34, 373-387.

Ogata, Y. and Katsura, K. (1986). Point Process Models with Linearly Parameterized Intensity for the Application to Earthquake Catalogue, *J. Applied Prob.*, 23A, 231-240.

Ogata, Y. and Katsura, K. (1988). Likelihood Analysis of Spatial Inhomogeneity of Marked Point Patterns, *Am. Inst. Stat. Math.*, 40, 29-39.

Ogata, Y. and Shimazaki, K. (1984). Transition from Aftershock to Normal Activity : The 1965 Rat Island Earthquake Aftershock Sequence, *Bull. Seisml. Soc. Am.*, 74, 1757-1765.

Ogata, Y. and Tanemura, M. (1984). Likelihood Analysis of Spatial Point Patterns, *J. Roy. Statist. Soc. B*, 496-518.

Ogata, Y. and Vere-Jones, D. (1984). Inference for Earthquake Models : A Self-Correcting Model, *Stoch. Proc. Appl.*, 17, 337-347.

Okubo, P. G. and Aki, K. (1987). Fractal Geometry in the San Andreas Fault System, *J. Geophys. Res.*, 92, 345-355.

Pisarenko, V. F. and Pisarenko, D. (1990). Spectral Properties of Singular Self-Similar Measures and Their Geophysical Applications. In: Analysis of Geophysical Fields, *Computational Seismology*, 23, 107-122.

Pisarenko, V. F. and Pisarenko, D. (1991). Spectral Properties of Multifractal Measures, *Phys. Letters A*, 153, 169-172.

Reid, H. Feilding (1910). The Mechanics of Earthquakes, *Univ. Calif. Publ. Bull. Dep. Geol. Sci.*, 6, 413-433.

Ripley, B. D. (1981). *Spatial Statistics*, Wiley, New York.

Sadovsky, M. A., Golubeva, T. V. et al. (1984). Characteristic Dimension of Rocks and Hierarchical Properties of Seismicity, *Izv. Acad. Sci. USSR Phys. Solid Earth*, 20, 87-95 (English trans.).

Saito, M., Kikuchi, M. and Kudo, M. (1973). Analytical Solution of "Gogame Model of Earthquakes", *Zishin*, 26(2), 19-25.

Scholz, C. H. (1990). *The Mechanics of Earthquakes and Faulting*, Cambridge University Press, New York.

Schwarz, D. P. and Coppersmith, K. (1984). Fault Behaviour and Characteristic Earthquakes : Examples from the Wasatch and San Andreas Faults, *J. Geophys. Res.*, 89, 5681-5698.

Shimazaki, K. and Nakata, T. (1980). Time-Predictable Recurrence Model for Large Earthquakes, *Geophys. Res. Letters*, 7, 279-282.

Sieh, K. (1981). A Review of Geological Evidence of Recurrence Times for Large Earthquakes. In: Earthquake Prediction, an International Review, (eds. D. Simpson and P. Richards), *M. Ewing Series 8*, *Am. Geophys. Union*, 209-216, Washington D.C.

Sieh, K., Stuiver, M. and Brillinger, D. (1988). A More Precise Chronology of Earthquakes Produced by the San Andreas Fault in Southern California, *J. Geophys. Res.*, 94, 603-623.

Utsu, T. (1961). A Statistical Study on the Properties of Aftershocks, *Geophys. Mag.*, 30, 521-605.

Utsu, T. and Seki, A. (1955). Relation Between the Area of Aftershock Region and the Energy of the Main Shock, *Zisin* (2nd series), 7, 233-240.

Vere-Jones, D. (1970). Stochastic Models for Earthquake Occurrence (with discussion), *J. Roy. Statist. Soc. B.*, 32, 1-62.

Vere-Jones, D. (1976). A Branching Model for Crack Propagation, *Pageoph.*, 114, 711-726.

Vere-Jones, D. (1977). Statistical Theories for Crack Propagation, *Math. Geol.*, 9, 455-481.

Vere-Jones, D. (1978a). Earthquake Prediction : A Statistician's View, *J. Phys. Earth*, 26, 129-146.

Vere-Jones, D. (1978b). Space-Time Correlations for Microearthquakes – A Pilot Study, *Adv. App. Prob. Supp.*, 10, 73-87.

Vere-Jones, D. (1982). On the Estimation of Frequency in Point Process Data. In: Essays in Statistical Science, (eds. J. Gani and E. J. Hannan), *J. App. Prob.*, 19A, 383-394.

Vere-Jones, D. (1988). On the Variance Properties of Stress Release Models, *Austral. J. Statist.*, 30A, 123-135.

Vere-Jones, D. and Davies, R. B. (1966). A Statistical Study of Earthquakes in the Main Seismic Region of New Zealand : Part II Time Series Analysis, *N.Z. J. Geol. Geophys.*, 9, 251-284.

Vere-Jones, D. and Deng, Y. L. (1988). A Point Process Analysis of Historical Earthquakes from North China, *Earthquake Research in China* (English trans.), 2(2), 165-181.

Vere-Jones, D. and Ogata, Y. (1984). On the Moments of a Self-Correcting Point Process, *J. Appl. Prob.*, 21, 315-342.

Vere-Jones, D. and Ozaki, T. (1982). Some Examples of Statistical Inference Applied to Point Process Data, *Am. Inst. Stat. Math.*, 34, 189-207 (Errata ******).

Vere-Jones, D. and Zheng, X. (1991). Application of Stress Release Models to Historical Earthquakes from North China, *Pageoph.*, 135, 559-576.

SEISMOLOGICAL APPLICATIONS OF STATISTICAL METHODS FOR POINT-PROCESS MODELLING

YOSIHIKO OGATA
The Institute of Statistical Mathematics
4-6-7 Minami-Azabu Minato-ku Tokyo

1. Introduction

Applications of point-process models seem to be needed in various fields such as engineering, neurophysiology, epidemiology, seismology, etc. On the other hand, unlike the case of ordinary time series, such process are sharply non-linear and non-Gaussian. Therefore almost no general models nor theories have been fully developed. Actually, modelling of point process data seems inherently specific and dependent on the field of application. Yet, statistical methods of point-process analyses are common among application fields, so that data analysts can learn them from case studies in other fields.

I have been concerned with the point-process models in application to seismology where the occurrence times of earthquakes can be considered to be a point process, and now believe that the modeling of point processes can be genuinely useful for the investigation of seismic activity in conjunction with its prediction problems. This paper summarizes a number of my papers which discuss statistical analyses of time series of events with application to earthquake occurrence data. Many references will be missed in the present review. Readers are referred to the books by Daley and Vere-Jones (1988) for the theory of point processes in general, Kutoyants (1984) and Karr (1991) for the asymptotic theory of the inferences, Cox and Lewis (1966) for statistical methods, and Snyder (1975) for various topics in engineering applications, besides many other books on the proportional hazard models such as Cox and Isham (1980).

2. A new look at the statistical methods for point-processes

This section describes a set of promising statistical methods for analyzing various point-process data. These are (1) efficient simulation method by the *thinning algorithm*, (2) model comparison and parameter estimation by the *AIC* and the *maximum likelihood method*, and (3) graphical diagnostic checks of models and data by the *residual point process.* These methods are closely related to each other and based on the idea of *conditional intensity* of a point process. Parametric models in terms of conditional intensity function is the key stone of the present approach.

H. Bozdogan (ed.), Proceedings of the First US/Japan Conference on the Frontiers of Statistical Modeling: An Informational Approach, 137–163.

2.1 STATISTICAL MODELLING OF POINT PROCESSES

2.1.1. *Conditional intensity function.* There have been two conventional approaches in describing a point processes, or time series of events. Consider a series of events $\{t_i;\ 0 < t_1 < t_2 < \cdots\}$ occurring at random on the half-line $(0, \infty)$. For describing this sequence we can take the time intervals, $X_i = t_i - t_{i-1}$, between the consecutive points, and we then consider $\{X_i\}$ as a positive valued stochastic process. When this is distributed independently and identically, we call this a *renewal process*. Further, if its marginal is exponential distribution, then this is the *stationary Poisson process*. A useful tool for analyzing renewal process may be the Laplace transform which were frequently used in the classic queueing or reliability theory (see Cox 1962, for example). *Wold process* is an extension of the renewal process to the case where $\{X_i\}$ is Markov process of any order.

Another conventional specification of point processes is based on counting of points on intervals. Let $N(a, b)$ be the number of points in an interval (a, b) on the real line so that this is a non-negative integer valued random variable. To specify a point process is to give consistently defined joint distributions of such random variables for any corresponding mutually disjoint intervals (see Daley and Vere-Jones, 1972). For example, a stationary Poisson process is defined by the following two conditions: 1) for any interval (a, b), $N(a, b)$ is the Poisson random variable with mean $\lambda(b - a)$ for a positive constant λ, and 2) random variables $N(a_k, b_k)$ are mutually independent if their corresponding intervals (a_k, b_k), $k = 1, 2, ..., K$ are disjoint. The probability generating functional is a tool to analyze point processes in this specification. This is particularly useful for clustering point processes whose parents are distributed according to Poisson processes, such as the Neyman-Scott cluster processes: see Vere-Jones (1970), and Daley and Vere-Jones (1972) for details.

The last specification of point processes is developed closely relating to the information theory in conjunction with the prediction of an event occurring on a small time interval. Suppose a point process on the half-line $(0, \infty)$, and divide it into small intervals of length δ. Then we get a stochastic process $\{\xi_k\}$, where $\xi_k = N[(k-1)\delta, k\delta)$ is k-th random variable on the subinterval $[(k-1)\delta, k\delta)$. For small enough δ we may assume that $\{\xi_k\}$ is a binary (i.e., 0 or 1 valued) process. If the considered point process is a stationary Poisson, then $\{\xi_k\}$ is identically and independently distributed, that is, a Bernoulli series. Otherwise, it is easily seen that, in general, the joint probability of the sequence is essentially determined by a sequence of conditional probabilities $P\{\xi_k = 1 \mid \xi_1, \ldots, \xi_{k-1}\}$, $k = 1, 2, \ldots$, on the history of events. Roughly speaking, as a derivative of the conditional probability with respect to the time, the *conditional intensity function* is defined by $\lambda(t \mid F_t)$ given by

$$P\{N(t, t+\delta) = 1 \mid F_t\} = \lambda(t \mid F_t)\delta + o(\delta);$$

in other words,

$$\lambda(t \mid F_t) = \lim_{\Delta \to 0} \mathrm{P}\{\text{an event in } [t, t+\Delta) \mid F_t\}/\Delta, \tag{1}$$

where F_t is a set of observations up to time t, including the history of the occurrence times of the events $H_t = \{t_i; t_i < t\}$ (see Rubin, 1972, for example). The conditional intensity completely characterizes the corresponding point process(see Liptzer and Shiryaev, 1978). For example, if the conditional intensity depends only on the elapsed time since the last occurrence t_L, then the corresponding point process is a renewal process, and the conditional intensity coincides with so-called *hazard rate*, which is familiar in the field of reliability theory, in such a way that

$$\lambda(t \mid H_t) = \nu(t - t_L) = \frac{f(t - t_L)}{1 - F(t - t_L)}, \tag{2}$$

where $F(x)$ is the probability distribution function of interval length of consecutive points and $f(x)$ is its density function. A conditional intensity function which is only dependent on lengths of the last n intervals determines the Wold process of $(n-1)th$ order. Also, if the conditional intensity is independent of the history but dependent only on the current time t, like $\lambda(t \mid H_t) = \nu(t)$, then this gives a *non-stationary Poisson process*. Of course, a constant conditional intensity provides a stationary Poisson process. There are many other interesting classes of point processes which are defined by certain conditional intensity functions such as the self-correcting processes, $\lambda(t \mid H_t) = \nu(\alpha t - N(0,t))$, (see Isham and Westcott, 1979; Ogata and Vere-Jones, 1983; Vere-Jones, 1978; and Vere-Jones and Ogata, 1983), and Hawkes' self-exciting process $\lambda(t \mid H_t) = \mu + \int_0^t g(t-s)dN_s = \mu + \sum_{t_i<t} g(t-t_i)$, (see Hawkes, 1971; Hawkes and Oakes, 1974). The last process is very interesting and useful from a statistical prediction viewpoint, because, like the AR time series, the expectation of an event occurring is given by a linear combination of past occurrences, where the so-called *impulse response function* $g(\cdot)$ measures the weights of such combinations (see Ogata and Akaike, 1982; and Ogata and Katsura, 1986, for example).

2.1.2. *Thinning simulation method.* It is most important that the conditional intensity function completely characterizes the corresponding point process. One practical application of this fact is in simulation of point process data using a conditional intensity function. In principle we can obtain a sample point process, or a series of events, by making use of the interval distributions. That is to say, consider a conditional distribution $F(t \mid t_1, \ldots, t_n)$ of an event occurring at time t, given a history of events occurred at times $t_1, t_2, \ldots, t_n$. Then the relation between this distribution and the corresponding conditional intensity is given by the equation,

$$\lambda(t \mid H_t) = \frac{f(t \mid t_1, \ldots, t_n)}{1 - F(t \mid t_1, \ldots, t_n)}, \tag{3}$$

where f is the density function of F (see Liptzer and Shiryaev, 1978, for example). This is an extension of equation (2) related to the hazard rate of renewal process. Solving this equation we have

$$F(t \mid t_1, \ldots, t_n) = 1 - exp\{-\int_{t_n}^{t} \lambda(s \mid t_1, \ldots, t_n)ds\}. \tag{4}$$

Now let me start with the simulation of a stationary Poisson process. Since the conditional intensity rate is constant, say λ, we get from (4) that distribution function of the intervals is exponential, and the sample events are then simulated, by the inversion of uniform random variable U_i, in such a way that $t_i = t_{i-1} - \log(U_i)/\lambda$. For general case, including renewal processes or non-stationary Poisson processes, we have to solve the equation

$$F(t \mid t_1, \ldots, t_n) = U_{n+1} \tag{5}$$

of t, given a uniform random variable U_{n+1} (see Ozaki, 1978). However, this simulation procedure is not efficient nor practical because of two technical reasons: firstly we sometimes have to use numerical integration in (4), and secondly we usually have to use numerical iteration algorithm to solve the equation (5).

Lewis and Shedler (1979) introduced the *thinning method* for the simulation of non-stationary Poisson process. The basic algorithm is very simple:

1. given a intensity function $\lambda(t)$, take a finite upper bound, say Λ, such that $\lambda(t) \leq \Lambda$, $0 \leq t \leq T$;
2. simulate trial events $\{t_i\}$ of the stationary Poisson with intensity Λ; and
3. for each event t_i we keep it with probability $\lambda(t_i)/\Lambda$, otherwise eliminate it.

Then the remaining events forms sample of the non-stationary Poisson process with intensity $\lambda(t)$. Actually this method is very similar to so-called the rejection method for simulating samples from a probability density function, $f(t) = \lambda(t)/\int_0^T \lambda(t)dt$, when the total number of accepted points is fixed. Nevertheless, the idea of thinning can be extended to general point processes. For example, a useful algorithm for simulation of events is as follows:

1. set $n = 0$, $t_n = 0$;
2. set $i = 1$, $r_1 = 0$;
3. take any $0 < r \leq \infty$ and find a constant Λ_i such that $\lambda(t \mid H_t) \leq \Lambda_i$ holds for all $t \in [r_i, r]$;
4. set $j = 0$, $\tau_0 = 0$;
5. set j equal to $j + 1$ and generate a uniform random variable U_j, and set $\tau_j = \tau_{j-1} - \log(U_j)/\Lambda_i$;
6. if $r_i + \tau_j > r$, set i equal to $i + 1$, set $r_i = r$, and go to step 3;
otherwise go to the next step;
7. set j equal to $j + 1$ and generate a uniform random variable U_j;
if $\lambda(t_n + r_i + \tau_j)/\Lambda_i > U_j$ then go to step 5, otherwise go to the next step;
8. set n equal to $n + 1$, then set $t_n = t_{n-1} + r_i + \tau_j$ and go to step 2.

This procedure is repeated until τ_j get out of the time interval $[0, T]$.

The thinning method can be further extended to the case of *multivariate* or *multi-channel* point processes as well as marked point processes with general state including space-time point processes (Musmeci and Vere-Jones, 1992). Also see Hawkes (1971) as an example of mutually interactive multivariate point process. For simplicity consider the bivariate point process $\{N_1(\cdot), N_2(\cdot)\}$ whose joint intensity rate is given by $\{\lambda_1(t|F_t), \lambda_2(t|F_t)\}$, where F_t is history of the both components of the process. Note that the *superposition* of the components, $N_1(\cdot)+N_2(\cdot)$, is specified by the direct sum

$$\lambda(t \mid F_t) = \lambda_1(t \mid F_t) + \lambda_2(t \mid F_t) \tag{6}$$

of the components of the corresponding conditional intensity. Then the simulation of the bivariate point process is carried out in a similar manner to the above algorithm except for the following remarks:

- A trial point $\tau = t_n + r_i + \tau_j$ is obtained as a univariate Poisson sample with intensity Λ_i which is an upper bound of the superposed intensity (6) on the determined interval.
- The trial point τ has three choices; that is, accepted as the first component $N_1(\cdot)$ with probability $\lambda_1(\tau \mid F_\tau)/\Lambda$, or accepted as the second component $N_2(\cdot)$ with probability $\lambda_2(\tau \mid F_\tau)/\Lambda$, otherwise deleted.

A mathematical justification of the thinning simulation is given in Ogata (1981) together with some details of the algorithms and numerical performances.

2.2 THE MAXIMUM LIKELIHOOD AND MINIMUM AIC PROCEDURES

Despite theoretical results advocating the maximum likelihood estimation procedure for point processes (Ogata, 1978, Ogata and Vere-Jones, 1984, and Kutoyants, 1979 and 1982), the modeling and its numerical implementation were not quite developed by the time. This is mainly due to that neither numerical algorithms of nonlinear optimization method were familiar nor any suitable criterion was available for the comparison of models besides the conventional likelihood ratio testing procedure for nested family of models. Now, with advancement of computing facilities, the major difficulties in the numerical aspect seem to be disappearing.

Given a set of occurrence data $t_1, t_2, ..., t_n$ in an observed time interval $[0, T]$ and a parameterized conditional intensity $\lambda_\theta(t|F_t)$, the likelihood is written down in the form

$$L_T(\theta \mid t_1, t_2, ..., t_n;\ 0, T) = \{\prod_{i=1}^{n} \lambda_\theta(t_i \mid F_{t_i})\} \exp\{-\int_0^T \lambda_\theta(t \mid F_t)dt\}.$$

The maximum likelihood estimate of θ is the value of the parameter vector which maximizes the likelihood or its logarithm

$$\log L_T(\theta \mid t_1, ..., t_n;\ 0, T) = \sum_{i=1}^{n} \log \lambda_\theta(t_i \mid F_{t_i}) - \int_0^T \lambda_\theta(t \mid F_t)dt. \tag{7}$$

If the integral in (7) can be expressed analytically in θ, then the gradients of the log-likelihood function can be easily obtained. In such case the maximization of the function can be carried out by using a standard non-linear optimization technique such as in Fletcher and Powell (1963). See Akaike et al. (1984) for the Fortran programs.

Suppose that we have to choose the best model among proposed competing models. The *Akaike Information Criterion* (Akaike, 1974, 1977),

$$\mathrm{AIC} = (-2)\ (\text{maximum log-likelihood}) + 2\ (\text{number of parameters})$$

is very suitable for such comparisons. Here "log" denotes natural logarithm, and a model with a smaller AIC is considered to be a better fit.

2.3 RESIDUAL ANALYSIS OF POINT PROCESSES

The AIC is useful for the comparison of competing models. However, having obtained the best model among proposed ones, there is still the possibility of the existence of a better fitting model. The *residual analysis* will be useful for checking such case (Ogata, 1988). Suppose that point-process data $t_1, ..., t_N$ are simulated from the conditional intensity $\lambda(t \mid H_t)$ (see Section 2.1 for the simulation algorithm by thinning). Then consider the integral

$$\Lambda(t) = \int_0^t \lambda(s \mid H_s) ds \tag{8}$$

of the conditional intensity function, which is a increasing function since $\lambda(s \mid H_s)$ is non-negative. Consider the transformation of time $\tau = \Lambda(t)$ from t to τ so that the original occurrence data $t_1, ..., t_N$ are transformed 1-to-1 into $\tau_1, ..., \tau_N$. Then, it is known that $\tau_1, ..., \tau_N$ are distributed as the standard stationary Poisson process (i.e., with the constant intensity 1; see Papangelou, 1972).

The similar transformation is considered using the estimated intensity $\lambda_{\hat{\theta}}(t \mid H_t)$ for the integrand, and we have the corresponding transformed data $\hat{\tau}_1, ..., \hat{\tau}_N$ which we call *residual process*. If the estimated conditional intensity is a good approximation to the true intensity, then the residual process is expected to behave like the standard stationary Poisson process. In other words, a deviation of any statistical property of the residual process $\hat{\tau}_1, ..., \hat{\tau}_N$ from that expected from the stationary Poisson process implies the existence of some features in the data $t_1, ..., t_N$ which were not captured by the model $\lambda_{\hat{\theta}}(t \mid H_t)$. Any conventional graphic tests for complete randomness, or the stationary Poisson, such as those in Cox and Lewis (1966) can be useful for such *residual analysis* (Ogata, 1988). When a model is reasonably good, the residual data can often provide new findings which were hard to see from the original data.

3. Application to earthquake occurrence data

The published lists of earthquakes usually consist of occurrence times, locations, depths and magnitudes. Such records from a self-contained seismic regions reveal time series of extremely complex structure. Large fluctuations in the numbers of shocks per month, complicated sequences of related shocks, dependence on activity from other seismic regions, fluctuation of seismicity on a larger time scale, and changes in the detection level of shocks, all appear to be characteristic features of such records. In our present analysis the origin times only are considered, other elements being largely ignored, except that the ETAS model discussed later includes magnitude data as covariates. Statistical application of point-process modeling for such data is pioneered by D. Vere-Jones (e.g., Vere-Jones and Davis, 1966; Vere-Jones, 1970; etc.). In the present review, emphasis is made on the intensity modeling of point process for earthquake series. Other important and frequently discussed data relates to magnitudes. Although I have written a number of works on such subject (Ogata, 1989; Ogata and Katsura, 1986; Ogata et al., 1991; and Ogata and Katsura, 1992), the methods are quite different from those given in the previous section, so that I will not review them.

The computer programs for the analyses discussed in this paper are being prepared for publication in the Computer Monographs of the Institute of Statistical mathematics, Tokyo, and also in the Software Library of the International Association of Seismology and Physics of the Earth's Interior (IASPEI) in collaboration with Seismological Society of America, accepting the invitation of the editorial committee.

3.1 DETECTION OF SEASONALITY FOR SHALLOW EARTHQUAKES AND CAUSAL RELATIONSHIP BETWEEN SEISMIC ACTIVITIES IN DISTINCTIVE AREAS

3.1.1. *Linearly parameterized intensity models.* Ogata (1978, 1983b) recommended a systematic use of the following parameterizations for the conditional intensity function

$$\lambda_\theta(t \mid F_t) = \sum_{k=1}^{K} \theta_k \cdot Q_k(t \mid F_t), \tag{9}$$

$$\log \lambda_\theta(t \mid F_t) = \sum_{k=1}^{K} \theta_k \cdot Q_k(t \mid F_t), \tag{10}$$

where $\theta = (\theta_1, ..., \theta_K)$ and each $Q_k(t \mid F_t)$ is independent of the parameter θ to be adjusted. A main advantage of such parameterization is that the log-likelihood function (7) has at most one maximum, no matter how the dimension of the parameter increases. This is because the Hessian matrix is everywhere negative-definite in θ. In the case of (9), if the analytic calculation of the integral $\int Q_k(t \mid F_t)dt$ is feasible, then the maximization algorithm for the log likelihood function (7) is implemented efficiently. However the major disadvantage of (9) is that the conditional intensity function may be negative for some values of t. The larger the number of the parameters, the more readily this occurs. This causes some difficulty in the numerical process of maximizing the log-likelihood, because the negative values in the conditional intensity function contributes to the seeming increase of log-likelihood due to the second term of (3). In order to retain non-negative conditional intensity function and to stabilize the non-linear optimization of such log-likelihood function, we impose restrictions on the parameters using a certain concave smooth penalty function $R(\theta)$ which takes the value 0 in the restricted region but takes large values on the outside, and then minimize

$$G(\theta) = -\log L(\theta) + R(\theta) \tag{11}$$

which is also concave since the intensity is linearly parameterized. The details are provided in Ogata and Katsura (1986).

In the log-linear case in (10), the analytic calculation of the integral is not generally feasible, except in some particular cases (e.g., Lewis, 1970). However, for the low-frequency variable intensity model such as the exponential polynomial rate for a trend (e.g., McLean, 1974), numerical integration well approximates the integral so that the maximum likelihood procedure is feasible. This parameterization has the advantage of non-negative valued intensity without any constraint within the whole parameter space. A FORTRAN program package TIMSAC-84 (Akaike et al., 1984) includes automatic and objective analyses for trend and cyclicity of point process data by the extensive use of the minimum AIC procedure. Berman and Turner (1992) recommend the use of GLIM for an alternative calculation procedure of the maximum likelihoods for the models with the form of (9) and (10).

Below, we apply the linearly parameterized intensity models in (9) to determine the approximate shapes of trend, periodic and clustering effects, and the relation to the activity in other seismic areas, if such exist.

3.1.2. *Trend/cyclic/clustering decomposition of earthquake risk.* In statistical seismology the analysis of cyclicity such as seasonal, monthly, diurnal and semidiurnal, has been attracting the interest of many investigators. The most common method is testing of uniformity of superposed point process on a unit circle. One may use statistics on the circle such as the periodogram by Schuster (1897). Vere-Jones and Ozaki (1982) and Vere-Jones (1985) discuss the significance level of periodicity in the periodogram on the basis of the extreme theory of the chi-square distributed statistics. Other method is use of histogram by a suitably categorized data. For instance, Oike (1977) suggests the seasohality of shallow seismicity in comparison with change of a precipitation pattern.

However, this kind of testing procedures must be carried out with declustered events and further without any trend whichever it is of real seismicity or of artificial due to observations. Carefully treated papers removes some portion of events somehow which usually reduces the data size significantly. For the effective use of data, Ogata (1983b) suggested a model which takes the form

$$\lambda_\theta(t \mid F_t) = a_0 + P_J(t) + C_K(t) + \sum_{t_i<t} g_M(t - t_i). \tag{12}$$

The first and second components in (12) represent the evolutionary trend where

$$P_J(t) = \sum_{i=1}^{J} a_j \phi_j(t/T), \qquad 0 < t < T, \tag{13}$$

T is the total length of the observed interval and $\phi_j(\cdot)$ is a polynomial of order j. We introduce these components either for the genuine seismic trend or the evolutionary change of the detection rate of shocks, or the both. The third component is the Fourier expansion

$$C_K(t) = \sum_{k=1}^{K} \{b_{2k-1} \cos(2k\pi t/T_0) + b_{2k} \sin(2k\pi t/T_0)\}, \tag{14}$$

for cyclic effects with a given fixed cycle length T_0. The last component in (12) stands for the clustering effects such as aftershocks and earthquake swarms. The function $g_M(x)$ measures the increase in risk of clustering due to a shock. We may call this function a *response function* of a shock, and parameterize it as

$$g_M(x) = \sum_{m=1}^{M} c_m x^{m-1} e^{-\alpha x}, \tag{15}$$

If the scaling parameter α is fixed in (15), then the model in (12) is linearly parameterized. This model is used for examining and establishing the existence of each component by the comparison of AIC values among a possible set of (J, K, M)-configurations. For instance, if the configuration with $K = 0$ is selected, that is to say, $C_0(t) = 0$, then this suggest no periodicity of T_0 existing in the seismic activity. Otherwise, its shape is estimated by the maximum likelihood method for the selected configuration of $(\hat{J}, \hat{K}, \hat{M})$.

Applying this model to shallow seismicity in the Inner Zone of the Southwest Japan, Ogata (1983b) showed the existence and the shape of seasonal component which is quite similar to the derivative of the averaged and smoothed precipitation curve in that area, where peaks are found at the typhoon season around early September and at the beginning of spring. As suggested by Oike (1977), it is understood that the drastic change of rainfall relates to the drastic change of underground water migration which can trigger shallow earthquakes with some probability. A similar feature is seen in the micro-seismicity around the Australian Capital Territory (ACT) including Canberra, Australia (Ogata and Katsura, 1986). The seasonality looks similar to the Sydney rainfall pattern rather than Canberra's. This might suggest some correlation between earthquake occurrence around ACT and the migration of underground water beneath the south-eastern highlands from Eastern Australia.

The present model is also included in TIMSAC-84. Matsumura (1986) extensively applied the program to whole-world earthquake data for some decades to find the seasonality in a number of areas of mid-latitude where there is clear seasonality in precipitation patterns.

3.1.3. *Detection of causal relations between two series of events.* It is also an interesting subject for the prediction of earthquakes to find a statistical causality of a seismic risk with some other observations. Conventionally, this problem have been investigated by examing the second order properties between point processes such as cross-covariances or cross-spectra (e.g., Brillinger, 1975). However, having known a cross-correlations between the point processes $N_1(\cdot)$ and $N_2(\cdot)$, say, this cannot discriminate the possibilities among the cases where (i) $N_1(\cdot)$ causes $N_2(\cdot)$, (ii) $N_2(\cdot)$ causes $N_1(\cdot)$, (iii) $N_1(\cdot)$ and $N_2(\cdot)$ cause each other, and (iv) some other factor causes the both $N_1(\cdot)$ and $N_2(\cdot)$. To discriminate (i), (ii) and (iii), Ogata and Akaike (1982) and Ogata et al. (1982) suggest a parametric model for the minimum AIC procedure based on Hawkes' mutually-exciting process (Hawkes, 1971). Further, for examining case (iv) as well, Ogata (1983b) extends the model in (12) to

$$\lambda_\theta(t \mid F_t) = a_0 + P_J(t) + C_K(t) + \sum_{t_i<t} g_M(t - t_i) + \sum_{u_j<t} h_N(t - u_j), \tag{16}$$

where $\{u_j\}$ is another series of events considered as inputs of the conditional intensity function. The response function $h_N(x)$ is parameterized by

$$h_N(x) = \sum_{n=1}^{N} d_n x^{n-1} e^{-\beta x}. \tag{17}$$

If there is no causal relation of $\{u_j\}$ to the conditional intensity function $\lambda_\theta(t \mid F_t)$, or to $\{t_j\}$, then $h_N(x) = 0$ is expected. Otherwise we are interested in knowing the influential scale and the approximate shape of the response function.

A parallel model to (16) for $\{u_j\}$ versus $\{t_i\}$ as the inputs is also considered. Since the joint density distribution of $\{t_i, u_j\}$ with bivariate intensity $(\lambda_\theta(t \mid F_t), \mu_\varsigma(u \mid F_u))$ is given by

$$L(\theta, \varsigma) = \prod_{t_i<t} \lambda_\theta(t_i) \prod_{u_j<t} \mu_\varsigma(u_j) \exp\{-\int_0^t (\lambda_\theta(s) + \mu_\varsigma(s) - 1)ds\},$$

it is noted that the maximization of the log-likelihood for the mutually exciting bivariate point process $\{t_i, u_j\}$ is equivalent to the maximization of both of the two partial log-likelihoods of the type in (7), as far as the sets of parameters between θ and ς of the two conditional intensity function are disjoint. Therefore the detection of mutual or one-way stimulation are reduced to the independent analysis of the model (16) using the partial likelihood. It is also noted that the existence of the second order correlation between the processes $\{t_i\}$ and $\{u_i\}$, or between their conditional intensity function, does not necessarily imply the existence of a stimulation. For example, a similar shape or periodicity of corresponding $P_J(t)$ or $C_K(t)$ can make some correlation even if $h_N(x) = 0$ holds which stands for the case (iv). There are also cases where some unknown factors which are approximated by $P_J(t) + C_K(t)$ stimulate both $\{t_i\}$ and $\{u_j\}$.

Utsu (1975) discussed the correlation between the deep earthquakes in Hida (the region around Takayama city, Gifu Prefecture, central Japan) and the shallow earthquakes in the central Kwanto (the region around Tokyo). He tested the independence of the Kwanto earthquakes from those of Hida area, assuming the former shocks to form a stationary Poisson process, and concluded that there was a significant dependence which could be attributed to the mechanical connection between the two seismic regions which belong to the same segment of the Pacific plate underthrusting at the northeastern Japan arc. The data compiled by Utsu is composed of 61 earthquakes with the magnitude $M \geq 5.5$ in central Kwanto and 16 earthquakes of $M \geq 5.0$ in Hida during the 51 years from 1924 through 1974.

Since the earthquakes in Kwanto District seem to have some slight clustering, Ogata et al. (1982) fitted the model (16) with $P_J(t) = C_K(t) = 0$ against the Hida shocks as inputs to estimate the scale and shape of the response functions $g_M(\cdot)$ and $h_N(\cdot)$. A similar analysis was made for Hida seismicity against Kwanto shocks as inputs. The results of the present analysis showed that the earthquakes in central Kwanto region are not only self-exciting but also receive significant one-way stimulation from earthquake occurrences in Hida region, and that Hida shocks are identified as a stationary Poisson process. Further, responding to a criticism that the data size of the Hida shocks may not be enough to conclude the independency from Kwanto shocks, Ogata and Katsura (1986) considered the similar problem extending the both areas with lower magnitude thresholds for the two areas: as a consequence, the time span had to be reduced for the completeness of the data. Except that some different trends exist in the two areas due to the increase of data sizes, the causality between the two area is the same as the case of Utsu data.

The model (16) is also applied to the shallow and deep earthquakes in a region covering the North Island area, New Zealand, from 1946 through 1980. It is clearly shown that the earthquake occurrences in the deep region significantly receive one-way stimulation from earthquake occurrences in the shallow region. This is different outcome from the above case of Japan where existence of the opposite one-way stimulation is concluded from the Utsu data (Utsu, 1975) or related data. Nevertheless, we learned after the present analysis that these two types of relationship between the shallow and deep seismicity in the eastern Pacific region was already discussed by Mogi (1973). Especially he found that the seismic activity in Mariana and Tonga areas gradually migrated from the shallow to the deep regions within the descending lithosphere, while the opposite migration is found in the Kurile-Kamchatka and northern Japan island-arc region. The migration rate or speed along the deep seismic zone of Tonga arc suggested by Mogi is about 45 km per year with which our estimated response impulse function is consistent. These may indicate that a similar tendency holds within the Tonga-Kermadec-New Zealand tectonic zone.

The present model was also used by De Natale et al. (1988) for the investigation of the causal relation between Calabrian and North-Agean earthquake sequence. This model is expected to be of wider use to investigate a causality in such a way that input process can be events of any type such as times when predictions are made.

3.2 ANALYSIS OF AFTERSHOCK OCCURRENCE

3.2.1. *Modified Omori formula.* The frequency of aftershocks per unit time interval (one day, one month, etc.) is well represented by the modified Omori formula (Utsu, 1961)

$$n(t) = K(t+c)^{-p} \quad (K, c, p: \text{ parameters}) \tag{18}$$

where t is the lapse time from the occurrence of the main shock, and K depends on the magnitude of the main shock and the lower bound of the magnitude of aftershocks counted, while p is known to be independent of these. According to Utsu (1969), for the case of Japanese aftershock sequences, the constant c is small, 1 day at most, and p varies in the range $0.9 \sim 1.9$ with values between 1.0 and 1.4 most frequent. In the case of $p = 1$, formula (18) is called the original Omori formula (Omori, 1894), while for general p, it is called the modified Omori formula (Utsu, 1957, 61). The value p is thought to reflect mechanical conditions of earth's crust. For example, Mogi (1962) demonstrate a certain systematic regional distribution of p values in Japan. Traditionally, estimates of the parameter p have been obtained in the following way. Plot $n(t)$ versus time t on a log-log scaled plane and then fit an asymptotic straight line; the slope of the line is an estimate for p. The values of c can be determined by another empirical technique.

Consider occurrence times of aftershock sequence $\{t_1, t_2, ..., t_N\}$ in a time interval $[S, T]$, where the origin of the time scale, $t = 0$, corresponds to the occurrence time of the main shock. Let us assume that the aftershock sequence is distributed according to a non-stationary Poisson process with the conditional intensity function

$$\lambda(t;\theta) = K(t+c)^{-p}, \quad \theta = (K, c, p), \tag{19}$$

which represent the modified Omori formula. Then from (7) the log-likelihood function of the aftershock sequence is written by

$$\log L(K, c, p;\ S, T) = N \log K - p \sum_{i=1}^{N} \log(t_i + c) - K\Lambda(c, p;\ S, T), \tag{20}$$

where

$$\Lambda(c, p;\ S, T) = \begin{cases} [(T+c)^{1-p} - (S+c)^{1-p}]/(1-p) & \text{for } p \neq 1 \\ \log(T+c) - \log(S+c) & \text{for } p = 1. \end{cases}$$

The maximum likelihood procedure has also the advantage of producing estimates for the asymptotic standard errors of parameters. In our case the Fisher information matrix is given by

$$\begin{aligned} J(\theta; S, T) &= \int_S^T \frac{1}{\lambda(t;\theta)} \frac{\partial\lambda(t;\theta)}{\partial\theta'} \frac{\partial\lambda(t;\theta)}{\partial\theta} dt \\ &= \int_S^T \begin{bmatrix} K^{-1}(t+c)^{-p} & -p(t+c)^{-p-1} & -(t+c)^{-p}\ln(t+c) \\ * & Kp^2(t+c)^{-p-2} & Kp(t+c)^{-p-1}\ln(t+c) \\ * & * & K(t+c)^{-p}\ln^2(t+c) \end{bmatrix} dt, \end{aligned} \tag{21}$$

and the inverse of the matrix (21) provides the asymptotic covariance matrix of the errors of the maximum likelihood estimates.

3.2.2. *Model selection for complex aftershock sequences.* First of all, we usually have the problem of determining the starting time S in the calculation of the log-likelihood (20) for aftershocks of a large earthquake. If the aftershocks were observed homogeneously (i.e., with a constant detectability) throughout the whole observation period $[0, T]$ then we can put $S = 0$. However, the rate of missing events immediately after the main shock can be large compared to the later parts. Another possibility may be that the beginning stages of some aftershock sequences are too complicated to be described by (19). In such cases, if the MLE is computed with putting $S = 0$, the estimates are very likely to be biased. Therefore we have to find a suitable S both to avoid significant bias and to get accurate estimation of p and c. One of the feasible methods for such a purpose is the comparison between

$$\mathrm{AIC} = (-2)\max \ln L(\theta; 0, T) + 2 \times 3$$

and

$$\mathrm{AIC} = (-2)\max \ln L(\theta; 0, S) + (-2)\max \ln L(\theta; S, T) + 2 \times 6 \tag{22}$$

for some candidates of S. Here we automatically take discrete values of S, say $S = i/8$ ($i = 1, 2, ..., 8$) days. Among the models we will adopt the one which has the least AIC value. Another possibility is just hypothesizing S subjectively by inspecting the log-log plots by eyes.

Next, if an earthquake sequence on an observed time interval does not consist of single pure aftershock sequence, the model (19) may not hold any more. The sequence of shocks may contain the secondary aftershocks (Utsu, 1970). another possibility may be that the modified Omori formula would not hold in some later stage [see Utsu (1957) and Mogi (1962) for example]. Testing these problems is reduced to the matter of model selection against alternative models such as

$$(i)\quad \lambda(t) = \begin{cases} K_1(t+c_1)^{-p_1} & for \quad 0 < t \le t_0, \\ K_1(t+c_1)^{-p_1} + K_2(t-t_0+c_2)^{-p_2} & for \quad t_0 < t, \end{cases} \tag{23}$$

and

$$(ii)\quad \lambda(t) = \begin{cases} K_1(t+c)^{-p} & for \quad 0 < t \le t_0, \\ K_2 e^{-\alpha t} & for \quad t_0 < t, \end{cases} \tag{24}$$

where $\theta = (K_1, K_2, c_1, c_2, p_1, p_2)$ for case (i) and $\theta = (K_1, K_2, c, p, \alpha)$ for case (ii).

Furthermore, when we are interested in examining whether a pair of aftershock sequences from distinctive areas or time periods have the same value p or not, we will calculate both

$$\mathrm{AIC}_0 = (-2)\max\{\ln L_1(K_1, p, c_1) + \ln L_2(K_2, p, c_2)\} + 2 \times 5$$

and

$$\mathrm{AIC}_1 = (-2)\{\max \ln L_1(K_1, p_1, c_1) + \max \ln L_2(K_2, p_2, c_2)\} + 2 \times 6, \tag{25}$$

where L_1 and L_2 stand for the likelihood of the first and second series of aftershocks, respectively. Then we choose the model which gives the smaller AIC values. See Ogata (1983a) for further details.

3.2.3. *Residual analysis of aftershock activity.* The integration in (8) gives a transformation to a frequency-linearized time τ where the occurrence of earthquakes becomes the standard stationary Poisson process (i.e. with unit intensity), if the intensity function $\lambda(t)$ in (8) is the true one for the data. For instance, the frequency-linearized time for a single aftershock sequence can be given by

$$\begin{aligned}\tau = \Lambda(t) &= \int_0^t K(s+c)^{-p} ds \\ &= \begin{cases} K \ln(t+c)/c & \text{for } p = 1.0 \\ K[(t+c)^{1-p} - c^{1-p}]/(1-p) & \text{otherwise} \end{cases}\end{aligned}$$

Using this transformation, it can be seen that the modified Omori formula holds as long as the occurrence rate of aftershocks surpasses that of background activity around the aftershock region (Ogata and Shimazaki, 1984). That is to say, cumulative numbers of aftershocks aginst the transformed time aligns on a straight line until the rate of occurrence is lower than the level of the background activity of the area. Therefore, if the level of the background seismicity around an aftershock region is very low, then the intensity trend of aftershocks of a major earthquake decays according to the formula in (19) for a long time span, even up to one century (Ogata, 1989). Utsu (1969) showed this fact directly by the log-log plot of frequency rate versus the elapsed years for the felt aftershock of Nobi earthquake of 1891.

It is often observed that a sequence of aftershocks contains secondary aftershocks, or aftershocks of a major aftershock (Utsu, 1970). If a secondary aftershock sequence starts at time $t = t_0$, then the intensity function $\lambda(t)$ can be given in (23), and the MLE of θ is also obtained by maximizing (7). For instance, Ogata and Shimazaki (1984) investigated the aftershock sequence of the 1965 Rat Islands earthquake of $M_w = 8.7$, for the time span about fifteen years. The selection of a statistical model based on AIC showed that the whole aftershock sequence is better modeled by the combination of an ordinary aftershock sequence and secondary aftershock activity associated with the largest aftershock, the 30 March 1965 event. The difference in both the parameters p and c between the main and the secondary aftershock sequences is insignificant, indicating that despite the contrast of the focal mechanism and tectonic location between the main shock and the largest aftershock, the decays in their aftershock sequences were very similar. The frequency-linearized time for a sequence including the secondary aftershock sequence which starts at time $t = t_0$, is given as

$$\tau = \int_0^t K_1(s+c_1)^{-p_1} ds + H(t-t_0) \int_{t_0}^t K_2(s-t_0+c_2)^{-p_2} ds, \tag{26}$$

where $H(t)$ is the Heaviside unit step function.

The transformation in (26) by substituting the MLEs shows that a transition from aftershock to normal activity is identified 2200 days since the main shock occurrence. After the transition time, we observe a significant upward deviation of the cumulative number curve against τ from the straight line, which means the departure of the seismicity in the focal region from the extrapolated trend of the aftershock activity predicted by the modified Omori formula.

On the other hand, Habermann (1983) concluded that a decrease of activity in the west Aleutians (170°E-180°E) occurred during early 1967 and that it marks the end of the Rat Islands aftershock sequence. This conclusion is different from the result of the present study which indicates the aftershock sequence lasted at least 6 years. In looking at time-dependent seismicity patterns, he used an algorithm which should have removed all the dependent events, i.e., aftershocks. However, since the algorithm does not work very well in removing all the aftershocks for the 1965 Rat Islands sequence, he interprets the decrease in seismicity around 1967 as the end of the aftershock sequence. In the present study, an algorithm which *can* predict the aftershock sequence is used to show that the sequence lasted much longer. The techniques proposed by Ogata and Shimazaki (1984) allow straightforward and objective detection of a transition from aftershock to normal activity. Defining the normal activity is essential in detection of a seismic quiescence which may take place prior to the next major event.

Furthermore, the cumulative number of such residual processes sometimes deviates downwards significantly in some stage, which means that the rate of aftershocks decays quicker in the corresponding stage than the expected rate according to the modified Omori formula. Using the methods and models described in in this section, Matsu'ura (1986) extensively investigated this phenomenon in aftershock sequence data of Japanese major earthquakes and found that largest aftershocks were very likely to take place in the recovery stage from the phenomenon in many cases. The reported phenomenon was nominated as one of the "Preliminary Candidate for Outstanding Geophysical Precursors to Earthquakes" by the IASPEI Sub-Commission on Earthquake Prediction (M. Wyss, Chairman) at the IASPEI Istanbul assembly at August 1989. The same results are confirmed in the aftershock sequences in China (Zhao et al. 1989).

3.3 THE ETAS MODEL AND RELATIVE QUIESCENCE

3.3.1. *Epidemic type aftershock-sequences model.* In many cases, aftershock activity appears multiple in such a way that

$$\lambda(t \mid H_t) = \mu + \sum_{\tau_m < t} \frac{K_m}{(t - \tau_m + c_m)^{p_m}}, \tag{27}$$

where μ stands for the rate of background activity, and $\{\tau_m\}$ are origin times of triggering events of secondary aftershocks, and the sum $\sum_{\tau_m<t}$ is taken over the trigger events $\{m\}$ which satisfy $\tau_m < t$. See previous subsection for the procedure of identifying the triggering events $\{\tau_m\}$ and the estimation of the parameters in (7). This model can be also applied to the case of general seismic activity. The trigger events of course include the main shocks. This is called by the restricted trigger model in Ogata (1988). The original trigger model was suggested by Vere-Jones and Davies (1966) where triggering events are defined without identifying them but, for each event, a probability of being triggering one is assumed for the calculation of the second-order moment statistics such as spectrum. Therefore, unless the so-called spectral log-likelihood (e.g., Hawkes and Adamopoulos, 1973; Ogata and Abe 1991; and Ogata and Katsura, 1991) is used, identification of trigger events is necessary for the ordinary maximum likelihood estimation procedure of the trigger models.

Extending these models further, Ogata (1985, 88, 89) demonstrated that the ordinary seismic activity of a wide region is remarkably well described, in terms of the conditional intensity, by the superposition of a constant rate for background seismicity and the modified Omori functions of *any* shocks i which occurred at time t_i, in such a way that

$$\lambda(t \mid H_t) = \mu + \sum_{t_i<t} \frac{K_i}{(t - t_i + c)^p}, \tag{28}$$

where μ is an occurrence rate for the background seismic activity. The sum $\sum_{t_i<t}$ is taken for *all* shocks i whose occurrence times satisfy the inequality $t_i < t$, and the parameter K_i for each shock i contributes to the size of the corresponding offspring, or aftershocks in a wide sense. The crucial point of the model here is that the parameter $\{K_i\}$ is dependent on its magnitude M_i as well as the cut-off magnitude M_0 of the data set according to the following exponential function form

$$K_i = K_0 e^{\alpha(M_i - M_0)}. \tag{29}$$

One justification for this form is the empirical formula obtained by Utsu and Seki (1955) on the linear relation between the logarithms of aftershock areas and the magnitudes M of the main shock (see Utsu, 1971, for some plotted data supporting the relation). This relation suggests that the number N of aftershocks with magnitudes over a threshold M_0 for a fixed time span is roughly estimated as

$$N \propto exp\{\alpha(M - M_0)\}, \tag{30}$$

for a constant α, where M is the magnitude of the main shock. The relation (30) appears to be true also for a set of the secondary aftershocks (Utsu, 1970, Section 8.1, and also Ogata, 1983a and 89). Yamanaka and Shimazaki (1990) discuss the distinction of the proportional constant in (30), or K_0 in (29), between intra- and interplate earthquakes.

Model (28) with (29) for the ordinary seismicity in terms of the occurrence rate of shocks is hereafter called *Epidemic Type Aftershock-Sequences (ETAS) model.* Note here that $H_t = \{(t_i, M_i); t_i < t\}$ is the history of occurrence times $\{t_i\}$ up to time t and their corresponding magnitudes $\{M_i\}$. The original epidemic model, called birth and death process, was considered by Kendall (1949) in application to population genetics in epidemiology. Hawkes (1971) modelled the birth process, allowing immigration at a constant rate μ per unit time, in terms of the conditional intensity rate depending on the history of occurrence times.

Compared to the restricted trigger model, the current ETAS model is extremely useful, especially when the identification of main shocks (triggering events) and aftershocks (triggered events) is not easy, as in earthquake swarms. Even with ordinary seismicity, the rigorous discrimination of these is impossible. Ogata (1985, 1988) compares the goodness-of-fit of ETAS model with any partially different alternative models including the restricted trigger models to show the better fit than all of the considered cases. Further, it appears that the ordinary seismicity is long-range

TABLE 1. Estimates of the Parameters of the ETAS Model

Area	T(days)	M_0	N	AIC	μ	K	c	α	p
1	30681.	7.0	823	7501.9	0.0188	0.0182	0.2121	1.224	1.027
2	6539.8	4.7	817	711.1	0.0083	0.0420	0.0298	1.421	1.130
3	35062.	6.0	483	4378.4	0.0054	0.9769	0.0196	1.614	1.000
4	229.4	4.8	236	1233.1	0.1415	0.0149	0.2270	2.259	1.280
5	150.	4.0	95	−182.6	0.0752	0.0005	0.1499	2.920	1.374
6	150.	2.5	1630	−11766.1	0.4655	0.0002	1.103	3.105	1.720
7	150.	2.9	583	−2483.8	0.0000	0.0319	0.0378	1.686	1.138
8	20208.	4.0	595	2589.6	0.0038	0.0616	0.0119	0.358	1.143
9	2676.	2.5	1097	−4881.3	0.0221	0.0350	0.0031	0.394	1.353
10	33.	2.9	255	1494.4	0.5872	0.0164	0.0089	0.703	1.726
11	7457.	3.9	344	1561.0	0.0006	0.0923	0.1286	0.848	1.138

The area numbers correspond as follows:
1) World's major shallow earthquakes (Ogata, 1987);
2) Central Aleutian aftershocks (Ogata, 1989);
3) Off eastern coasts of Tohoku area (Ogata, 1988, 1989);
4) 1968 Tokachi-Oki aftershocks (Ogata, 1987);
5) Niigata aftershocks (Ogata, 1987);
6) Niigata aftershocks; felt shocks (Ogata, 1987);
7) Western Nagano Prefecture aftershocks (Ogata, 1989);
8) Izu Islands area (Ogata, 1987);
9) Izu Peninsula area (Ogata, 1987);
10) Off east coast of Izu swarms and aftershocks (Ogata, 1987); and
11) Matsushiro Swarm (Ogata, 1987).

*) N is the number of shocks in the time interval T.
*) M_0 is the cut-off magnitude in equation (29) except for the Niigata aftershock sequences 5 and 6.

correlated and approximately self-similar for various ranges of time and space (e.g., Ogata, 1987; Ogata and Katsura, 1991; and Ogata and Abe, 1991), and ETAS model is shown to be self-similar in an approximate sense (Ramselaar, 1990).

Among the parameters $\theta = (\mu, K, c, \alpha, p)$ of ETAS model in (28) with (29), the last two parameters α and p are extremely useful for characterizing the temporal pattern of seismicity. The p value in (28) indicates the decay rate of aftershocks, and the α value measures an efficiency of magnitude of an earthquake in generating its offspring, or aftershocks in a wide sense. For example, Ogata (1987) shows that swarm-type activity has a smaller α value than that of ordinary main shock and aftershock activity, and that α is likely to be large if there are no conspicuous secondary aftershocks in an aftershock sequence. To illustrate this, Table 1 lists the maximum likelihood estimates of the ETAS model for the data from a variety of local and global seismic regions. The first four cases are typical seismicity, mostly at plate boundaries, and the next three cases are intraplate seismic activity, while the last four are the swarm type activity.

3.3.2. *Change point problem and model selection.* The question of whether or not the temporal patterns of seismicity changed before and after a time T_0 in a given

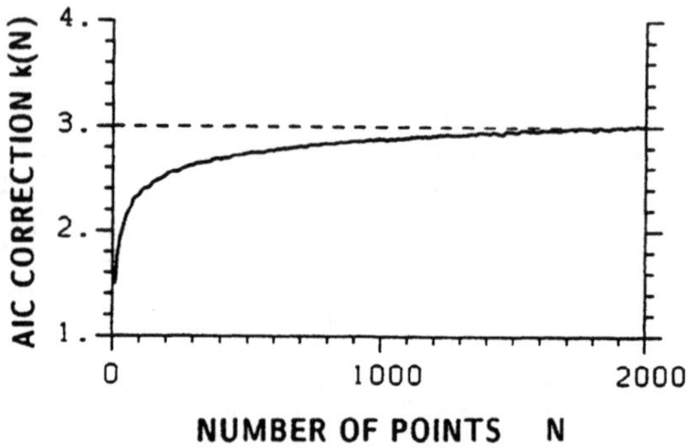

Figure 1. Penalty factor of the AIC for the change-point parameter. The factor $k(N)$ versus sample size N is obtained by a Monte Carlo simulation experiment: see Ogata (1992).

data set on a time interval $[0, T]$ can be reduced to a problem of model selection. First, we calculate the following AICs, taking it into account that each ETAS model usually has five parameters to be adjusted,

$$\begin{aligned} AIC_0 &= (-2)\max_\theta \ln L(\theta; 0, T) \ + 2 \times 5, \\ AIC_1 &= (-2)\max_\theta \ln L(\theta; 0, T_0) \ + 2 \times 5, \\ AIC_2 &= (-2)\max_\theta \ln L(\theta; T_0, T) \ + 2 \times 5, \end{aligned}$$

where the log-likelihood is given in (7). We should note here that the log-likelihood function $\ln L(\theta; T_0, T)$ includes the data in the previous time span $[0, T_0]$, as well as these in $[T_0, T]$, since $\lambda_\theta(t \mid H_t)$ is the function of the history from the time origin. Also, in order to minimize the effect of the unknown data before the observed time span $[0, T]$, it is recommended to set the starting time $t = 0$ to be the occurrence time of the last great earthquake in the considered region.

If a potential change point T_0 is set based on other information than the data, such as a scientific reason or a systematic setting, AIC_0 is compared with the following AIC_{12} of the alternative model for the changed seismicity at the time T_0 where

$$AIC_{12} = AIC_1 + AIC_2$$

(see Ozaki and Tong, 1975; and Kitagawa and Akaike, 1978). Otherwise, AIC_0 is compared with

$$AIC_{12} = AIC_1 + AIC_2 + 2k(N),$$

where the quantity $k(N)$ is the contribution of T_0 as an adjusted parameter and is dependent on the number of events in the interval $[0, T]$ as shown in Figure 1. The $k(N)$ is obtained by Monte Carlo experiments, and accurately given by the Padé approximant

$$k(N) = 1 + \frac{0.0770N + 0.000143N^2}{1 + 0.0511N + 0.0000650N^2}$$

for $10 \le N \le 2000$. Adjusting the parameter T_0 belongs to the so-called change-point problem, and the contribution of $k(N)$ is extraordinary because the ordinary contribution to the AIC is unity for a parameter which satisfies standard large sample theory. See Ogata (1992) for the estimation procedure of $k(N)$.

$AIC_{12} < AIC_0$ indicates a significant distinction between the seismicity patterns in the two divided time spans. This procedure can be extended for detection of further possible seismicity changes, but notice that number of parameters to be adjusted increases as model becomes complicated. See Ogata (1981, 1983a, b, 1985 and 1988) and Ogata et al. (1982, 1984 and 1986) for a variety of comparisons among statistical models for earthquake occurrence including (18) and (27).

3.3.3. *Residual point process and relative quiescence.* Suppose we have the maximum likelihood estimates of the parameters $\theta = (\mu, K, c, \alpha, p)$ of the ETAS model for general seismicity, including complex aftershock activity and swarms, assuming that the temporal seismic pattern is broadly identical throughout the time span and region. Now, consider the time-change, using the increasing function in (8), such that $\tau_i = \Lambda(t_i)$. Then origin times $\{t_i\}$ are transformed 1-to-1 into the Residual Point Process (RPP) $\{\tau_i\}$. Since we do not know the true model nor its parameter value, we use the maximum likelihood estimate $\hat{\theta}$ for the parameter value of the assumed conditional intensity model in (8). If the model provides a good fit to the seismicity, then the RPP is well approximated by the standard stationary Poisson process. If, on the contrary, we find a significant deviation of any characteristic property of the RPP from that expected from the stationary Poisson, this suggests some discrepancy between the model and the data, such as non-homogeneity of the data or the existence of seismic quiescence, which is not captured by the estimated ETAS model (28) with (29).

To demonstrate that RPP obeys very closely the standard Poisson process, we apply the following asymptotic theory: for the cumulative number of the standard Poisson N_τ defined on the interval $(0, \tau_0)$, the process $(N_{\sigma\tau_0} - \sigma\tau_0)/\sqrt{\tau_0}$ with σ converges to the standard *Brownian Bridge* $\{B_\sigma^0; 0 < \sigma < 1\}$ as τ_0 increases (Billingsley, 1961, p.65). Therefore, the mean and standard deviation of the cumulative number N_τ at fixed τ, $(0 < \tau < \tau_0)$, are τ and $\sqrt{\tau(1 - \tau/\tau_0)}$, respectively. Hereafter, in the figures showing the significance level of cumulative curve of RPP, we draw lines of 3-fold standard deviations at each τ. Among the significant deviations of RPP from what would be expected from the standard Poisson process, the *relative quiescence* is shown by the significant decrease of the occurrence rate of the RPP. Such quantitative analysis relates to the detection of statistical change-points. For instance, Habermann (1988), Reasenberg and Matthews (1988) and Wyss and Habermann (1988) apply their respective test statistics for detecting a change-point in the background seismicity and evaluating the significance of the naive quiescence against the stationary Poisson processes. Ogata (1992) likewise investigate the RPP events versus the transformed time. If the seismic activity is well described by an ETAS model, the cumulative curve of the RPP data is nearly a straight line like that of the stationary Poisson process. By contrast, a significant decrease of the slope shows the integrated effect of seismicity decreases from the expected rate, no matter how high or low the original seismicity is.

How can we see the significance of the seismicity decrease? We could apply the statistics suggested by the above authors to the RPP data. Instead, however, we adopt the following procedures. First of all, by inspection of the cumulative number curve of either the ordinary data or the corresponding RPP for the whole time interval $(0, S)$, we hypothesize a change-point t_0 where onset time of the relative quiescence is defined. This point is seen clearly in some cases, as described in Wyss and Habermann (1988). Otherwise, we can use the procedure described in the last paragraph in Section 3.3.2 to find the candidates of t_0. We then calculate the maximum of log-likelihoods to confirm whether $AIC_{12} < AIC_0$ or not, where the AICs stand for the data sets on the time intervals $(0, t_0) \cup (t_0, S)$ and $(0, S)$, respectively (see Section 3.3.2). If the above inequality holds, then the estimated ETAS model in the former time interval $(0, t_0)$ is identified to be the standard seismic activity of the focal region, and the time transformation in (8) from $(0, t_0)$ onto $(0, \tau_0)$ is made, by which the cumulative curve of RPP is expected to be nearly the straight line with unit slope on $(0, \tau_0)$. Then, the RPP and its cumulative curve is extended for the latter interval, using the same transformation from (t_0, S) onto the time interval (τ_0, σ). We then examine the deviation of the cumulative number of events on the time interval (τ_0, τ) from the corresponding expected number of events, that is to say, $\Delta N(\tau_0, \tau) \equiv N_\tau - N_{\tau_0} - (\tau - \tau_0)$, to measure the significance of the decrease after the suspected change-point τ_0 (see Figure 2a). If no statistical change has occurred, the extended RPP is exactly the standard Poisson process with the same unit intensity, and the deviation $\Delta N(\tau_0, \tau)$ of the cumulative curve is approximately distributed according to the normal distribution $\mathcal{N}(0, \Delta\tau)$, where $\Delta\tau = \tau - \tau_0$.

However, when the number of events N_{τ_0} in the interval $(0, \tau_0)$ is not large enough, the significance actually depends on the sample size, due to the estimation accuracy of the parameter for the transformation. Thus, taking such accuracy into consideration, the error distribution of $\Delta N(\tau_0, \tau)$ is modified to be $\mathcal{N}(0, \Delta\tau + (\Delta\tau)^2/N_{\tau_0})$. (See Ogata, 1992, Appendix 2). Based on the error distribution, as a change-point problem, we examine the significance level of the maximum downward deviation of $\Delta N(\tau_0, \tau_1)$ with respect to τ_0 and τ_1. Ten thousand standard Poisson sequences were simulated on the interval $(0, 1)$ with number $N = N(0, 1)$ of events (see Figure 2a), and for each sequence, the optimal times τ_0 and τ_1 are adjusted to reach the minimum

$$\xi(\hat{\tau}_0, \hat{\tau}_1) = \min_{\tau_0, \tau_1} \frac{N_{\tau_1} - N_{\tau_0} - \hat{\lambda}(\tau_1 - \tau_0)}{\sqrt{\hat{\lambda}(\tau_1 - \tau_0) + \hat{\lambda}^2(\tau_1 - \tau_0)^2/N_{\tau_0}}}, \tag{31}$$

where $\hat{\lambda} = N_{\tau_0}/\tau_0$, and the denominator in (31) corresponds to the standard deviation of the above-mentioned error distribution. For instance, in fewer than 1% of these sets, $\xi(\hat{\tau}_0; \hat{\tau}_1) \leq -2.4$,$-2.8$ and -3.1 for the case of simulated numbers $N = 200$, 500 and 1000, respectively (see Figure 2b). Thus, these levels indicate significance level

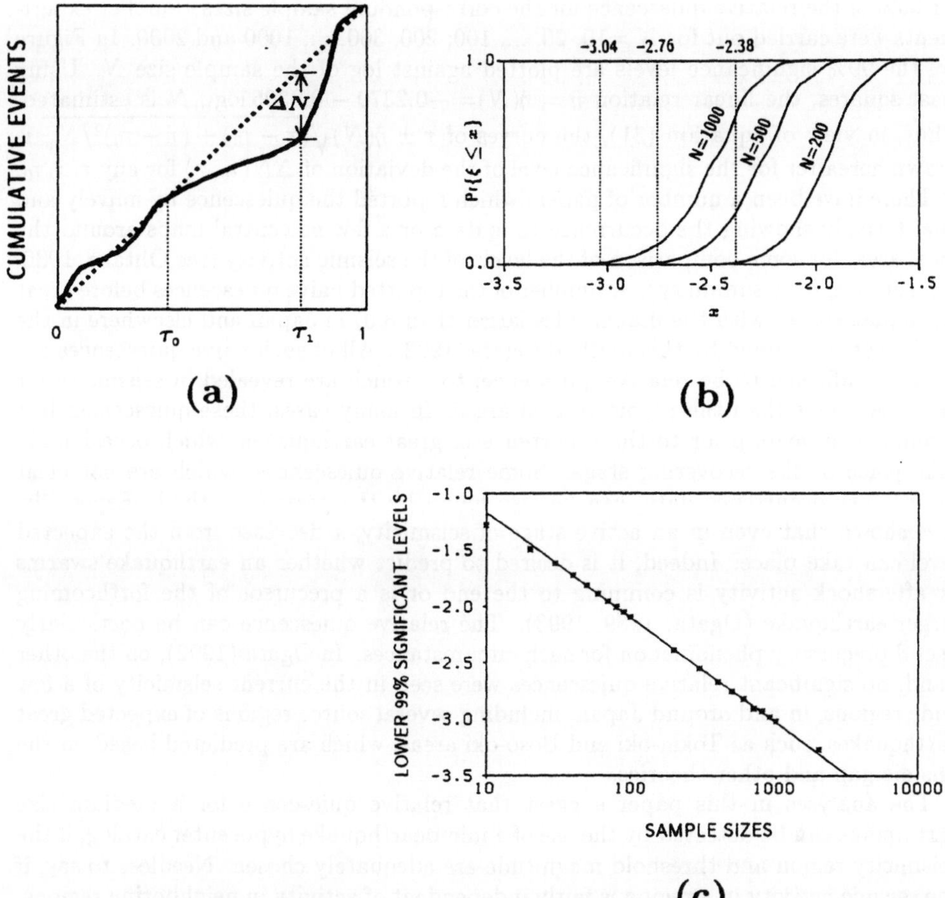

Figure 2. Illustration of experiment for evaluating the significance level of relative quiescence. (*a*) A schematic diagram of the cumulative number curve of a stationary Poisson process and the calculation of the ξ-statistic. Here, $\Delta N = N_{\tau_1} - N_{\tau_0} - \hat{\lambda}(\tau_1 - \tau_0)$ with $\hat{\lambda} = N_{\tau_0}/\tau_0$. See text for details. (*b*) Empirical distribution of the ξ-statistic for 10,000 repeated experiments with sample sizes N = 200, 500, and 1000, and their 99% significance levels (the vertical dotted lines). (*c*) Diagram of the lower 99% significant level of ξ-statistic against the sample size N in the logarithmic scale. Asterisks indicate data obtained by the simulated experiment, and the linear relation $\hat{\eta} = -0.2370 - 0.9305 \log_{10} N$ is obtained by the least squares fitted to asterisked data.

for 99% of the relative quiescence for the corresponding sample sizes. Further experiments were carried out for N =10, 20, ..., 100; 200, 300, ..., 1000 and 2000. In Figure 2c, the 99% significance levels are plotted against log of the sample size N. Using least squares, the linear relation $\hat{\eta} = \hat{\eta}(N) = -0.2370 - 0.9305 \log_{10} N$ is estimated. Thus, in view of equation (31), the curves of $\tau \pm \hat{\eta}(N)\sqrt{(\tau - \tau_0) + (\tau - \tau_0)^2/N_{\tau_0}}$ is drawn hereafter for the significance level of the deviation of $\Delta N(\tau_0, \tau)$ for any $\tau > \tau_0$.

There have been a number of papers which reported the quiescence by naively and qualitatively showing the occurrence time data or a few epicentral maps around the focal area, for some comparison of the levels of the seismic activity (see Ohtake, 1980, for instance, for a summary). A number of the reported naive quiescences before great earthquakes (i.e. with the magnitudes larger than 8.0) in Japan and elsewhere in the world were examined by this method (ogata, 1992). All of such naive quiescences are clearly confirmed to be relative quiescence, too, which are revealed in seismicity for far wider areas than the reported focal areas. In many cases, these quiescences last a number of years prior to the occurrence of great earthquakes, which occasionally take place in the recovering stage. Some relative quiescences, which are not clear in ordinary occurrence data, can be also found by the present method. Especially, it is shown that even in an active stage of seismicity, a decrease from the expected level can take place. Indeed, it is desired to predict whether an earthquake swarms or aftershock activity is comming to the end or is a precursor of the forthcoming larger earthquake (Ogata, 1989, 1993). The relative quiescence can be particularly useful precursory phenomenon for such circumstances. In Ogata (1992), on the other hand, no significant relative quiescences were seen in the current seismicity of a few wide regions, in and around Japan, including several source regions of expected great earthquakes, such as Tokai-oki and Boso-oki areas, which are predicted based on the seismic gap and other theories.

The analyses in this paper suggest that relative quiescence for a medium size earthquake can be detected by the use of a microearthquake hypocenter catalog, if the seismicity region and threshold magnitude are adequately chosen. Needless to say, if the seismic activity in a region is fairly independent of activity in neighboring regions, such as aftershocks of large earthquakes or very local swarm activity, selection of the data is usually easy. Otherwise, selection of a suitable region for analysis becomes increasingly critical as either the size of the data set increases or the cut-off level of magnitude decreases.

Acknowledgements

It is my great preasure to note that Professor Hiro Akaike organized the research project on statistical inference for point processes at the Institute of Statistical Mathematics to join Professor David Vere-Jones in 1976. Without such a fortunate opportunity and without their warmest encouragement and valuable suggestions since then, a series of my works as referred in the present paper could not exist. For all of these I am very grateful. Also, careful review of the referees were useful for clarification of the present paper.

References

Akaike, H.(1974). A new look at the statistical model identification. *IEEE Trans. Autom. Control* **19**, pp. 716-723.

Akaike, H. (1977). On entropy maximization principle, in *Applications of Statistics*, P.R. Krishnaiah (ed.), North-Holland, Amsterdam, pp. 27-41.

Akaike, H., Ozaki, T., Ishiguro, M., Ogata, Y., Kitagawa, G., Tamura, Y.H., Arahata, E., Katsura, K. & Tamura, Y.(1984). *Time Series Analysis and Control Program Package, TIMSAC-84*, The Institute of Statistical Mathematics, Tokyo.

Berman, M. and Turner, T.R. (1992). Approximating point process likelihoods *Applied Statist. (J. Roy. Statist. Soc. C)* **41**, pp. 31-38.

Brillinger, D. (1988). Some statistical methods for random process data from seismology and neurophysiology, The 1983 Wold Memorial Lectures. *Ann. Statist.* **16**, pp. 1-54.

Cox, D. (1962). *Renewal Theory*, Wiley, New York.

Cox, D. and Isham, V. (1980). *Point Processes*, Chapman & Hall, London.

Cox, D. and Lewis, P.A.W. (1966). *The Statistical Analysis of Series of Events*, Methuen, London.

Daley, D.J. and Vere-Jones, (1972). A summary of the theory of point processes, *Stochastic Point Processes: statistical analysis, theory and applications*, Lewis, P.A.W. ed., Wiley, New York.

Daley, D.J. and Vere-Jones, (1988). *An Introduction to the Theory of Point Processes*, Springer-Verlag, New York.

De Natale, G., Musmeci, F. and Zollo, A. (1988) A linear intensity model to investigate the causal relation between Calabirian and North-Agean earthquake sequences, *Geophys. J.*, **95**, pp. 285-293.

Fletcher, R., and M.J.D. Powell (1963). A rapidly convergent method for minimization, *Comput. J.*, **6**, pp. 163-168.

Habermann, R.E. (1983). Teleseismic detection in the Aleutian Island Arc, *J. Geophys. Res.*, **88**, pp. 5056-5064.

Habermann, R.E. (1988). Precursory seismic quiescence: past, present, and future, *Pure Apple. Geophys.* **126**, pp. 279-318.

Hawkes, A.G. (1971). Point spectra of some mutually exciting point processes, *J. R. Statist. Soc.* **B 33**, pp. 438-443.

Hawkes, A.G. and Adamopoulos, L. (1973), Cluster Models for Earthquakes–Regional Comparisons, *Bulletin of the International Statistical Institute*, **45**, Book 3, pp. 454-461.

Hawkes, A.G. and Oakes, D.A. (1974). A cluster process representation of self-exciting process, *J. Appl. Probab.* **11**, pp. 493-503.

Isham, V. and Westcott, M. (1979). A self-correcting point process, *Stoc. Proc. Appl.* **8**, pp. 335-348.

Karr, A.F. (1991). *Point Processes and Their Statistical Inference*, 2nd ed., Marcel Dekker, New York.

Kendall, D. G. (1949), Stochastic Processes and Population Growth, *Journal of the Royal Statistical Society*, Ser. B, **11**, pp. 230-264.

Kitagawa, G. and Akaike, H. (1978). A procedure for the modeling of non-stationary time series, *Ann. Inst. Statist. Math.*, **30**, pp. 351-363.

Kutoyants, Yu. A. (1979). Local asymptotic normality for processes of Poisson type, *Izvest. Akad. Arm. Nauk. Ser. Matematika* **14**, pp. 3-20.

Kutoyants, Yu. A., (1982). Multidimensional parameter estimation of the intensity function of inhomogeneous Poisson processes, *Probl. Control Inf. Theory*, **11(4)**, pp. 325-334.

Kutoyants, Yu. A., (1984). *Parameter Estimation of Stochastic Processes*, Helderman Verlag, Berlin.

Lewis, P. A. W. (1970) Remarks on the theory, computation and application of the spectral analysis of series of events, *J. Sound Vib.*, **12**, pp. 353-75.

Lewis, P. A. W. and Shedler, G. S. (1976) Statistical analysis of non-stationary series of events in a data base system, *IBM J. Res. Develop*, **5**, pp. 429-528.

Lewis, P.A.W. and Shedler, G.S. (1979). Simulation of non-homogeneous Poisson processes by thinning, *Naval Res. Logistics Quart.* **26**, pp. 403-413.

Liptzel, R.S. and Shiryaev, A.N. (1978). *Statistics of Random Processes, part II*, Springer, Berlin.

Maclean, C. J. (1974) Estimation and testing of an exponential polynomial rate function within the non-stationary Poisson process, *Biometrika*, **61**, pp. 81-86.

Matsumura, K. (1986). On regional characteristics of seasonal variation of shallow earthquake activities in the world, em Bull. Disas. Prev. Inst. of Kyoto Univ. **36**, pp. 43-98.

Matsu'ura R. S. (1986). Precursory quiescence and recovery of aftershock activities before some large aftershocks, *Bull. Earthq. Res. Inst., Univ. Tokyo* **61** pp. 1-65.

Mogi, K. (1962) On the time distribution of aftershocks accompanying the recent major earthquakes in and near Japan, *Bull. Earthquake Res. Inst.*, Univ. of Tokyo, 40, 107-124.

Mogi, K. (1973) Relationship between deep and shallow seismicity in the Western Pacific region, *Tectonophysics*, **17**, pp. 1-22.

Musmeci, F. and Vere-Jones D. (1992) A space-time clustering model for historical earthquakes, *Ann. Inst. Statist. Math.*, **44**, pp. 1-11.

Ohtake, M. (1980) Earthquake prediction based on the seismic gap with special reference to the Oaxaca, Mexico earthquake (in Japanese with English summary), *Report of the National Reseach Center for Disaster Prevention*, **23**, pp. 65-110.

Ogata, Y. (1978) Asymptotic behaviour of the maximum likelihood estimators for the stationary point processes, *Ann. Inst. Statist. Math.*, **A 30**, pp. 243-261.

Ogata, Y.(1981). On Lewis' simulation method for point processes. *IEEE Trans. Inform. Theory* **IT-30**, pp. 23-31.

Ogata, Y. (1983a). Estimation of the parameters in the modified Omori formula for aftershock frequencies by the maximum likelihood procedure, *J. Phys. Earth* **31**, pp. 115-124.

Ogata, Y. (1983b). Likelihood analysis of point processes and its applications to seismological data, *Bull. Int. Statist. Inst.* **50**, Book 2, pp. 943-961.

Ogata, Y. (1985). Statistical models for earthquake occurrences and residual analysis for point processes, *Research Memo. (Technical report)*, No. 288, Inst. Statist. Math., Tokyo.

Ogata, Y.(1987). Long term dependence of earthquake occurrences and statistical models for standard seismic activity. *Suri Zisin Gaku (Mathematical Seismology) II*, Ed. M. Saito, pp. 115-124, Cooperative Research Report 3, Inst. Statist. Math., Tokyo. (In Japanese).

Ogata, Y. (1988). Statistical models for earthquake occurrences and residual analysis for point processes, *J. Amer. Statist. Assoc.*, **83**, No. 401, pp. 9-27.

Ogata, Y. (1989). Statistical model for standard seismicity and detection of anomalies by residual analysis for point process, *Tectonophysics* **169**, pp. 1-16.

Ogata, Y. (1992). Detection of precursory relative quiescence before great earthquakes through a statistical model, *J. Geophys. Res.*, **97**, pp. 19845-19871.

Ogata, Y. (1993). Precursory relative quiescence in swarm and aftershock activities before great earthquakes, preprint (in Japanese).

Ogata, Y. and Akaike, H. (1982). On linear intensity models for mixed doubly stochastic Poisson and self-exciting point processes, *J. Royal Statist. Soc. B* **44**, pp. 102-107.

Ogata, Y., Akaike, H. and Katsura, K. (1982). The Application of linear intensity models to the investigation of causal relations between a point process and another stochastic process, *Ann. Inst. Statist. Math.* **34B**, pp. 373-387.

Ogata, Y. and Shimazaki, K. (1984). Transition from aftershock to normal activity, *Bull. Seismol. Soc. Amer.* **74**, pp. 1757-1765.

Ogata, Y. and Vere-Jones, D. (1984). Inference for Earthquake Models: A Self-Correcting Model, *Stoch. Processes Appl.*, **17**, pp. 337-347.

Ogata, Y. and Katsura, K. (1986). Point process model with linearly parameterized intensity for the application to earthquake data, *J. Appl. Probab.* **23A**, pp. 291-310.

Ogata, Y. and Yamashina, K. (1986). Unbiased Estimate for b-Value of Magnitude Frequency, *Journal of Physics of the Earth* **34**, pp. 187-194.

Ogata, Y. and Katsura, K. (1988). Likelihood Analysis of Spatial Inhomogeneity for Marked Point Patterns, *Ann. Inst. Statist. Math.* **40**, pp. 29-40.

Ogata, Y. and Abe, K. (1991). Some statistical features of long term variation of the global and regional seismic activity, *Int. Statist. Rev.*, **59**, pp. 139-161.

Ogata, Y., Imoto, M. and Katsura, K. (1991). Three-dimensional spatial variation of b-values of magnitude frequency distribution beneath the Kanto District, Japan. *Geophys. J. Int.*, **10**, pp. 135-146.

Ogata, Y. and Katsura, K. (1991). Maximum likelihood estimates of the Fractal dimension for Spatial Patterns, *Biometrika*, **78**, pp. 463-467.

Oike, K. (1977) On the relation between rainfall and the occurrence of earthquakes, *Bull Disas. Prev. Res. Inst.*, **20 B1**, pp. 35-45 (in Japanese).

Omori, F. (1894). On the aftershocks of earthquake, *J. Coll. Sci. Imp. Univ. Tokyo*, **7**: pp. 111-200.

Ozaki, T. (1978). Maximum likelihood estimation of Hawkes' self-exciting point processes, *Ann. Inst. Stat. Math.* **30**, pp. 145-155.

Ozaki, T. and Tong, H. (1975). On the fitting of non-stationary autoregressive models in time series analysis, *Proceeding of the 8-th Hawaii International Conference on System Science*, Western Periodical Company.

Papangelou, F. (1972). Integrability of expected increments of point processes and related random change of scale, *Trans. Amer. Math. Soc.* **165**, pp. 483-506.

Ramselaar, P.A. (1990). The mean behaviour of the Ogata earthquake process, *Master's Thesis, Dept. Math., Univ. Utrecht.*

Reasenberg, P.A. and Matthews, M.V. (1988). Precursory seismic quiescence: a preliminary assessment of the hypothesis,*Pure Apple. Geophys.* **126**, pp. 373-406.

Rubin, I. (1972). Regular point processes and their detection, *IEEE Trans. Inform. Theory* **IT-18**, pp. 547-557.

Shuster, A. (1897). On lunar and solar periodicities of earthquakes, *Proc. Roy. Soc.* **61**, pp. 455-465.

Snyder, D.L. (1975). *Random Point Processes*, Wiley-Interscience, New York.

Utsu, T. (1957). Magnitude of earthquakes and occurrence of their aftershocks, *Zisin (J. Seismol. Soc. Japan), Ser 2*, **10**, pp. 35-45, in Japanese.

Utsu, T. (1961) A statistical study on the occurrence of aftershocks, *Geophys. Mag.*, **30**, pp. 521-605.

Utsu, T. (1968). Seismic activity in Hokkaido and its vicinity, *Geophys. Bull. Hokkaido Univ.*, **13**, pp. 99-103 (in Japanese).

Utsu, T. (1969). Aftershocks and earthquake statistics (I): some parameters which characterize an aftershock sequence and their interaction, *J. Faculty Sci., Hokkaido Univ., Ser. VII*, **3**: pp. 129-195.

Utsu, T. (1970). Aftershocks and earthquake statistics (II): Further investigation of aftershocks and other earthquakes sequence based on a new classification of earthquake sequences, *J. Faculty Sci., Hokkaido Univ., Ser. VII*, **3**: pp. 379-441.

Utsu, T. (1971). Aftershocks and earthquake statistics (III): Analyses of the distribution of earthquakes in magnitude, time, and space with special consideration to clustering characteristics to earthquake occurrence (1), *J. Faculty Sci., Hokkaido Univ., Ser. VII*, **3**: pp. 379-441.

Utsu, T. (1972). Aftershocks and earthquake statistics (IV): Analyses of the distribution of earthquakes in magnitude, time, and space with special consideration to clustering characteristics to earthquake occurrence (2), *J. Faculty Sci., Hokkaido Univ., Ser. VII*, **3**: pp. 379-441.

Utsu, T. (1975) Correlation between shallow earthquakes in Kwanto region and intermediate earthquakes in Hida region, central Japan, *Zisin (J. Seism. Soc., Japan)* 2nd Ser, **28**, pp. 303-311 (in Japanese).

Utsu, T. and Seki, A. (1955). Relation between the area of aftershock region and the energy of the main shock, *Zisin (J. Seismol. Soc. Japan), 2nd Ser., ii*, **7**, pp. 233-240, (in Japanese).

Vere-Jones, D. (1970), Stochastic Models for Earthquake Occurrence (with discussion), *Journal of the Royal Statistical Society*, Ser. B, **32**, pp. 1-62.

Vere-Jones, D. (1978). Earthquake prediction - a statistician's view, *J. Phys. Earth* **26**, pp. 129-146.

Vere-Jones, D. and Ozaki, T. (1982), Some examples of statistical estimation applied to earthquake data, 1. Cyclic Poisson and self-exciting models, *Ann. Inst. Statist. Math.* **34B**, pp. 189-207.

Vere-Jones, D. (1985), The detection and estimation of periodicities in point process data, *Technical report*, ISOR, Victoria Univ. of wellington.

Vere-Jones, D. and Davies, R.B. (1966). A statistical survey of earthquakes in the main seismic region of New Zealand, Part 2, Time series analyses, *New Zealand J. Geology and Geophys.* **9**, pp. 251-284.

Vere-Jones, D. and Ogata, Y. (1984). On the Moments of a Self-Correcting Process, *J. Appl. Probab.* **21**, pp. 335-352.

Wyss, M. and Habermann, R.E. (1988). Precursory seismic quiescence, *Pure Apple. Geophys.* **126**, pp. 319-332.

Yamanaka, Y. and Shimazaki, K. (1990). Scaling relationship between the number of aftershocks and the size of the main shock, *J. Phys. Earth* **38** pp. 305-324.

Zhao, Z., Matsumura, K. and Oike, K. (1989). Precursory change of Aftershock activity before large aftershock: a case study for recent earthquakes in China, *J. Phys. Earth* **37**, pp. 155-177.

ONE CHANNEL AT-A-TIME MULTICHANNEL AUTOREGRESSIVE MODELING OF STATIONARY AND NONSTATIONARY TIME SERIES

Will GERSCH and David STONE
Department of Information & Computer Sciences
University of Hawaii
Honolulu, HI 96822 U.S.A.

1. Introduction

This paper exploits a known but insufficiently attended to idea for the autogressive modeling of multivariate stationary and nonstationary covariance time series. (The term multichannel is common in the engineering literature. Here, we use both terminologies, multivariate and multichannel.) The idea is to model multichannel things one channel at-a-time as an instantaneous response-orthogonal innovations autoregressive model. Such modeling can be realized using only scalar computations. That idea was implicit in Pagano (1978) and was explicit in a program MULMAR by Kitagawa (Akaike et al. 1979), in Newton (1982) and also Sakai (1982). Each of those papers only treated multichannel stationary time series AR modeling. Here, we exploit that idea to address time series analysis problems via the realization of autoregressive models of both multichannel stationary and multichannel nonstationary covariance times series data.

Three well known and important multivariate modeling problems are addressed here via the one channel at-a-time paradigm. In the order in which they are presented: The first is an attempt to achieve a more parsimonious modeling than conventional Yule-Walker modeling of stationary time series for the purpose of decreasing one-step-ahead prediction variance. The second is an attempt to enhance the estimation of the power spectral density matrices for stationary covariance time series as compared to what can be achieved by conventional Yule-Walker VAR modeling. The third is an attempt to achieve a multivariate time varying AR coefficient modeling of multivariate nonstationary covariance times series. In addition to the the one channel at-a-time paradigm, the application of smoothness priors concepts in the analysis of time series, (reviewed in Gersch and Kitagawa 1988) is also exploited in the second and third problems. In what follows, in order we elaborate on each of those problems.

1.1 Parsimonious MCAR Modeling: Covariance Stationary Data

Our objective is to achieve a relatively parsimonious representation of the autoregressive modeling of multichannel stationary covariance time series, (MCAR), The motivation for such modeling of stationary time series is to reduce the inherent tendency to overparameterize that appears in conventional MCAR model fitting to enhance mean square prediction error performance as compared to conventional Yule-Walker MCAR stationary time series modeling.

By virtue of the one channel at-a-time instantaneous response orthogonal innovations representation, each orthogonal innovations channel, is modeled as a scalar autoregressive linear combination of itself and the other data channels. The computations are based on instantaneous data using only scalar computations. By expressing each of the individual scalar AR models in the form of a linear regression model and employing a subsequent subset selection procedure, a parsimonious MCAR model is realized. The one channel at-a-time instantaneous response orthogonal innovations variance model is related to the usual MCAR model with general innovations variance by an algebraic transformation.

H. Bozdogan (ed.), Proceedings of the First US/Japan Conference on the Frontiers of Statistical Modeling: An Informational Approach, 165–192.

In the one channel at-a-time MCAR modeling of stationary time series the order of each of the scalar AR models is determined using Akaike's AIC, (Akaike 1973,1974), (see Shibata 1976, 1980, 1981). Also here and wherever else the AIC is referred to we refer to it generically and actually consider it as well as the Hurvich and Tsai small sample corrections for scalar and multivariate series, (Hurvich and Tsai 1989, 1992). Then, representing each scalar AR model in a linear regression model the subset selection is achieved also using Akaike's AIC. The subsetted instantaneous MCAR model with orthogonal innovations may be algebraically transformed to the conventional MCAR model type.

The classical method of fitting MCAR models to data is by solving the multichannel Yule-Walker equations, using estimates of the data's covariance matrix function. This method is described for example in Whittle (1963), Marple (1987), Kay (1988) and Lutkepohl (1991). Fitting a d-channel autoregressive (AR) model of order M to data implies the estimation of $d \times d \times M$ AR model parameters and $d \times (d + 1)/2$ innovations parameters. Whittle (1963), was a seminal development in MCAR modeling. Only the inversion of block Toeplitz $d \times d \times M$ matrices is required in the Whittle-Levinson-type recursion algorithm solution of the Yule-walker equations, (also using covariance data). Neither the direct Yule-Walker nor the Whittle-Levinson type algorithms address the inherent tendency to overparameterize in fitting the MCAR model. Hsiao (1979) and Kitagawa (the program MULMAR in Akaike et al. 1979), were attempts to achieve parsimonious multichannel AR modeling. We demonstrate that our method is more parsimonious than Hsiao's method. (Kitagawa's method was a one-channel at-a-time exploratory method. It differs from ours primarily in the subset selection procedure.) In examples and by simulations we also demonstrate superior mean square tracking and means square prediction error performance compared to that achieved by Yule-Walker equation MCAR modeling.

1.2 Smoothness Priors-long MCAR Modeling For Spectrum Estimation

Here we combine the one channel at-a-time methodology and the scalar case smoothness priors long AR model for spectral estimation (Kitagawa and Gersch, 1985a), and apply those ideas to the problem of multichannel stationary time series spectral estimation.

In the scalar case the smoothness priors long AR method of estimating power spectral density is an alternative to the now conventional method of estimating power spectral density as a functional of a fitted AR model. That the computation of the spectral density for any particular AR model order can not possibly be optimum can be understood from the following Bayesian style argument: Independent of which parametric model order is selected, there is information in the data to select models of different orders. A proper Bayesian estimate of power spectral density requires that the spectral density of parametric models of different model orders be weighted in accordance with the likelihood and the prior of the model order of different models. On the other hand, the smoothness priors long AR model for spectral estimation, (Kitagawa and Gersch 1985a), does achieve a quasi-Bayesian solution to the problem of parametric power spectral density modeling. A particular class of frequency domain smoothness priors is assumed for the (normal) distribution of the coefficients of AR model order M, with M relatively large. Under the assumption that the conditional data distribution is also normal, the likelihood of the hyperparameters that characterize the class of smoothness priors is maximized to yield the best AR model of order M with the best data dependent priors. (A more complete treatment of the smoothness priors time series modeling is in Gersch and Kitagawa 1988.)

It should be noted that this procedure of computing the likelihood of a small number of hyperparameters under the conditions of normal priors and conditionally normal data distributions was introduced in Akaike (1980) for the quasi-Bayesian modeling of a variety of time series problems.

The extension of the smoothness priors long AR model to the multivariate case via the one channel at-a-time paradigm is straightforward. Examples are given in which the expected cross-entropy of the multichannel smoothness priors model spectral density is compared the the spectral density computed via the MCAR model fitted by solution of the Yule-Walker equations.

1.3 Time Varying MCAR Modeling: Nonstationary Covariance Data.

Here the problem of modeling multichannel nonstationary covariance time series data with a multichannel time varying autoregressive model is achieved with a one channel at-a-time smoothness priors, (SPMCTVAR), paradigm. Modeling of multichannel nonstationary covariance time series is relevant to many econometric, human electroencephalogram, oceanographic, meteorological and seismic data analysis situations.

Wiggins and Robinson (1965), a development similar to Whittle (1963) Levinson type algorithm for fitting MCAR models to data of the was a multichannel lattice-form algorithm. Subsequent multichannel realizations in the engineering literature by Lee et al. (1981), Friedlander (1982), and Porat et al. (1982) involved matrix computations. Scalar implementations of multichannel recursions, which required no matrix processing, were realized by Sakai (1982), Lev-Ari (1983),(1987), and Ling and Proakis (1984). Motivated by Pagano (1978), Sakai implemented a multichannel AR models into scalar one channel at-a-time AR models. (This was referred to by Sakai as the "circular lattice filter".) The realization by Lev-Ari emphasizes the modularity of the architecture and is equivalent to Sakai's. Sakai's algorithm is based on sample covariance function computations.

Our own SPMCTVAR modeling can be seen to have evolved from the scalar SPTVAR modeling efforts in Kitagawa (1983), Kitagawa and Gersch (1985b) and Kitagawa (1988). Here in our one channel at-a-time varying AR coefficient modeling, we exploit a random walk model in the regression equation for the partial correlation coefficients, (PARCORS). Instantaneous PARCORS are computed for each channel which in turn yields an instantaneous multichannel AR model. The use of a stochastic regression model for the PARCORS derives from the concept of smoothness priors time series modeling. Here the smoothness priors specify a prior distribution on the autoregressive time series PARCOR parameters in the form of stochastic difference equation constraints. The estimate of the time series model parameters, in this case the estimate of the instantaneous PARCORS, is the mean of the posterior distribution of the PARCORS. The algorithm for computing the instantaneous PARCORS is a version of that in Kitagawa (1988). There it was used in the AR modeling of scalar non-Gaussian nonstationary covariance time series.

The development shown here is a superior way of modeling multichannel nonstationary covariance time series as compared to that done in Gersch and Kitagawa (1983), and Gersch (1985), Gersch (1987). In those earlier papers we exploited an orthogonal polynomial-least squares MCTVAR method of modeling of multichannel nonstationary covariance data. That was applied to the analysis of econometric data, (Gersch and Kitagawa 1983), and multichannel human epileptic episode data, (Gersch 1985, Gersch 1987).

1.4 Organization Of The Paper

The multichannel to one channel at-a-time development appears in Section 2. Bivariate examples comparing results of our subset selection approach with the MCAR modeling approach in Hsiao (1979), and a 3-variate data example of the modeling of bathymetry (depth of the ocean) data are shown in Section 3. Also in Section 3 is a comparison of the one-step-ahead prediction performance of the Whittle-AIC and MCAR modeling methods. The SPMCAR model for enhancing spectral estimation and examples are in Section 4. The SPMCTVAR modeling and examples of its application appear in Section 5. A list of references is included.

2. Multichannel One Channel At-a-time AR Modeling

An instantaneous response orthogonal innovations MCAR model is modeled one autoregressive channel at-a-time with computations on instantaneous data. The conventional MCAR model is related to the instantaneous response orthogonal innovations model by an invertible linear transformation.

2.1 The Instantaneous Response-Orthogonal Innovations and the Conventional MCAR Models.

From Theorem 1, Pagano (1978) an instantaneous response-orthogonal innovations MCAR model of the stationary d-vector random process $\{x(n)\}$, is in the form

$$\sum_{m=0}^{M} B_m x(n-m) = \tilde{\varepsilon}(n) \tag{2.1}$$

with $\tilde{\varepsilon}$ a zero mean uncorrelated process with diagonal innovations variance matrix D

$$D = \begin{bmatrix} \sigma_{11} & . & . & . & . & . \\ . & \sigma_{22} & . & . & . & . \\ . & . & . & . & . & . \\ . & . & . & . & . & . \\ . & . & . & . & . & \sigma_{dd} \end{bmatrix}. \tag{2.2}$$

and B_0 a unit (ones on the diagonal) lower triangular matrix

$$B_0 = \begin{bmatrix} 1 & 0 & 0 & . & 0 & 0 \\ b_{210} & 1 & 0 & . & . & . \\ b_{310} & b_{320} & 1 & . & . & . \\ . & . & . & . & 0 & . \\ . & . & . & . & 1 & 0 \\ b_{d10} & b_{d20} & b_{d30} & . & b_{d(d-1)0} & 1 \end{bmatrix} \tag{2.3}$$

The model in (2.1) can be re-expressed in the conventional MCAR form

$$x(n) + \sum_{m=1}^{M} A_m x(n-m) = \varepsilon(n) . \tag{2.4}$$

with innovations matrix V and MCAR parameter matrices, $A_m = B_0^{-1}B_m$. Also, $\varepsilon(n) = B_0^{-1}\tilde{\varepsilon}(n)$. The relation between the innovations variance matrix V of the model in (2.4) and D the diagonal innovations matrix in (2.2) is

$$V = E[\varepsilon(n)\varepsilon(n)'] = B_0^{-1}DB_0^{-T\prime}. \tag{2.5}$$

Conversely, consider the conventional MCAR model in (2.4). The unique modified Cholesky decomposition of $V = LDL^T$ where L is unit lower triangular and D is as in (2.3). Then $L^{-1} = B_0^{-1}$ resulting in the instantaneous response model in (2.1) with orthogonal innovations.

The motivation for this paper is based on the key observation that the instantaneous response-orthogonal innovations MCAR model permits multichannel data to be modeled one AR channel at-a-time.

Another useful way of interpreting the relationship between the conventional MCAR representation in (2.4) and the instantaneous response model in (2.1) is that the latter can be thought of an interlacing of the data in the former. That is, first consider interlacing the d-channel time series $x(t)$ into a scalar series $y(t)$ using,

$$y(j + d(t-1)) = x_j(t), \ j = 1, \ldots, d. \tag{2.6}$$

Then, re-express the $\{y(t)\}$ sequence into a d-vector AR model according to

$$y(k+d(t-1)) + \sum_{j=1}^{p_k} \alpha_k(j) y(k+d(t-1)-j) = \tilde{\varepsilon}\,,\ (k+d(t-1)) \quad k = 1,\ldots,d. \tag{2.7}$$

with $\tilde{\varepsilon}(k+d(t-1)) \sim N(0,\sigma_k^2)$ i.i.d. and with p_k the order of the k-th channel AR model. For example for $d = 3$, (2.7) can be written,

$$\begin{bmatrix} 1 & 0 & 0 \\ \alpha_2(1) & 1 & 0 \\ \alpha_3(2) & \alpha_3(1) & 1 \end{bmatrix} \begin{bmatrix} y(1+3(t-1)) \\ y(2+3(t-1)) \\ y(3+3(t-1)) \end{bmatrix} + \begin{bmatrix} \alpha_1(3) & \alpha_1(2) & \alpha_1(1) \\ \alpha_2(4) & \alpha_2(3) & \alpha_2(2) \\ \alpha_3(5) & \alpha_3(4) & \alpha_3(3) \end{bmatrix} \begin{bmatrix} y(1+3(t-2)) \\ y(2+3(t-2)) \\ y(3+3(t-2)) \end{bmatrix} \tag{2.8}$$

$$+ \quad . \; . \; . \quad = \begin{bmatrix} \varepsilon(1+3(t-1)) \\ \varepsilon(2+3(t-1)) \\ \varepsilon(3+3(t-1)) \end{bmatrix}.$$

Using (2.7), and $(t-1) = n$, the $\{x(t)\}$ sequence can be written in terms of the $\{y(t)\}$ of (2.8) in the matrix form

$$\sum_{m=0}^{M} \tilde{A}_m y(n-m) = \tilde{\varepsilon}(n) \tag{2.9}$$

where the $\tilde{A}_0$ is lower unit triangular and the d vector sequence $\tilde{\varepsilon}(n)$ has diagonal covariance matrix $D = diag(\sigma_1^2, \ldots, \sigma_d^2)$.

That is, the model in (2.9) is precisely the model in (2.1) with $\tilde{A}_m = B_m$, and $y(n) = x(n)$.

For completeness here and later use, the power spectral density matrix of the conventional MCAR model (2.4)-(2.5) is given by

$$S(f) = A(f)^{-1} V A^*(f)^{-T} \tag{2.10}$$

where

$$A(f) = \sum_{k=0}^{p} A_k \exp[-2\pi f k],$$

$A_0 = I$, and $A^*(f)$ is the complex conjugate matrix of A(f) where A(f) is the whitening filtering or inverse filter of the AR process.

The power spectral density matrix for the system expressed in the instantaneous response orthogonal innovations form is given by

$$S(f) = B(f)^{-1} D B^*(f)^{-T} \tag{2.11}$$

where

$$B(f) = \sum_{k=0}^{p} B_k \exp[-2\pi f k].$$

For example consider a $d = 2$ AR_4 model of the Box-Jenkins Series J input-output physical plant data as in Newton (1988), p. 342. The conventional MCAR model form (2.4) with parameter matrix $A_0 = I$ and (to within four decimal digits),

$$A_1=\begin{bmatrix}-1.9259 & 0.0012\\ -0.0505 & -1.2998\end{bmatrix},\quad A_2=\begin{bmatrix}1.2017 & -0.0042\\ 0.0205 & 0.3277\end{bmatrix},\quad A_3=\begin{bmatrix}-0.1169 & 0.0087\\ 0.7118 & 0.2570\end{bmatrix},\quad A_4=\begin{bmatrix}-0.1042 & -0.0033\\ -0.1954 & -0.1334\end{bmatrix}.$$

$$V=\begin{bmatrix}0.3516 & -0.0731\\ -0.0731 & 0.9724\end{bmatrix}.$$

Correspondingly the equivalent instantaneous response-orthogonal innovations matrix representation may be computed. By simple algebraic manipulation, from (2.3)-(2.5), the matrix B_0 is

$$B_0=\begin{bmatrix}1.0000 & 0.0000\\ 0.2081 & 1.0000\end{bmatrix}$$

where $b_{210}=-V(1,2)/V(1,1)$. Then computing $B_m=B_0A_m, m=1,\ldots,4$ and $D=B_0VB_0^T$, we arrive at the equivalent theoretical instantaneous response orthogonal innovations model. The B_i and D matrices are then

$$B_1=\begin{bmatrix}-1.9259 & 0.0012\\ -0.4512 & -1.2995\end{bmatrix},\quad B_2=\begin{bmatrix}1.2016 & -0.0042\\ 0.2705 & 0.3268\end{bmatrix},\quad B_3=\begin{bmatrix}-0.1169 & 0.0087\\ 0.6875 & 0.2588\end{bmatrix}.\quad B_4=\begin{bmatrix}-0.1042 & -0.0033\\ -0.2171 & -0.1341\end{bmatrix}.$$

$$D=\begin{bmatrix}0.3516 & 0.0000\\ 0.0000 & 0.9572\end{bmatrix}.$$

3. Multichannel Autoregressive Modeling of Stationary and Nonstationary Mean Time Series"

In this section the MCAR modeling of nonstationary mean time series and the tracking and one-step-ahead prediction performance of stationary MCAR models are considered. The application of the one channel at-a-time paradigm and subsequent subset selection of the individual scalar AR modeled channels achieves a "thinning" of what are in general overparameterized MCAR models. A version of this approach of thinning the one channel at-a-time scalar AR models was done by Kitagawa in the TIMSAC-78 program MULMAR, (Akaike et al. 1979, Kitagawa and Akaike 1981).

For completeness we begin in Sec. 3.1 by reviewing the Yule-Walker-AIC criterion method of fitting the stationary MCAR model to observed data. The subset selection procedure is in Sec. 3.2 and examples are shown in Sec. 3.3.

3.1 The Yule-Walker Akaike-AIC approach to MCAR modeling.

Consider the MCAR model

$$\sum_{m=0}^{p} A_m x(n-m)=\varepsilon(n) \tag{3.1}$$

with $A_0=I_d$ and $\{\varepsilon(n)\}$ a zero mean uncorrelated d-vector sequence with covariance matrix V. Define the matrix covariance function $\{\Gamma_k, k=0,\pm1,\pm2,,,,p\}$ of the stationary stochastic process $\{X(n)\}$, where $\Gamma_k=E[X(n+k)-\bar{X}][X(n)-\bar{X}]^T$, with $\bar{X}=\frac{1}{N}\sum_{n=1}^{N}(n)$. Let the d-variate observed data be $\{x(n), n=1,2,\ldots,N\}$, where $x(n)=[x_1(n),\ldots x_d(n)]^T$ and A^T denotes the transpose of A. Then let $C_k=\hat{\Gamma}_k$ be an estimate of Γ_k with $C_k=\frac{1}{N}\sum_{n=k+1}^{N}[x(n)-\bar{x}][x(n-k)-\bar{x}]^T$ with $\bar{x}=\frac{1}{N}\sum_{n=1}^{N}x(n)$. The Yule-Walker equations may be derived by post-multiplying (3.1) by $x(n-j)^T$ and summing over n. The resulting Yule-Walker equations satisfy:

$$C_j = \sum_{i=1}^{p} A_i C_{j-i} + \hat{V}\delta_{j,0} \qquad j = 0, 1, \ldots p \tag{3.2}$$

where $C_{-j} = C_j^T$.

Evaluating (3.1) for $j = 1, , , , p$, the resulting equation that must be satisfied for fitting the order p MCAR model to the data is

$$[\,\hat{A}_1, \ldots, \hat{A}_p\,] \begin{bmatrix} C_0 & C_1 & \cdots & C_{p-1} \\ C_1^T & C_0 & \cdots & C_{p-2} \\ \cdot & \cdot & \cdots & \cdot \\ \cdot & \cdot & \cdots & \cdot \\ \cdot & \cdot & \cdots & \cdot \\ C_{p-1}^T & C_{p-2}^T & \cdots & C_0 \end{bmatrix} = -[\,C_1, \ldots, C_p\,]. \tag{3.3}$$

An estimate of the innovations matrix V is given by $\hat{V} = \sum_{m=0}^{p} \hat{A}_m C_m^T$.

For large values of d and p, inversion of the $d \times d \times p$ matrix in (3.2) can be a prohibitive numerical computation. The Whittle (1963) Levinson-type algorithm achieves recursively increasing AR model order fits to the data requiring inversion of only $d \times d$ matrices. Given a set of AR models of increasing order fitted to the data, the selection of the statistically 'best' AR order model fitted to the data can be determined via the Akaike (1974) AIC criterion. Using that criterion, the quantity

$$AIC(p) = N \times d\ \ln(\hat{V}) + 2 \times d(1 + dp), \tag{3.4}$$

is computed for each of the fitted AR models of orders $p = 0,1,2, , , pmax$. The AIC best model is the one for which the AIC is smallest.

3.2 The Individual Channel AR modeling and Subset Selection Procedure.

The one channel at-a-time modeling of d-channel multichannel stationary time series yield d individual AR modeled data channels. To each individual channel consider fitting an autoregressive model of order p

$$y(n) = \alpha_1 y(n-1) + \ldots + \alpha_p y(n-p) + \varepsilon(n) \tag{3.5}$$

to the time series data $\{y(n), n=1, , , , N\}$. Define the design matrix X and the vectors y, ε and α by

$$X = \begin{bmatrix} y(p) & y(p-1) & \cdots & y(1) \\ y(p+1) & y(p) & \cdots & y(2) \\ \cdot & \cdot & & \cdot \\ \cdot & \cdot & & \cdot \\ \cdot & \cdot & & \cdot \\ y(N-1) & y(N-2) & \cdots & y(N-p) \end{bmatrix} \quad y = \begin{bmatrix} y(p+1) \\ y(p+2) \\ \cdot \\ \cdot \\ \cdot \\ y(N) \end{bmatrix} \quad \varepsilon = \begin{bmatrix} \varepsilon(p+1) \\ \varepsilon(p+2) \\ \cdot \\ \cdot \\ \cdot \\ \varepsilon(N) \end{bmatrix} \quad \alpha = \begin{bmatrix} \alpha_1 \\ \cdot \\ \cdot \\ \cdot \\ \alpha_p \end{bmatrix}. \tag{3.6}$$

Then the scalar autoregressive time series model fitting problem is in the linear model form

$$y = X\alpha + \varepsilon. \tag{3.7}$$

In detail the structure of the of the linear model regression equation components corresponding to the individual channels for $d = 3$, and MCAR model order 3 associated with (2.8), (2.9) and (3.7) are

$$y_1 = \begin{bmatrix} x_1(10) \\ x_1(11) \\ \cdot \\ \cdot \\ x_1(N) \end{bmatrix}, \quad X_1 = \begin{bmatrix} x_3(9) & x_2(9) & x_1(9) & \cdot\cdot & x_3(1) & x_2(1) & x_1(1) \\ x_3(10) & x_2(10) & x_1(10) & \cdot\cdot & x_3(2) & x_2(2) & x_1(2) \\ \cdot & \cdot & \cdot & \cdot\cdot & \cdot & \cdot & \cdot \\ \cdot & \cdot & \cdot & \cdot\cdot & \cdot & \cdot & \cdot \\ x_3(N-1) & x_2(N-1) & x_1(N-1) & \cdot\cdot & x_3(N-9) & x_2(N-9) & x_1(N-9) \end{bmatrix}, \quad \alpha_1 = \begin{bmatrix} \alpha_1(1) \\ \alpha_1(2) \\ \cdot \\ \alpha_1(9) \end{bmatrix}. \tag{3.8}$$

$$y_2 = \begin{bmatrix} x_2(10) \\ x_2(11) \\ \cdot \\ \cdot \\ x_2(N) \end{bmatrix}, \quad X_2 = \begin{bmatrix} x_1(10) & x_3(9) & x_2(9) & x_1(9) & \cdots & x_3(1) & x_2(1) & x_1(1) \\ x_1(11) & x_3(10) & x_2(10) & x_1(10) & \cdots & x_3(2) & x_2(2) & x_1(2) \\ \cdot & \cdot & \cdot & \cdot & \cdots & \cdot & \cdot & \cdot \\ \cdot & \cdot & \cdot & \cdot & \cdots & \cdot & \cdot & \cdot \\ x_1(N) & x_3(N-1) & x_2(N-1) & x_1(N-1) & \cdots & x_3(N-9) & x_2(N-9) & x_1(N-9) \end{bmatrix}, \quad \alpha_2 = \begin{bmatrix} \alpha_2(1) \\ \alpha_2(2) \\ \cdot \\ \alpha_2(9) \\ \alpha_2(10) \end{bmatrix}.$$

$$y_3 = \begin{bmatrix} x_3(10) \\ x_3(11) \\ \cdot \\ \cdot \\ x_3(N) \end{bmatrix}, \quad X_2 = \begin{bmatrix} x_2(10) & x_1(10) & x_3(9) & x_2(9) & x_1(9) & \cdot\cdot & x_3(1) & x_2(1) & x_1(1) \\ x_2(11) & x_1(11) & x_3(10) & x_2(10) & x_1(10) & \cdot\cdot & x_3(2) & x_2(2) & x_1(2) \\ \cdot & \cdot & \cdot & \cdot & \cdot & \cdot\cdot & \cdot & \cdot & \cdot \\ \cdot & \cdot & \cdot & \cdot & \cdot & \cdot\cdot & \cdot & \cdot & \cdot \\ x_2(N) & x_1(N) & x_3(N-1) & x_2(N-1) & x_1(N-1) & \cdot\cdot & x_3(N-9) & x_2(N-9) & x_1(N-9) \end{bmatrix}, \quad \alpha_3 = \begin{bmatrix} \alpha_3(1) \\ \alpha_3(2) \\ \cdot \\ \cdot \\ \alpha_3(9) \\ \alpha_3(10) \\ \alpha_3(11) \end{bmatrix}.$$

In (3.8) the notation y_1, y_2, y_3 refer respectively to the dependent variable associated with the individual channels, X_i, $i = 1,2,3$ refers to the design matrix of the i-th channel and $\alpha_1, \alpha_2, \alpha_3$ refer to the estimated i-th channel AR parameters. Also note that if p is the order of the MCAR model that $p_k = d \times p + (k - 1)$, $k = 1,2,3$ where k is the channel number and p_k is the scalar AR model order in the k-th channel.

With the problem cast into an ordinary regression model in the familiar linear model form, we exploit a modified S language (Becker et al. 1988), function "leaps" and the C_p subset selection algorithm to achieve subset selection using the AIC_C (Hurvich and Tsai 1989), criterion. The AIC_C is a small sample modification of the Akaike AIC, (Akaike 1974). The C_p criterion is due to Mallows (1973) and the S language C_p subset selection algorithm, "leaps" is due to Furnival (1954). (We also note that it is only a nominal exercise to implement an alternative model selection criterion such as PRESS, Allen 1974, in the S environment without resorting to the leaps algorithm.)

The subset selection on the individual orthogonal innovations channels yields an instantaneous model in the form (2.8). Application of (2.9) and (2.10) to (2.8) returns the MVAR model form (2.1). The MCAR forms (2.1) and (2.8) are used to evaluate the tracking performance of the full and subset selection models. Tracking performance may be used as a diagnostic to compare the compatibility and relative performance of the subset selection model and the Yule-Walker models.

3.3 Examples.

Instantaneous response-orthogonal variance-subset procedure modeling is first applied to the nonstationary mean data examined in Tsaio and Tsay (1989).

Secondly tracking performance of $d = 2$ and $d = 3$ channel data are computed for the subsetted one channel at-a-time orthogonal innovations model and competitive model alternatives. The $d = 2$ channel data is from an attempt at parameter thinning in MCAR modeling from an example considered in Hsiao (1979). The $d = 3$ channel data is from an a bathymetry (depth of the ocean)example from Goff and Jordan (1988).

Finally, the one-step-ahead prediction performance of the Whittle-AIC_C criterion method of fitting MCAR models and the subsetted one channel at-a-time orthogonal innovations MCAR model for $d = 2$ examples are considered. The situations considered are simulations of the Newton (1988) AR_4 model, an AR_2 model and an AR_2MA_1 model.

3.3.1 Modeling the Tsaio-Tsay flour price data.

The first example that illustrates the parsimonious property of our AR model instantaneous response orthogonal innovation-subset selection procedure is an application to the $d = 3$, $n = 100$ flour price data considered by Tsaio and Tsay (1985), (1989). Respectively Tsaio and Tsay (1985), (1989) consider transformations of the original series, cannonical correlation analyses, AR_2 and AR_1MA_1 models. The fitted AR_2 model was a full model with 18 AR parameters and 5 innovation variance parameters. An objective of the 1989 paper is to achieve parsimonious modeling. The 1989 AR_1MA_1 model conceptually can have as many as 18 AR-MA parameters and 6 innovation matrix parameters. The Tsaio-Tsay model fitted to the flour price data consists of 9 AR-MA parameters and 5 innovations variance parameters. In comparison with that latter model our AR_2 instantaneous response orthogonal innovation-subset selection procedure model fitted 11 AR parameter, 2 parameters in the B_0 matrix, and 3 instantaneous variance matrix parameters. That is a total of 16 parameters or 2 parameters more than the Tsaio-Tsay model. Of course we should take into account that fact that the Tsaio-Tsay model also requires computation, via an fairly complex analysis procedure, of a 3x3 transformation matrix. Consequently, feel that our own rather simple computational procedure (which incidentally yields excellent tracking results) is at least a reasonable parsimonious model contender if not superior to the Tsaio-Tsay modeling of the flour price data. For completeness the B_i, $i=0,1,2$ parameter matrices fitted to the flour price data are

$$B_0 = \begin{bmatrix} 0.0000 & 0.0000 & 0.0000 \\ 1.0393 & 0.0000 & 0.0000 \\ 0.9832 & 0.0000 & 0.0000 \end{bmatrix}, \; B_1 = \begin{bmatrix} 0.0000 & 1.0902 & 0.0000 \\ -0.5466 & 0.5259 & 0.0000 \\ 0.0000 & -0.8081 & 0.8956 \end{bmatrix},$$

$$B_2 = \begin{bmatrix} 1.1132 & -1.4888 & 0.2326 \\ -0.4569 & 0.4357 & 0.0000 \\ 0.0000 & -0.0865 & 0.0000 \end{bmatrix}. \; D = \begin{bmatrix} 44.0000 & 00.0000 & 00.0000 \\ 00.0000 & 11.8192 & 00.0000 \\ 00.0000 & 00.0000 & 12.1898 \end{bmatrix}.$$

As a result of the non-negligible feedthrough terms B_{120} and B_{130} the innovations variance in the first channel, $D[1,1]$, is considerably larger than in channels 2 and 3, $D[2,2]$,$D[3,3]$.

3.3.2 Tracking performance: Hsiao's 2 channel data example.

Here we compare results of our one channel at-a-time subset selection approach with the attempt in Hsiao(1979) to achieve parsimonious MCAR modeling. Hsiao's model for Canadian post war money (M1) and income (GNP) series is in the form:

$$\begin{bmatrix} (1-L)^2\log GNP \\ (1-L)^2\log M1 \end{bmatrix} = \begin{bmatrix} \psi_{11}^{m_{11}}(L) & \psi_{12}^{m_{12}}(L) \\ \psi_{21}^{m_{21}}(L) & \psi_{22}^{m_{22}}(L) \end{bmatrix} \begin{bmatrix} (1-L)^2\log GNP \\ (1-L)^2\log M1 \end{bmatrix} + \begin{bmatrix} a \\ b \end{bmatrix} + \begin{bmatrix} u \\ v \end{bmatrix}, \tag{3.9}$$

where $\psi_{ij}(L) = \sum_{l=1}^{M_{ij}} \psi_{ijl} L^l$, L is a lag operator, $Ly_t = y_{t-1}$, and u, v are zero mean white noise with

$cov(u,v)_{s,t} = \delta_{t,s}\Omega$ and a and b are constants (mean values).

To determine the orders m_{11}, m_{12}, m_{21}, and m_{22}, the Hsiao procedure is as follows:
First consider modeling the GNP series.
1) Determine the order of GNP series as a scalar autoregressive process using Akaike's FPE criterion. Assume that order is p and refer to this as the FPE(p,0) model.
2) With the order of GNP fixed at p, add components of ψ_{12}, (the $M1$ process) to the p components of the GNP series, one-component at-a-time and use the FPE to determine q the FPE best number of $M1$ components to be added. This yields a $FPE(p,q)$ component model.
3) Now "reverse" the process and fixing the order of the $M1$ components at q, starting at zero components, add components of ψ_{11}, the GNP series, one lag component at-a-time and using the FPE, determine FPE best FPE(p',q) model.
4) Finally, using the FPE criterion select the FPE best of the FPE(p',q) and FPE(p,0) models.
Repeat the process for the M1 series.

Using the procedure specified above, Hsiao identified the model orders m_{11}, m_{12}, m_{21} and m_{22} to be 6, 8, 4 and 9, respectively. The resulting total number of AR parameters estimated is 27. In addition, 3 parameters were estimated to obtain the innovation variance matrix.

In our approach, because ours is an instantaneous response model, the choice of which series is modeled first is significant. We chose the $M1$ series first and allow that series to have an instantaneous effect on GNP. Here we use a notation that is similar to that of Hsiao's and the model is

$$\begin{bmatrix} (1-L)^2 log(M1) \\ (1-L)^2 log(GNP) \end{bmatrix} = \begin{bmatrix} \psi_{11}^{m_{11}}(L) & \psi_{12}^{m_{12}}(L) \\ \psi_{21}^{m_{21}}(L) & \psi_{22}^{m_{22}}(L) \end{bmatrix} \begin{bmatrix} (1-L)^2 log(M1) \\ (1-L)^2 log(GNP) \end{bmatrix} + \begin{bmatrix} a \\ b \end{bmatrix} + \begin{bmatrix} u \\ v \end{bmatrix}, \tag{3.10}$$

where $\psi_{21}(L) = \sum_{l=0}^{M_{21}} \psi_{21l} L^l$, and everything else remains the same as defined in Hsiao's model.

Computational experiments suggested that the maximum value of 10 for m_{11}, m_{12}, m_{21} and m_{22} was sufficient. This maximum value is also compatible with Hsiao's model. We then apply our procedure and we found that only 4 of the 10 $\psi_{11}'s$, only 4 of the 10 $\psi_{12}'s$, 6 of the 11 $\psi_{21}'s$, and 7 of the 10 $\psi_{22}'s$ were nonzero values. That is, a total of 21 AR parameters were estimated using our approach as compared to 27 AR parameters using Hsiao's approach. Also only 2 additional parameters are required to identify the innovations variance matrix in our model.

We do not consider the M1, GNP series to be a very good example of AR model fitting. Nevertheless, the mean square errors for the M1 and GNP series respectively for Hsiao's model are 0.000181 and 0.000123. Correspondingly the means square errors for the M1 and GNP series for our model are 0.000197 and 0.000110.

This kind of model fit or tracking performance result revealed in this example is typical of performance that can be expected from our procedure in comparison with other procedures. The first channel modeled generally will have a tracking performance comparable with that of other procedures, however subsequent channels will be better fitted by our procedure in the sense of smaller means square error than other procedures. Very likely this is a consequence of the "feedthrough" of the instantaneous response. This property is more dramaticlly illustrated in the next example.

3.3.3 Tracking performance: Modeling 3 channels of bathymetric data.

In this example we consider a 3-variate bathymetry data. Bathymetric data is a measure of the depth of the ocean. The data reported on here is a 3 channel subset of the 15 channel data in Figure 18 in Goff and Jordan (1988). The length of the (reduced) data set, N, is 60. The MCAR model in (3.1) was first fitted by solving Yule-walker equations. The order of the model, M= 2, was determined using the Hurvich-Tsai AIC_C. That is, 3 x 3 x 2 AR parameters and 3 x (3+1) / 2 innovations matrix parameters or a total of 24 parameters were estimated for this model. On the other hand, using our procedure only 8 AR parameters and 3 parameters of the matrix D, or a total of 11 parameters were required to fit the instantaneous response-orthogonal innovations model.

The tracking performance of both the Yule-Walker method of modeling and the one channel at-a-time subset procedure are shown in Figure 3.1.

Respectively for channels 1,2 and 3 the full model mean square tracking errors were 3722.416, 3763.362 and 4764.678. Correspondingly for our procedure the mean square tracking errors were 3092.006, 535.204 and 238.055. The overall improved tracking performance or superior model fit using our procedure as compared with that of the Yule-Walker procedure is quite apparent.

3.3.4 Prediction performance.

Lewis and Reinsel (1985) is a a well known paper on the prediction performance of multivariate *AR* model fitting. n that paper, a $d = 2$ AR_1MA_1 model was used and the empirical results were compared to theoretical expectations. In the particular model the zeros of the *MA* polynomial were close to the unit circle making a statistical criterion selected *AR* model order quite unstable for the $n = 100$ simulated data points. Apparantly a fixed AR_4 model order was fitted to each of the 100 and 2500 replications described in that paper. Instead, a statistical criterion should have been used to determine the AR model order for each model fitted to the simulated data.

We did not make that mistake. Several $d = 2$ prediction performance experiments were conducted. In one experiment the one step-ahead prediction performance of the Whittle-AIC criterion MCAR modeling method and our subset selection one channel at-a-time orthogonal innovations MCAR method was conducted on simulations of a model of the (two channel) Box-Jenkins Series J, gas-furnace data. The AR_4 model described in Newton (1988), is taken to be the "true" AR model. Because it a Yule-Walker method model of a physical data problem, this model can be expected to be overparametrized. 100 trials of 299 data points each were modeled and a 300-th data point was predicted. The AR_4 model has 16 AR model parameters. Typically the subset selection model required only 12 parameters to be fitted to that data. On each channel the prediction variance of the subset selection method was approximately 85% of that achieved by the Yule-Walker solution method. In particular the one step-ahead prediction variances for the Yule-Walker method were 0.04305 and 0.13784 for channels one and two respectively. The corresponding prediction variances for the subset selection method were 0.03719 and 0.11722.

In two other Monte Carlo simulation studies of one-step-ahead prediction performance, the data from simulated models was fitted and predicted by the two different modeling methods, the Whittle-AIC MCAR method and the and our subset selection one channel at-a-time orthogonal innovations MCAR method. Two different bivariate time series models respectively an AR_2 model and an AR_2MA_1 model were simulated. In the conventional MCAR form the (highly oscillatory) AR_2 model considered was

$$x(n) = \sum_{m=0}^{2} A_m x(n-m) = \varepsilon(n), \quad E[\varepsilon(n)] = 0, \quad COV[\varepsilon(n), \varepsilon(n')] = V\delta_{n,n'} \tag{3.11}$$

with parameters

$$A_0 = I, \quad V = \begin{bmatrix} 1.000 & 0.500 \\ 0.500 & 1.250 \end{bmatrix}, \quad A_1 = \begin{bmatrix} 0.130 & -0.330 \\ -0.830 & -1.130 \end{bmatrix}, \quad A_2 = \begin{bmatrix} 0.750 & 0.085 \\ -0.185 & 0.850 \end{bmatrix}.$$

In the AR_2MA_1 model the innovation variance of the model and the AR_2 part of the model were identical to that in (3.11) The *MA* parameters were $B_0 = I$ and

$$B_1 = \begin{bmatrix} 0.900 & -0.900 \\ -0.600 & 0.900 \end{bmatrix} .$$

Data from 100 simulations of each model were computed with $n = 200$. For the AR_2 model, the prediction performance using the Whittle-AIC and our subset selection instantaneous response-orthogonal innovations subset selection method respectively was

$$\begin{bmatrix} 1.20052 & 0.56857 \\ 0.56857 & 1.34850 \end{bmatrix} , \begin{bmatrix} 1.22846 & 0.54305 \\ 0.54305 & 1.28969 \end{bmatrix} .$$

Typically an AR_2 model was fitted to the simulated AR_2 model.

For the AR_2MA_1 model, The prediction performance using the Whittle-AIC and our subset selection instantaneous response-orthogonal innovations subset selection method respectively was

$$\begin{bmatrix} 1.12177 & 0.58894 \\ 0.58894 & 1.60493 \end{bmatrix} , \begin{bmatrix} 1.08432 & 0.58672 \\ 0.58672 & 1.48887 \end{bmatrix} .$$

Typically the AR model order fitted to the simulated AR_2MA_1 model was 7.

The performance observed was anticipated. The highly oscillatory AR_2 model was expected to be "tight" without extraneous parameters. Thus with data length $n = 200$, the one-step-ahead prediction variance performance for the first channel of the AR_2 model orthogonal innovations-subset selection model is not as good as that achieved by the Whittle-AIC model. That is similar to the performance observed in the first channel in the tracking experiments. We note however, as anticipated because of the feedthrough from the first channel to the second channel, the prediction performance of our method is superior for the second channel. Also as expected, with respect the the simulation of the AR_2MA_1 model, the prediction performance of our model is superior to that achieved by the Whittle-AIC model in both channels. This is a consequence of the fact that the *AR* model is only an approximation to the *ARMA* model and that the Yule-Walker modeling of data from the *ARMA* model can be expected to be overparameterized.

Several research issues remain to be explored in comparing the prediction performance of the conventional Whittle-AIC and our innovations variance-subset selection MCAR modeling. For one thing, it is known that even for normally distributed data, the one-step-ahead prediction is itself not normally distributed. It remains to explore the distributions of those prediction errors. That may be accomplished empirically with a bootstrap analysis. A second topic is to explore the relationship between the measure of information from the first channel to the second channel, (Gewke 1982), to the expected one-step ahead prediction variance improvement

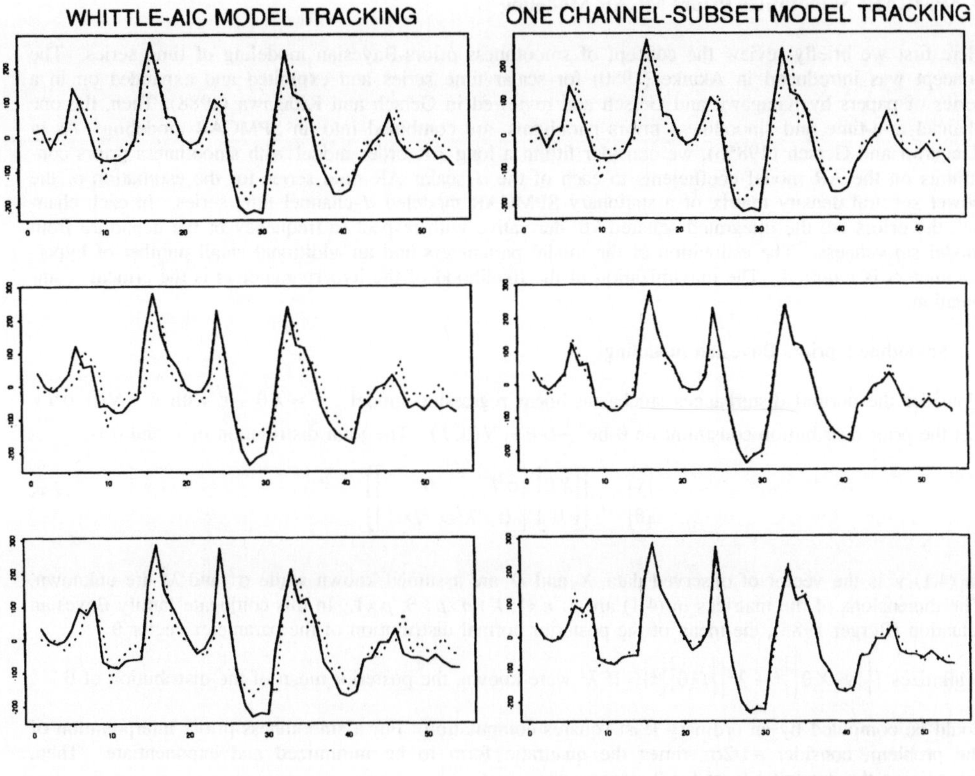

Figure 3.1 Bathymetry data tracking performance Whittle-AIC and one channel at-a-time MCAR models.

4. SPMCAR: Smoothness Priors MCAR Modeling

Here first we briefly review the concept of smoothness priors-Bayesian modeling of time series. The concept was introduced in Akaike (1980) for scalar time series and exploited and expanded on in a series of papers by Kitagawa and Gersch and reviewed in Gersch and Kitagawa (1988). Then, the one channel at-a-time and smoothness priors paradigms, are combined into an SPMCAR modeling. As in Kitagawa and Gersch (1985b), we consider fitting a long AR order model with smoothness priors constraints on the AR model coefficients to each of the d scalar AR time series for the estimation of the power spectral density matrix of a stationary SPMCAR modeled d-channel time series. In each channel, the priors are the integrated squared kth derivative with respect to frequency of the departure from model smoothness. The estimation of the model parameters and an additional small number of hyperparameters is required. The maximization of the likelihood of the hyperparameters is the critical computation.

4.1 Smoothness priors Bayesian modeling.

Consider the normal disturbances stochastic linear regression model , $y = X\theta + \varepsilon$ with $\varepsilon \sim N(0, \sigma^2 I)$. Let the prior distribution constraint on θ be $\frac{1}{\lambda} D\theta \sim N(0, I)$. The joint distribution of y and θ is,

$$\begin{bmatrix} y \\ \theta \end{bmatrix} \sim N\left(\begin{bmatrix} X\theta \\ 0 \end{bmatrix}, \begin{bmatrix} \sigma^2 I & 0 \\ 0 & \lambda^2 D^{-1} D^{-T} \end{bmatrix}\right). \tag{4.1}$$

In (4.1) y is the vector of observed data, X and D are assumed known while σ^2 and λ^2 are unknown. The dimensions of the matrices in (4.1) are y: $n \times 1$; X: $n \times p$; θ: $p \times 1$. In this conjugate family Bayesian situation (Berger 1985), the mean of the posterior normal distribution of the parameter vector θ

minimizes $\{\left\|y - X\theta\right\|^2 + \lambda^2 \left\|D\theta\right\|^2\}$. If λ^2 were known, the posterior mean of the distribution of θ

could be computed by an ordinary least squares computation. For a smoothness priors interpretation of the problem, consider $-1/2\sigma^2$ times the quadratic form to be minimized and exponentiate. Then, minimizing the quadratic form for θ maximizes

$$\exp\left\{-\frac{1}{2\sigma^2}\left\|y - X\theta\right\|^2\right\}\exp\left\{-\frac{\lambda^2}{2\sigma^2}\left\|D\theta\right\|^2\right\}. \tag{4.2}$$

In (4.2), the posterior distribution interpretation of the parameter vector θ is that it is proportional to the product of the conditional data distribution (likelihood), $p(y|X,\theta,\sigma^2)$, and a prior distribution, $\pi(\theta|\lambda^2,\sigma^2)$ on θ,

$$\pi(\theta|y,\lambda^2,\sigma^2) \;\alpha\; p(y|X,\theta,\sigma^2)\pi(\theta|\lambda^2,\sigma^2)\,. \tag{4.3}$$

Integration of (4.3) yields $L(\lambda^2,\sigma^2)$, the likelihood for the unknown parameters λ^2 and σ^2,

$$L(\lambda^2,\sigma^2) = \int_{-\infty}^{\infty} \pi(\theta|y,\lambda^2,\sigma^2)d\theta\,. \tag{4.4}$$

Since $\pi(\theta|y,\lambda^2,\sigma^2)$ is normally distributed, (4.4) can be expressed in the closed form, (Akaike 1980),

$$L(\lambda^2,\sigma^2) = (2\pi\sigma^2)^{-N/2}|\lambda^2 D^T D|^{1/2}|X^T X + \lambda^2 D^T D|^{-1/2}\exp\left\{\frac{-1}{2\sigma^2}SSE(\hat{\theta},\lambda^2)\right\}. \tag{4.5}$$

In (4.5), the maximum likelihood estimator of σ^2 is $\hat{\sigma}^2 = SSE(\hat{\theta},\lambda^2)/N$ where SSE denotes the sum of squares of errors of the linear regression problem. Then -2 log likelihood is,

$$-2logL(\lambda^2,\hat{\sigma}^2) = \tag{4.6}$$

$$Nlog\,2\pi + NlogSSE\,((\hat{\theta}^2,\lambda^2)/N) + \log|X^TX + \lambda^2D^TD| - \log|\lambda^2D^TD| + N$$

A practical way to determine the value of λ^2 for which the -2log-likelihood is minimized, is to compute the likelihood for discrete values of λ^2 and search the discrete -2log likelihood-hyperparameter space for the minimum. The Akaike (1980) paper was very likely the first practical application of the likelihood of the Bayesian model which used the likelihood of the hyperparameters as a measure of the goodness of fit of a model to data.

4.2 A Smoothness Priors Long AR Model For Spectral Estimation.

Consider the scalar autoregressive model of order p,

$$y_n = \sum_{m=1}^{p} a_m y_{n-m} + \varepsilon_n \tag{4.7}$$

In (4.7) $\{\varepsilon_n\}$ is a Gaussian white noise with mean zero and variance σ^2. A least squares fit of the AR model to the data, $y_1,...,y_N$, with the first p observations $y_{1-p}, y_{2-p},...,y_0$ treated as given constants, leads to the minimization of $\sum_{n=1}^{N}[y_n - \sum_{m=1}^{p} a_m y_{n-m}]^2$.

The frequency response function of the whitening filter of the AR process is

$$A(f) = 1 - \sum_{m=1}^{p} a_m \exp\,[-2\pi imf]\,. \tag{4.8}$$

Consider a frequency domain smoothness priors constraint on the AR model parameters. Let measures of the unsmoothness of the frequency response function be the integrated squared kth and zero-th derivative with respect to frequency of the AR process whitening filter. That is,

$$R_k = \int_{-1/2}^{1/2} \left|\frac{d^kA(f)}{df^k}\right|^2 df = (2\pi)^{2k}\sum_{m=1}^{p} m^{2k}a_m^2, \quad R_0 = \int_{-1/2}^{1/2}|A(f)|^2df = 1 + \sum_{m=1}^{p} a_m^2 \tag{4.9}$$

are the terms used as penalties to the whitening filter.

With these constraints, and with λ^2 and v^2 fixed, the AR model coefficients $\{a_m, m=1,...,p\}$, minimize

$$\sum_{n=1}^{N}[y_n - \sum_{m=1}^{p} a_m y_{n-m}]^2 + \lambda^2\sum_{m=1}^{p} m^{2k}a_m^2 + v^2\sum_{m=1}^{p} a_m^2. \tag{4.10}$$

In (4.10), λ^2 and v^2 are hyperparameters, (Lindley and Smith 1972). By a proper choice of these parameters, the estimates of the AR model coefficients balance the tradeoff between the infidelity of the model to the data and the infidelity of the model to the frequency domain smoothness constraints. To within a constant, the Gaussian priors on the AR model coefficients corresponding to the R_0 and R_k constraints are

$$\exp\frac{-\lambda^2}{2\sigma^2}\sum_{m=1}^{p} m^{2k}a_m^2 \quad \exp\frac{-v^2}{2\sigma^2}\sum_{m=1}^{p} a_m^2. \tag{4.11}$$

Following our earlier discussion, define the matrices D and a and the matrices X and y by

$$D = \begin{bmatrix} (\nu^2+\lambda^2)^{1/2} & & & \\ & (\nu^2+2^{2k}\lambda^2)^{1/2} & & \\ & & \cdot & \\ & & & \cdot \\ & & & & (\nu^2+p^{2k}\lambda^2)^{1/2} \end{bmatrix}, \quad a = \begin{bmatrix} a_1 \\ a_2 \\ \cdot \\ \cdot \\ \cdot \\ a_p \end{bmatrix}, \quad X = \begin{bmatrix} y_0 & \cdots & y_{1-p} \\ y_1 & \cdots & y_{2-p} \\ \cdot & & \cdot \\ \cdot & & \cdot \\ \cdot & & \cdot \\ y_{N-1} & \cdots & y_{N-p} \end{bmatrix}, \quad y = \begin{bmatrix} y_1 \\ y_2 \\ \cdot \\ \cdot \\ y_N \end{bmatrix} \tag{4.12}$$

Then respectively, the fitted AR model coefficients and residual sum of squares satisfy

$$\hat{a} = (X^TX + D^TD)^{-1}X^Ty, \qquad S(\lambda^2,\nu^2) = y^Ty - \hat{a}^T(X^TX + D^TD)\hat{a}. \tag{4.13}$$

The likelihood of the hyperparameter model is

$$L(\lambda^2,\nu^2,\sigma^2) = \left(\frac{1}{2\pi\sigma^2}\right)^{N/2}|D^TD|^{1/2}|X^TX + D^TD|^{-1/2}\exp\left\{\frac{-1}{2\sigma^2}S(\lambda^2,\nu^2)\right\}. \tag{4.14}$$

Given λ^2 and ν^2, the maximum likelihood estimate of σ^2 is, $\hat{\sigma}^2=S(\lambda^2,\nu^2)/N$. The maximum likelihood estimates of λ^2 and ν^2 are obtained by minimizing

$$-2logL(y\,|\lambda^2,\nu^2,\sigma^2) = N\log 2\pi\sigma^2 + \log|D^TD| - log\,|X^TX + D^TD| + N \tag{4.15}$$

with respect to λ^2 and ν^2. Computation of the likelihood over a discrete k,λ^2,ν^2 parameter grid and searching over the resulting discrete likelihood-hyperparameter space for the minimum of -2 log likelihood yields the desired smoothness priors long AR model.

The frequency domain smoothness priors constraints used here have interpretations as constraints on the smoothness of the whitening filter of the AR model. The 0th derivative has an energy constraint interpretation. These computations allow us to remain within the framework of the general linear model and are therefore computationally tractable.

Kitagawa and Gersch (1985a) show smoothness priors- long AR model results that were superior to the minimum AIC-AR model method for estimating the power spectral density. Two problem situations were studied. In one case, the simulation model was in the AR model class. The other was a two sine waves in noise case, in which the data actually correspond to an ARMA model. These results support the soundness of the empirical frequency domain smoothness priors constraint approach long AR modeling method to the estimation of the power spectral density matrix for the multichannel case by exploiting the one channel at-a-time paradigm.

4.3 Extension to the multichannel case, SPMCAR.

Our objective here is is to extend the smoothness priors long AR modeling method to the estimation of the power density matrix for the multichannel case by exploiting the one channel at-a-time paradigm. As in Section 3, the data from the individual channels is interlaced to permit an instantaneous response orthogonal innovations representation of the multichannel AR model. In that way the data can be modeled by a succession of scalar AR models to which the frequency domain smoothness priors can be applied.

4.3.1 Examples.

In a Monte Carlo simulation study of power spectrum estimation performance, the data from simulated models was fitted by two different modeling methods, the Whittle-AIC MCAR method and the smoothness priors multichannel AR, (SPMCAR) modeling method. Two different bivariate time series models respectively an AR model and an ARMA model were simulated. In the conventional MCAR form the AR model considered was

$$x(n) = \sum_{m=0}^{M} A_m x(n-m) = \varepsilon(n), \quad E[\varepsilon(n)] = 0, \quad COV[\varepsilon(n),\varepsilon(n')] = V\delta_{n,n'} \tag{4.14}$$

with parameters $A_0 = I$, $V = I$ and

$$A_1 = \begin{bmatrix} 0.130 & 0.130 \\ -0.130 & -1.130 \end{bmatrix}, \quad A_2 = \begin{bmatrix} 0.850 & 0.085 \\ -0.085 & 0.850 \end{bmatrix}.$$

The AR_1MA_1 model considered was

$$x(n) = \sum_{m=0}^{1} A_m x(n-m) = \sum_{m=0}^{1} B_m \varepsilon(n-m), \quad E[\varepsilon(n)] = 0, \quad COV[\varepsilon(n),\varepsilon(n')] = V\delta_{n,n'} \tag{4.16}$$

with parameters $A_0 = I$, $B_0 = I$ and

$$A_1 = \begin{bmatrix} -1.200 & -0.600 \\ 0.500 & -0.300 \end{bmatrix}, \quad B_1 = \begin{bmatrix} 0.900 & -0.900 \\ -0.600 & 0.900 \end{bmatrix}, \quad V = \begin{bmatrix} 1.000 & 0.500 \\ 0.500 & 1.250 \end{bmatrix}.$$

Data from thirty simulations of each model were computed with $N = 100$. Three different SPMCAR models were fitted one with effective MCAR model orders 2,4 and 8.

The theoretical MCAR order 2 model and typical Whittle-AIC algorithm (MCAR order 2), and SPMCAR model (SPMCAR model order 4) fits to an $n = 100$ simulation of the MCAR model appear in order in (4.17).

$$M_{theoretical} = \begin{bmatrix} 1.00 & -0.50 \\ -0.50 & 1.00 \\ \text{---} & \text{---} \\ 1.00 & 0.00 \\ 0.00 & 1.00 \\ \text{---} & \text{---} \\ 0.13 & -0.33 \\ -0.83 & -1.13 \\ 0.85 & 0.09 \\ -0.19 & 0.95 \end{bmatrix}, \quad MCAR_{Whittle-AIC} = \begin{bmatrix} 0.96 & -0.46 \\ -0.46 & 1.69 \\ \text{---} & \text{---} \\ 1.00 & 0.00 \\ 0.00 & 1.00 \\ \text{---} & \text{---} \\ 0.22 & -0.34 \\ -0.88 & -1.09 \\ 0.78 & 0.09 \\ -0.12 & 0.93 \end{bmatrix}, \quad MCAR_{onechannel-subset} = \begin{bmatrix} 0.963 & -0.41 \\ -0.41 & 1.15 \\ \text{---} & \text{---} \\ 1.00 & 0.00 \\ 0.00 & 1.00 \\ \text{---} & \text{---} \\ 0.23 & -0.34 \\ -0.82 & -1.04 \\ 0.85 & 0.15 \\ -0.07 & 0.85 \\ -0.02 & -0.11 \\ 0.03 & 0.04 \\ 0.04 & 0.07 \\ 0.11 & 0.01 \end{bmatrix}. \tag{4.17}$$

In (4.17) in each model, the first two rows correspond to the innovations matrix and the next two rows correspond to the $d = 2$, A_0 term.

The Whittle-AIC MCAR order 2 model fitted to the $n = 100$ simulation of the theoretical MCAR order

2 model appear to be a reasonable fit. In the SPMCAR order 4 we can see the smoothing effects of the priors on the Whittle-AIC model parameters. As expected we note that the successive AR coefficients tend to diminish in magnitude.

A sum of relative squares of the difference between the estimated model function and the true model function was computed as a goodness of fit criterion. Results were computed for the four functions, power spectral density in channels one and two, coherence squared and transfer function as functions of frequency. Numerical results are shown in Tables 1 and 2.

Table 1: AR_2 MCAR Model: Comparative spectral function estimate performances: Whittle-AIC MCAR and SPMCAR modeling.

Simulation Results: Relative MSE, 100 Trials			
	Whittle	SP Order 3	SP Order 4
Channel 1 Mean	12.907	11.683	13.975
S.D.	9.976	8.643	9.606
Channel 2 Mean	27.637	11.930	14.022
S.D.	46.416	8.436	8.473
Coherence Mean	5833.624	2907.954	3795.207
S.D.	18845.650	7165.363	8375.217
T.funct. Mean	180.296	88.628	119.158
S.D.	401.906	138.082	172.563

Table 2: AR_2MA_1 MCAR Model: Comparative spectral function estimate performances: Whittle-AIC MCAR and SPMCAR modeling.

Simulation Results: Relative MSE, 100 Trials			
	Whittle	SP Order 10	SP Order 13
Channel 1 Mean	37.768	18.369	21.060
S.D.	73.077	6.974	10.460
Channel 2 Mean	28.935	7.732	8.102
S.D.	50.428	2.929	3.851
Coherence Mean	14.396	14.168	14.396
S.D.	8.647	6.251	6.589
T.funct. Mean	10.735	7.556	7.814
S.D.	9.261	3.101	3.211

Figure 4.1 and Figure 4.2 respectively the theoretical spectral densities, coherence and transfer function (dark lines) and the empirical mean and plus and minus one standard deviation (dotted lines) for the Whittle-AIC and SPMCAR modeled data. Figure 4.1 corresponding to the simulated AR_2 model, shows the results for SPMCAR model order 4. The graphical results for SPMCAR model orders 3 and 4 were visually almost indistinguishable. Similarly, Figure 4.2 corresponding to the simulated AR_2MA_1 model shows the results for SPMCAR model order 13. The graphical results for SPMCAR model orders 10 and 13 were visually almost indistinguishable.

The results of our experiments suggest that the SPMCAR method of power spectrum estimation is superior to the Whittle-AIC method in a mean square sense and that it is more stable in the sense that the estimated function variances are smaller. Furthermore and quite important, the SPMCAR method is relatively insensitive to SPMCAR model order and it can realize good model estimation results with relatively short duration data sets.

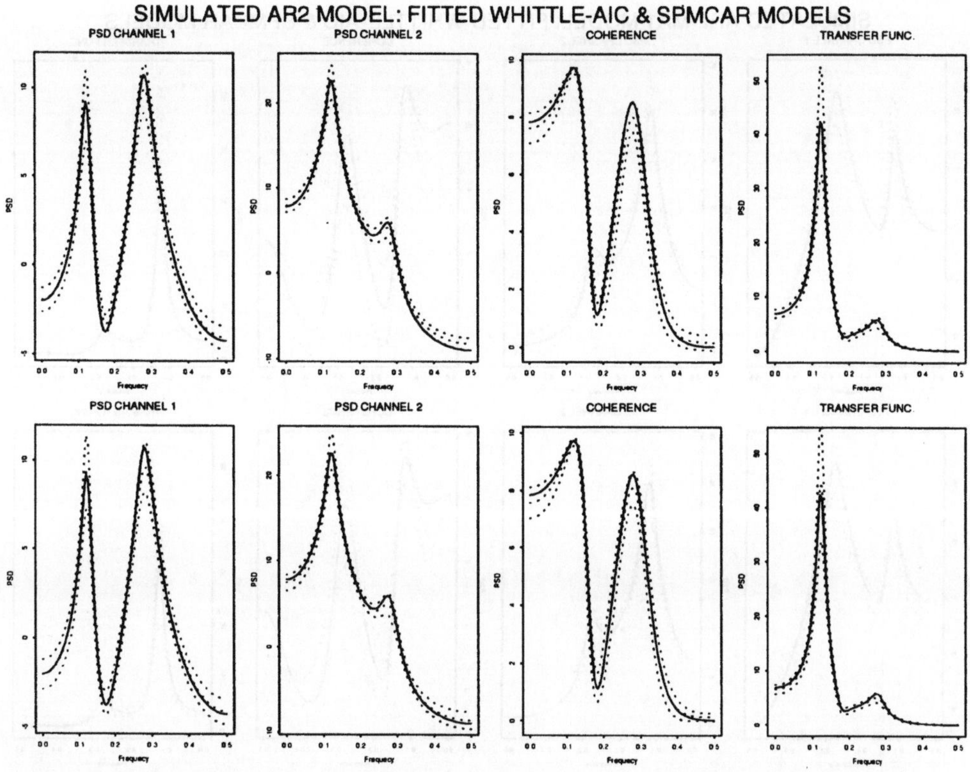

Figure 4.1 Simuated AR_2 model spectral density, coherence and transfer function (dark lines) and the empirical mean and plus and minus one standard deviation (dotted lines) for the Whittle-AIC and SPMCAR modeled data.

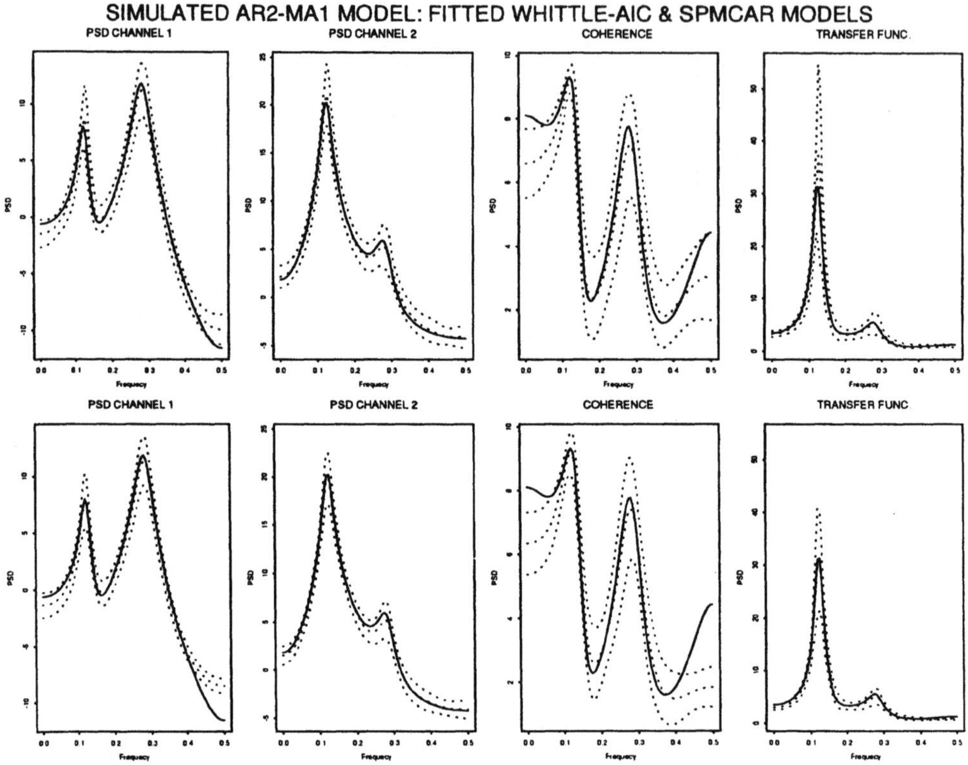

Figure 4.2 Simuated AR_2MA_1 model spectral density, coherence and transfer function (dark lines) and the empirical mean and plus and minus one standard deviation (dotted lines) for the Whittle-AIC and SPMCAR modeled data.

5. SPMCTVAR MODELING

The problem in modeling nonstationary covariance time series is to achieve an efficient parametrization to capture the local and global statistical properties of the time series. That is achieved here using the one channel at-a-time smoothness priors method. As in the preceding sections, the multichannel AR model can be realized one-channel at-a-time using only scalar computations on instantaneous data. The smoothness priors modeling used here is a stochastic linear regression of PARCORS that captures the time varying behavior of the individual AR model channels. A p-th order MCTVAR model of d channels each of duration N time series, (consisting of $d \times N$ observations), is in general specified by $p \times d^2 \times N$ AR parameters. In our algorithm, those parameters are expressed implicitly in terms of $2 \times d$ explicitly computed hyperparameters.

5.1 Development.

A multichannel time varying AR coefficient model for modeling the d-channel N duration data, $\{x(1), \ldots, x(N)\}$, is in the form

$$x(n) = \sum_{m=1}^{p} A_{n,m} x(n-m) + \varepsilon(n). \tag{5.1}$$

In (5.1) the $x(n)$'s are d-dimensional vectors and the $\{\varepsilon(n), n = 1, \ldots, N\}$ is a zero-mean uncorrelated normally distributed d-dimensional vector sequence with instantaneous covariance matrix, V. The number of AR coefficients fitted in this model is $p \times d^2 \times N$ and there are only $d \times N$ observations.

In the practical analysis of nonstationary covariance data, we are usually interested in the evolution with time of transfer functions, coherences etc. between channels as a function of frequency. Such quantities can be readily computed from the evolutionary power spectral density matrix, (Gersch and Kitagawa 1983, Gersch 1985, Gersch 1987),

$$S(f, n) = A(f, n)^{-1} V(n) A(f, n)^{-T} \tag{5.2}$$

where the multichannel AR model instantaneous system function $A(f, n)$ is given by

$$A(f, n) = I - \sum_{m=1}^{p} A_{n,m} \exp[-2\pi i m f]. \tag{5.3}$$

In order to simplify the presentation, much of the development is shown in terms of multichannel stationary AR time series. Correspondingly, the presentation follows substantially from Sakai (1982). Nonstationarity is re-invoked in computing the instantaneous partial correlation coefficients.

5.1.1 Forward and Backward PARCORS, Multichannel Stationary Time Series.

The Levinson algorithm is crucial here. It is invoked to obtain a recursive order updating algorithm in terms of a reflection coefficient parameterization, (Morf et al. 1978a,b). Toward that end, define the j-th order k-th channel forward and backward prediction errors,

$$\varepsilon(j, k + nd) = y(k + nd) + \sum_{i=1}^{j} \alpha_k(j, i) y(k + nd - i) \tag{5.4}$$

$$\eta(j, k + nd) = y(k + nd - j) + \sum_{i=1}^{j} \beta_k(j, j + 1 - i) y(k + nd - i + 1)$$

for $k = 1, \ldots, d$ and integer n respectively in terms of the forward model $\{\alpha_k(j,1), \ldots, \alpha_k(j,j)\}$ parameters where k is the channel number and j is the AR model order and backward model parameters $\{\beta_k(j,1), \ldots, \beta_k(j,j)\}$.

Invoking the recursive order update relations yields

$$\varepsilon(j+1,k+nd) = \varepsilon(j,k+nd) + \alpha_k(j+1,j+1)\eta(j,k-1+nd) \tag{5.5}$$

$$\eta(j+1,k+nd) = \eta(j,k-1+nd) + \beta_k(j+1,j+1)\varepsilon(j,k+nd)$$

with

$$\sigma^2_{k(j)} = E[\varepsilon^2(j,k+nd)], \quad \tau^2_{k(j)} = E[\eta^2(j,k+nd)]. \tag{5.6}$$

The parameter update equations then become

$$\alpha_k(j+1,i) = \alpha_k(j,i) + \alpha_k(j+1,j+1)\beta_{k-1}(j,j+1-i) \tag{5.7}$$

$$\beta_k(j+1,i) = \beta_{k-1}(j,i) + \beta_k(j+1,j+1)\alpha_k(j,j+1-i)$$

$$\sigma_k(j+1) = \sigma^2_k(j)[1-\alpha_k(j+1,j+1)\beta_k(j+1,j+1)]$$

$$\tau_k(j+1) = \tau^2_{k-1}(j)[1-\alpha_k(j+1,j+1)\beta_k(j+1,j+1)]$$

$$\sigma^2_k(0) = E[\varepsilon(0,k+nd]^2, \quad \tau^2_k(0) = E[\eta(0,k+nd]^2,$$

with the constraint $0 \le \alpha_k(j+1,j+1)\beta_k(j+1,j+1) < 1$.

5.1.2 Forward and Backward Instantaneous PARCORS.

In our algorithm, the k th channel zero-th order forward and backward innovations are

$$\varepsilon(0,k+nd) = y(k+nd), \quad \eta(0,k+nd) = y(k-1+nd). \tag{5.8}$$

From (5.5), imposing the nonstationarity of the PARCORS, the relations between the order update innovations and the instantaneous PARCORS are given by

$$\varepsilon(j+1,k+nd) = \varepsilon(j,k+nd) + \alpha_{n,k}(j+1,j+1)\eta(j,k-1+nd) \tag{5.9}$$

$$\nabla\alpha_{n,k} = v_{n,k} \quad v_{n,k} \sim N(0,\tau_1^2)$$

$$\eta(j+1,k+nd) = \eta(j,k-1+nd) + \beta_{n,k}(j+1,j+1)\varepsilon(j,k+nd)$$

$$\nabla\beta_{n,k} = u_{n,k} \quad u_{n,k} \sim N(0,\tau_2^2)$$

The instantaneous updated forward and backward model parameters become,

$$\alpha_{n,k}(j+1,i) = \alpha_{n,k}(j,i) + \alpha_{n,k}(j+1,j+1)\beta_{n,k-1}(j,j+1-i) \tag{5.10}$$

$$\beta_{n,k}(j+1,i) = \beta_{n,k-1}(j,i) + \beta_{n,k}(j+1,j+1)\alpha_{n,k}(j,j+1-i)$$

From (5.9) the order updated instantaneous PARCORS, $\alpha_{n,k}(j+1,j+1)$, $\beta_{n,k}(j+1,j+1)$ are scalar regression coefficients and the updated forward and backward innovations are the residuals of the regressions. The key innovation in (5.9) is the stochastic difference equation representation for the order updated instantaneous PARCORS. That development is related to our previous work on smoothness priors, (reviewed in Gersch and Kitagawa 1988).

To clarify this point, consider the stochastic time-varying linear regression model:

$$y_n = \phi_n^T \theta_n + \eta_n, \quad n = 1,...,N \tag{5.11}$$

where y_n, η_n, ϕ_n, and θ_n are respectively, the observation and the (unobserved) observation noise, a d-dimensional stochastic regression vector and an unknown time-varying parameter. For example, consider the parameter variation at time n to be governed by the stochastic difference equation,

$$\nabla^k \theta_n = u_n \tag{5.12}$$

where $\nabla\theta_n = \theta_n - \theta_{n-1}$, $\nabla^2\theta = \nabla(\nabla\theta)$ etc. and $\{\eta_n\}$ and $\{u_n\}$ are normally distributed uncorrelated sequences with variances σ^2 and τ^2 respectively. For example, if $\phi_n = 1$ and $\nabla^k \theta_n = u_n$, (i.e. ϕ_n and θ_n are scalars), then $\{y_n,\ n = 1,...,N\}$ is a nonstationary mean time series. Let a general model be $y_n = \sum_{i=1}^{p} \phi_{i,n}\theta_{i,n} + \eta_n$, $n = 1,...,N$ and $\nabla^k \theta_{i,n} = u_{i,n}$, with the $\phi_{i,n} = 1$. Then $y_n = \theta_{1,n} + \theta_{2,n} + \cdots \theta_{p,n} + \eta_n$, is a nonstationary mean time series-variance component model. That model has proven to be useful for econometric time series modeling, (Gersch and Kitagawa 1983, Kitagawa and Gersch 1984). Alternatively, the model $y_n = \sum_{i=1}^{M} \theta_{i,n} y_{n-i} + \eta_n$, with $\nabla^k \theta_{i,n} = u_{i,n}$, (i.e. the time varying AR coefficient model, with stochastically varying AR coefficients), is useful for modeling scalar nonstationary covariance time series. Note that in both the variance component and time varying AR coefficient model, the number of parameters exceeds the number of data. The key idea in fitting such models is to express the parameters implicitly in a prior distribution as functions of a hyperparameter, (Lindley and Smith 1972). Only the likelihood for the hyperparameter needs to be computed explicitly.

Returning to (5.11) and (5.12), in the model $y_n = \theta_n + \eta_n$, $\nabla\theta_n = u_n$, $\eta_n \sim N(0,\sigma^2)$, $u_n \sim N(0,\tau^2)$, the normal distribution for the $\{\theta_n\}$ sequence is the prior and τ^2 is the hyperparameter of the prior distribution. Gersch and Kitagawa (1988) exhibit Kalman (Kalman 1960), filter algorithm-likelihood computations for the hyperparameter τ^2 and the implicit estimation of the time-varying parameter sequence $\{\theta_n,\ n = 1,...,N\}$. Kitagawa (1988) shows a variation of the time-varying stochastic linear regression model for the computation of PARCORS for a scalar time varying AR coefficient model. A related approach is adapted here. Our earlier treatment of time varying stochastic linear regression has attracted recent research attention. The stability, convergence and tracking properties of the time varying stochastic linear regression model, are treated for example in Guo (1990), Guo et al (1991), Ljung (1991).

In (5.11), the innovations variances τ_1^2 and τ_2^2 are hyperparameters of a smoothness priors regression-trend model for the forward and backward PARCORS. In our algorithm, they are computed via a maximized likelihood of the Bayesian-smoothness priors model. We compute the sum of squares of forward prediction errors of each channel, one channel at-a-time, for each model order for use with Akaike's AIC statistic, (Akaike 1973,1974), in order to determine the AIC best order PARCOR model. Following that model order selection, the instantaneous multichannel AR model parameters are computed from the instantaneous PARCORS.

5.2 Examples.

Figure 5.1 illustrates the results of a simulation example for a $d = 2$ AR model order 2 channel quasi-periodic time series with $n = 200$. The theoretical model is

$$x(n) + \begin{bmatrix} 1.73 - 1.05*sin(\pi*n/200) & 0.2 \\ 0.2 & -1.55 + 1.25*sin(\pi*n/200) \end{bmatrix} x(n-1)$$

$$+ \begin{bmatrix} 0.85 & 0.2 \\ 0.2 & 0.85 \end{bmatrix} x(n-2) = \varepsilon(n),$$

where the $\varepsilon(n)$, $n = 1, \ldots 200$ are independent normal $d = 2$ vectors with identity covariance matrix. The two channels of simulated data are shown. Also shown are the theoretical and fitted evolution of the the instantaneous power spectral density in each channel, and the coherence between the channels for every 10 th data point. Following the development in Section 3, the simulation model corresponds to AR orders 4 and 5 for channels one and two respectively. The channel model order selected in the fitted model were four and five respectively.

Next we show an example of the data analysis of a three-channel data episode of an "aura", a common pre-epileptic seizure EEG event. Some of our previous work in the study of human epileptic EEGs is in Gersch (1985, 1987). The epileptic episode analyzed here is the same as that analyzed in those papers.

Three channels of the aura event are shown in Figure 5.2. (The data analyzed was low pass filtered and data-rate reduced to 50 samples per second.) The electrical event is the incidence of a spontaneous change of a "background" EEG with no apparent structure to one which has a quasi-periodic structure that is reminiscent of the motor response of people during the convulsion stage of an epileptic seizure. The quasi-periodicity is apparent in each channel. Also, after approximately three seconds, the appearance of the electrical activity in each channel changes. The concept of the event analyzed here is that it was was provoked by the electrical activity at a single anatomical site (an epileptic focus), which "drives" the electrical activity of the rest of the brain. The technical problem in the analysis of this data is the identification of the driving channel. This can be done by a partial spectral coherence analysis, (Gersch and Goddard 1970, Gersch 1985, 1987).

The evolution of the power spectral density, the instantaneous power spectral density, in each channel, the coherences between channel pairs and the partial coherence between channel pairs, with the influence of the excluded channel removed are also shown for every 10 th data point in Figure 5.2.

A complete discussion of the clinical significance of the analysis of this data, is out of place here. However, it might be of interest here to note the depressed partial coherence between channels 2 and 3, (with the influence of channel 1 removed both from channel 2 and channel 3). That phenomenon suggests that channel 1 accounts for the coherence or is "driving" channels 2 and 3. Another detail of clinical significance is that the temporal duration of driving is short. After the driving or initiating episode, the aura or epileptic event evolves to become a very different process.

The covariance structure and hence the ability to identify the location of the focal driving site of the epileptic episode changes over a very short time interval. That fact makes a proper method of analysis of multichannel nonstationary covariance data critical for the analysis of human focal epilepsy EEG data.

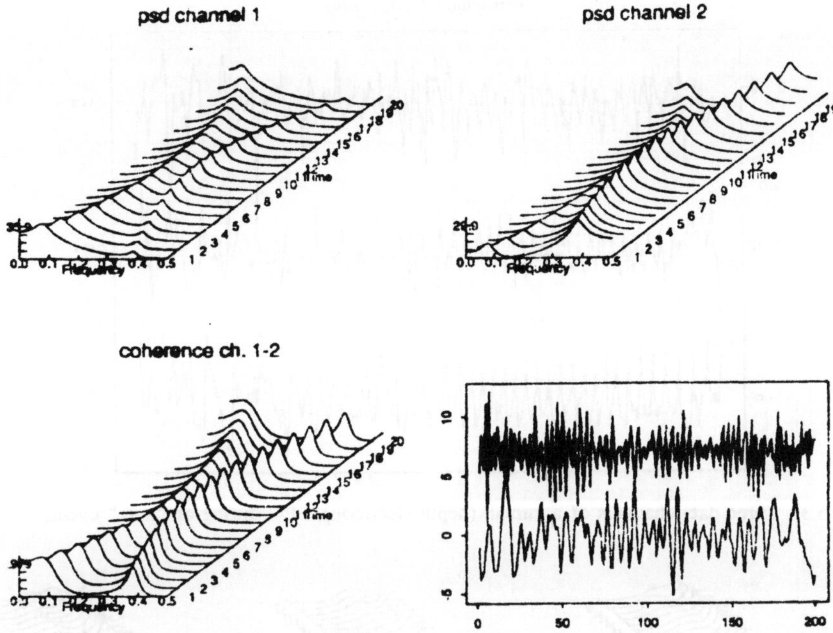

Figure 5.1: A Top Row: The theoretical evolutionary spectral densities versus frequency for channels 1 and 2 of a quasiperiodic nonstationary covariance SPMCTVAR model.
B Second Row: The theoretical evolutionary spectral coherences versus frequency between channels 1 & 2 and two channels of simulated data.

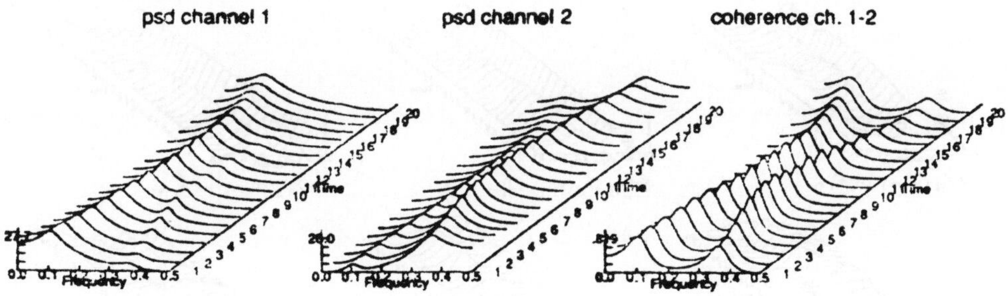

Figure 5.2: A Top Row: The fitted model evolutionary spectral densities versus frequency for channels 1 and 2, SPMVTVAR model.
B Second Row: The fitted model evolutionary spectral coherences versus frequency between channels 1 & 2.

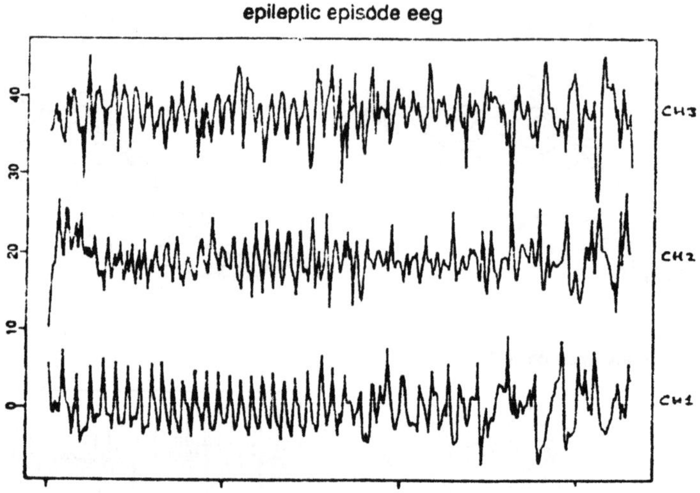

Figure 5.3: Three data channels of a human (depth electrode), EEG during an "aura" event.

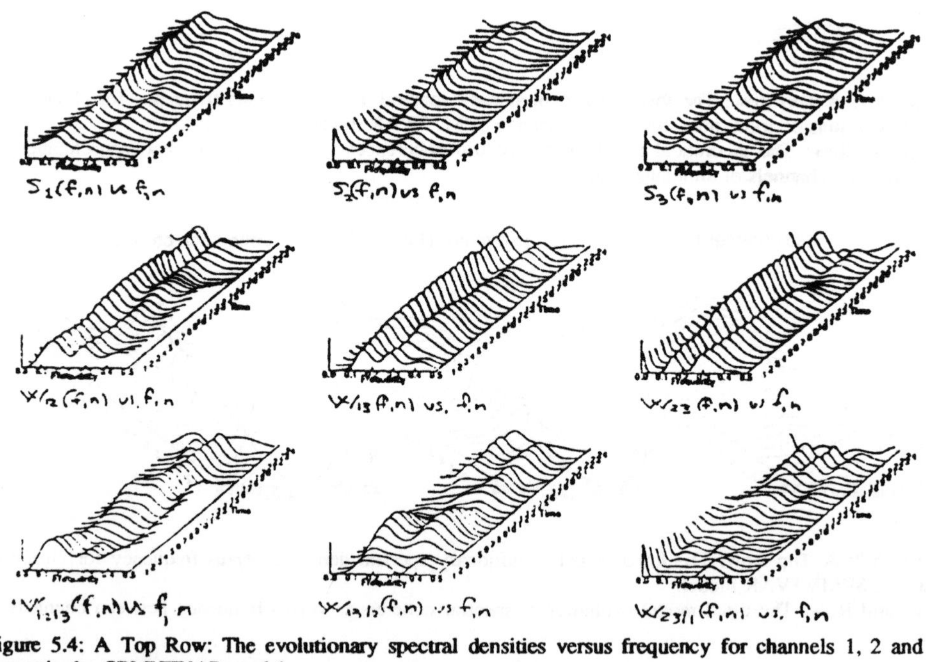

Figure 5.4: A Top Row: The evolutionary spectral densities versus frequency for channels 1, 2 and 3 respectively, SPMVTVAR model.

B Second Row: The evolutionary spectral coherences versus frequency between channels 1 & 2, channels 1 & 3, and channels 2 & 3, respectively.

C Third Row: The evolutionary partial spectral coherences versus frequency between channels 1 & 2, channels 1 & 3, and channels 2 & 3, (regressed on the excluded channel), respectively.

REFERENCES

Akaike H. (1973), Information theory and an extension of the maximum likelihood principle, *2nd International Symposium on Information Theory*, (eds. B.N. Petrov and F. Csaki), Akademia Kiado, Budapest, pp 267-281.

Akaike H. (1974), A new look at the statistical model identification, *IEEE Transactions on Automatic Control*, AC-19, 716-723.

Akaike H., Kitagawa G, Arahata E. and Tada F. (1979), *TIMSAC-78, Computer Science Monograph No. 10*, Institute of Statistical Mathematics, Tokyo

Akaike, H. (1980), Likelihood and the Bayes procedure, *Bayesian Statistics*, (eds. J.M. Bernardo, M.H. De Groot, D.V. Lindley and A.F.M. Smith), University Press, Valencia , Spain, 143-166.

Allen D.M. (1971), Mean square error of prediction as a criterion for selecting variables, *Technometrics*, 13, 469-475.

Becker, R.A. , Chambers J.M. and Wilks A.R. (1988), *The New S Language, A Programming Environment for Data Analysis and Graphics*, Wadsworth & Brooks/Cole, Pacific Grove.

Berger J.O. (1980), *Statistical Decision Theory*, Springer-Verlag, New York.

Friedlander B. (1982), Lattice filters for adaptive processing, *Proceedings IEEE*, 70, 829-867.

Furnival , G. M. and Wilson R. W. Jr. (1974), Regressions by leaps and bounds, *Technometrics*, 16, 499-511.

Gersch W. (1985), Modeling nonstationary time series and inferring instantaneous dependency, feedback and causality: An application to human epileptic seizure event data, Identification and System Parameter Estimation 1985, *7 th IFAC/IFORS Symposium*, York, UK, 737-742.

Gersch W. (1987), Non-Stationary Multichannel Time Series Analysis, Chapter 10, *EEG Handbook, Revised Series*, vol 1, (ed. A. Gevins), Academic Press, 261-296, 1987.

Gersch W. (1991), Nonstationary covariance time series: Modeling by smoothness priors methods, *Systems and Control*, (eds. T.Ono and F. Kozin), Mita Press, Tokyo, 73-84.

Gersch W. and Goddard G.V. (1970), Epileptic focus location, spectral analysis method, *Science*, 150, 701-702.

Gersch, W. and Kitagawa, G. (1983), The Prediction of Time Series with Trends and Seasonalities, *Journal of Business and Economic Statistics*,1,253-264.

Gersch W. and Kitagawa G. (1983), A multivariate time varying autoregressive modeling of nonstationary econometric time series, *Proceedings of the Business and Economics Statistics Section, American Statistical Association*, Toronto Canada, August 1983.

Gersch W. and Stone D. (1990), Multichannel time varying autoregressive modeling: A circular lattice-smoothness priors realization, *Proceedings 29th IEEE Conference on Decision and Control*, Honolulu, Hawaii. A longer version is in the *Proceedings of the 9th IFAC/IFORS Symposium on Identification and System Parameter Estimation*, Budapest, Hungary, (1991).

Geweke J. (1982), The measurement of linear dependence and feedback between multiple time series, (with discussion), *Journal of the American Statistical Association*, 77, 304-324.

Goff J.A. and Jordan T.H. (1988) Stochastic modeling of seafloor morphology: Inversion of Sea Beam data for second-order statistics, *Journal of Geophysical Research*, 93, 13,589-13,608.

Guo L. (1990), Estimating time-varying parameters by the Kalman filter based algorithm: Stability and convergence, *IEEE Transactions on Automatic Control*, AC-35, 141-147.

Guo L. , L. Xia and J.B. Moore (1991), Tracking randomly varying parameters: Analysis of a standard algorithm, *MCSS, Mathematics of Control, Signals and Systems*.

Hsiao C. (1979), Autoregressive modeling of Canadian Money and Income Data, *Journal of the American Statistical Association*, 74, 553-560.

Hurvich C. M. and Tsai C. L. (1989), Regression and time series model selection in small samples, *Biometrika* 76, 297-307.

Hurvich C. M. and Tsai C. L. (1992), A corrected Akaike information criterion for vector autoregressive model selection, Report SOR-92-10 New York University L. N. Stern School of Business.

Kalman R.E.K. (1960), A New Approach to Linear Filtering and Prediction Problems, *Transactions of ASME, Journal of Basic Engineering*, 82D,35-45.

Kay(1988), *Modern Spectral Estimation: Theory and Applications*, Prentice Hall.

Kitagawa, G. (1981), A Nonstationary Time Series Model and its Fitting by a Recursive Filter, *Journal*

of Time Series Analysis, 2,103-116.
Kitagawa G. (1983), Changing spectrum estimation, *Journal of Sound and Vibration*, 89, 433-445.
Kitagawa G. (1988), Numerical Approach to non-Gaussian smoothing and its applications, *20th Interface Symposium Computer Science and Statistics.*
Kitagawa G. and Akaike H. (1981), On TIMSAC-78 in D. F. Findley ed. *Applied Time Series Analysis II*, Academic Press, 499-547.
Kitagawa G. and Gersch W. (1984), A Smoothness Priors-State Space Modeling of Time Series with Trend and Seasonality, *Journal of the American Statistical Association*, 79, 378-389.
Kitagawa G. and Gersch W. (1985a), A Smoothness Priors Long AR Model Method for Spectral Estimation, *IEEE Transactions on Automatic Control*, AC-30,57-65.
Kitagawa G. and Gersch W. (1985b), A Smoothness Priors Time Varying AR Coefficient Modeling of Nonstationary Time Series, *IEEE Transactions on Automatic Control*, AC-30,48-56.
Lewis R and Reinsel G. C. (1985), Prediction of multivariate time series by autoregressive model fitting, Journal of Multivariate Analysis 16, 393-411.
Mallows C.L. (1973) Some comments on the C_p, *Technometrics* 15, 661-675.
Marple (1987), *Digital Spectral Analysis*, Prentice Hall.
Lindley D.V. and Smith A.F.M. (1972), Bayes estimate for the linear model, *Journal of the Royal Statistical Society, B*,34,1-41.
Lev-Ari H. (1983), Modular architectures for adaptive multichannel lattice algorithms, *Proc. 1983 IEEE Int. Conf. Acoust., Speech, Signal Processing*, Boston MA, Apr 1983, 455-458
Lee D.T.L., Morf M. and Friedlander B. (1981), Recursive least squares ladder estimation algorithms, *IEEE Transactions Acoust., Speech, Signal Processing*, ASSP-29, 627-641.
Lev-Ari H. (1987), Modular architectures for adaptive multichannel lattice algorithms, *IEEE Transactions Acoust. Speech & Signal Process.*, ASSP-35, 543-552.
Ling F and Proakis, (1984), A generalized multichannel least squares lattice algorithm based on sequential processing stages, *IEEE Trans. Acoust. Speech & Signal Process.*, ASSP-32, 381-390.
Ljung L. (1991), An exact formula for the tracking ability of adaptive algorithms: Implications for the general case, *Proc. 9th IFAC/IFORS Symp. on Identification and System Parameter estimation*, Budapest, Hungary.
Lutkepohl H. (1991), *Introduction To Multiple Time Series Analysis*, Springer-Verlag, Berlin.
Morf M., Viera A., and Kailath T. (1978a), Covariance characterization by partial autocorrelation matrices, *Annals of Statistics*, 6, 643-648.
Morf M., Viera A., Lee D.T.L. and Kailath T. (1978b), Recursive multichannel maximum entropy spectral estimation, *IEEE Trans. Geosci.. Electron.*, GE-16, 85-94.
Newton J.J. (1982), Using periodic autoregressions for multiple spectral estimation, *Technometrics*, 24, 109-116.
Pagano M. (1978), Periodic and multiple autoregressions, *Annals of Statistics* 6, 1310-1317.
Porat B., Friedlander B. and Morf M. (1982), Square-root covariance ladder estimation algorithms, *IEEE Transactions Automat. Control*, AC-17, 813-829.
Sakai H. (1982), Circular lattice filtering using Pagano's method; *IEEE Trans. Autom. Control*, AC-19, 716-723.
Shibata R (1976), Selection of the order of an autoregressive model by Akaike's information criterion, *Biometrika*, 63, 117-126.
Shibata R (1980), Asymptotically efficient selection of the order of the model for estimating the parameters of a linear process, *Annals of Statistics*, 8 147-164.
Shibata R (1981), An optimal selection of regression variables, *Biometrika*, 68, 47-54.
Tsaio G.C. and Tsay R.S. (1985), A canonical correlation approach to modeling multivariate time series, *Proceedings Econ. Bus. Stat. Sec. Amer. Stat. Assoc.*, 112-120.
Tsaio G.C. and Tsay R.S. (1989), Model specification in multivariate time series, *Jour. Royal Statistical Society*, (with discussion) 157-213
Whittle P. (1963), On the fitting of multivariable autoregressions and the approximate canonical factorization of a spectral density matrix, *Biometrika* 50, 129-134.
Wiggins R.A. and Robinson E.A. (1965), Recursive solution to the multichannel filtering problem, *Journal of Geophys. Research*, 70, 1885-1891.

SEPARATION OF SPIN SYNCHRONIZED SIGNALS USING A BAYESIAN APPROACH

T. HIGUCHI
The Institute of Statistical Mathematics
4–6–7 Minami–Azabu, Minato–ku
Tokyo 106, Japan

1. Introduction

Data taken aboard a spinning spacecraft frequently suffers from an unexpected modulation synchronized with the rotation of the spacecraft. If there is a strong noise source along some view direction, a resulting strong periodic noise appears in the measurements. Consequently, this spin–synchronized signals often dominate over the real signal. In other words, the observed real signals are masked by periodic noise due to the spacecraft spin. Furthermore, we frequently face situations such that we cannot study the observed data without eliminating this periodic noise from the original data. It is therefore important to develop methods that can sufficiently separate the noise closely associated with the rotation of the spacecraft.

The importance of an elimination of the spin–synchronized noise is reinforced with the fact that the periodic noise is crucial in inverse problems which often arise in geophysics (Jackson and Matsu'ura (1985)). For example, the quasi–periodic noise due to the spacecraft rotation is easily amplified by a simple inversion technique such as the direct differentiation of the observed data (Higuchi *et al.* (1988)). Hence we cannot avoid the problem of noise reduction when we examine data measured by a spacecraft.

The spin–synchronized noise can be easily removed from the observed data by using a simple conventional digital signal procedure only if it repeats the same pattern in time domain. However, in general cases, it shows non–stationary behavior which requires more complicated treatment inevitably involving human interpretation and interaction under the direction of a specialist of algorithm development to remove the background noise. To treat nonstationarity of time series, Akaike (1980) has introduced a Bayesian approach which allows us to construct a flexible model. This approach facilitates fitting a large amount of the time series model parameters (Gersch and Kitagawa (1988), Sakamoto (1991)). For example, the number of parameters generally exceeds the number of observations. Although this type of model with too many parameters causes nonuniqueness in solutions for estimating parameters, this problem can be solved by imposing the smoothness constraint on the time series model parameters in Akaike's (1980) Bayesian approach. Namely, Akaike considers the *prior distribution* of parameters, $p(\Theta)$, as well as the *conditional data distribution*, $p(Y|\Theta)$, where Θ is a parameter vector and Y is the observation

H. Bozdogan (ed.), Proceedings of the First US/Japan Conference on the Frontiers of Statistical Modeling: An Informational Approach, 193–215.

vector $Y = [y(1), y(2), \ldots, y(N)]$. Smoothness prior is closely related to the ill–posed (or ill–conditioned) problems (Tikhonov (1965)) and the penalized likelihood methods (Good (1965), Good and Gaskin (1971, 1980), Leonard (1978), Ogata (1990)). In addition, when we consider a normal disturbance–linear–stochastic regression prior for the prior distribution as well as an linearly additional normal observation noise for the conditional data distribution, an estimation of Θ can be reduced to the well-known constrained least squares problem (Lindley and Smith (1972), Kitagawa and Gersch (1984), Titterington (1985), Silverman (1985), Gersch and Kitagawa (1988)). The basic principles of the Bayesian approach is illustrated in Higuchi (1991b) by using a simple problem.

As mentioned above, Akaike's Bayesian approach can be characterized by considering two noise distributions. One is the observation noise with which the observation vector, Y, is expressed in an *observational model* as a function of parameters, Θ. Accordingly, this noise determines a goodness of fit of parameters to the observed data. The other is the system noise which controls through a *system model* as to how the parameters obey the given constraints. In modeling along Akaike's Bayesian approach, the smoothness constraint is usually adopted as given constraints in an attempt to get a unique solution for parameters. However, we need not adhere to this smoothness constraint and will consider various constraints by taking account of any useful prior information about parameters as much as possible. Here prior information means kind of information about Θ such as physical law, well–established results, analyst's insight.

Although it is important to explicitly incorporate various prior information into the analysis process, we cannot justify this procedure without an information criterion which allows us to objectively determine a tradeoff between the "goodness" of fit of parameters to the observed data and that to the given constraints. This tradeoff parameter plays an important role in Bayesian statistics and is referred to as the hyper–parameter (Lindley and Smith (1972)). A key point of the modern Bayesian approach along Akaike's model is to introduce Akaike Bayesian Information Criterion, abbreviated as **ABIC**, which enables us to objectively define a value of this hyper–parameter.

As previously mentioned, the Bayesian approach along the lines of Akaike (1980) is a constrained least squares computational approach. Thus an estimation of Θ can be reduced to the well-known constrained least squares problem in the simplest case where prior– and conditional data distributions are the normal disturbance–linear–stochastic regression prior and linearly additional normal noise, respectively. Since any model in this simplest case also involves the hyper–parameter, its value should be given beforehand in a process of estimation of Θ. Of course, the optimal value of hyper–parameter must be determined according to **ABIC**.

The Bayesian approach along the lines of Akaike (1980) has computational complexity of $O(N^3)$. Since the constraints between the estimated parameters are usually local in the Bayesian approach to model a time series, we can reduce this complexity to be $O(N)$ by the efficient scheme (Ishiguro (1984)). There is another approach to realize the Bayesian time series modeling along Akaike's idea (Kitagawa, (1981)).

This approach is called a state space model (**SSM**) approach. and is also designed to have the computational complexity of $O(N)$ by the computationally efficient recursive algorithm based on the Kalman filtering (Anderson and Moore (1979)). The **SSM** representation is very flexible enough to rewrite any constrained least squares computational approach only if the relationships between the estimated parameters are linear and local. Several examples for representations by the **SSM** help us to understand the equivalence between the **SSM** and constrained least squares approaches (Kitagawa (1981), Kitagawa and Gersch, (1984)). The estimation of parameters by the constrained least squares approach can be in the **SSM** approach achieved by the Kalman smoother algorithm. For detail descriptions of the procedure based on the **SSM** approach, the reader is referred to Kitagawa (1981, 1983) and Kitagawa and Gersch (1984, 1985).

Although the Bayesian approach forms a general scheme, the solutions must be tailored to each application of the actual data. Our purpose in this paper is not to explain the general scheme, but to summarize several useful models to extract the background noise which is closely associated with the spin of the satellite. We will show that these methods enable us to study the nature in situations in which we cannot analyze the data because of the instrument noise levels. We expect they would allow the nonspecialists in the data analysis to interpret large quantities of satellite data without relinquishing control of the noise reduction technique.

2. Model Interpreted as a Bandpass Filter

A bandpass filter is generally used to extract a spin synchronized signal which repeats in time domain with almost same pattern and a characteristic frequency f_c, such as a rotation of a spacecraft. However, a general bandpass filter does not work well for a quasi–periodic wave which shows a gradual change in its amplitude and/or an abrupt shift in its phase. Such situations often occur in data taken aboard a rotational body such as a spacecraft, rocket, and balloon. In this section, we present several models which can be interpreted as a band pass filter in a framework of a linear filter. Higuchi (1991a) has studied the frequency domain properties of these models and shown the performance of the models in frequency domain parametrically as a function of the hyperparameters.

2.1 QUASI–SINUSOIDAL WAVE MODEL

To extract a quasi–sinusoidal signals in the observations, a Bayesian approach has been proposed by Higuchi *et al.* (1988). In our procedure, a component with quasi–sinusoidal wave behavior, $s(n)$, is extracted from the observation, $y(n)$. We assume that $y(n)$ can be linearly decomposed as

$$y(n) = t(n) + s(n) + \varepsilon(n), \tag{2.1}$$

where $t(n)$, $s(n)$, and $\varepsilon(n)$ are a trend, quasi–sinusoidal wave, and random noise components, respectively. A smoothness constraint for prior information of the trend component is imposed and realized by minimizing the sum of the squares of its second difference.

As for the prior information for $s(n)$, we proposed the following constraint to minimize

$$\sum_n \Big(s(n) - 2c\ s(n-1) + s(n-2)\Big)^2, \tag{2.2}$$

where c is given by $c = \cos(2\pi f_c)$. If $s(n)$ exactly follows a sinusoidal variation, the residual of $s(n) - 2c\ s(n-1) + s(n-2)$ should be zero. Accordingly, to minimize (2.2) is appropriate for the prior information to describe a signal which was assumed as a locally well–behaved sinusoidal wave. Since this model for $s(n)$ specifies a local relationship around $s(n)$ and the dependence of $s(n)$ on $s(n+h)$ for the larger value of h becomes smaller, this model satisfactorily represents a decaying (or growing) sinusoid. This model can describe a sinusoid whose amplitude gradually changes as a function of time (Higuchi *et al.* (1988)).

In an actual application, there sometimes occurs a situation such that f_c is unknown. In this case, we simply initiate an estimation of its value by a visual inspection into data, and get its optimal value by minimizing **ABIC** with respect to f_c. Namely, we treat f_c as a hyperparameter within a framework of the Bayesian approach. A more complete treatment of the modeling discussed above is in Higuchi *et al.* (1988).

The procedure based on the model described here has been first applied to the airglow data obtained with a photometer on board the sounding rocket as a function of an altitude (Higuchi *et al.* (1988)). A view angle was simultaneously measured by means of both a geometric aspectmeter and a star sensor. This allows us to subtract an effect from modulations due to extraterrestrial background emissions and geometrical change in view direction to the object. However, even if we eliminate these modulations, there still remains an unexpected spurious modulation in conjunction with a periodic change in the rocket attitude. A further complication is that the amplitude of modulations strongly depends on an altitude and that there is no simple analytical function to describe this dependency. The proposed procedure based on the Bayesian approach enables us to remove this spurious modulation as well as a random noise. The high performance of this procedure to recover each component has been demonstrated by applying it to synthetic data sets (Higuchi *et al.* (1988)).

Kita *et al.* (1989) has compared Higuchi *et al.*'s procedure with two conventional methods which has been frequently used in estimating an altitude profile of the airglow: the Fourier filtering and incremental straight line fitting methods (Murtagh *et al.* (1984)). They emphasized two major advantages of the procedure based on the Bayesian approach. One is the usage of **ABIC** by which we can objectively select an optimal model to decompose the observations. The other one is that we can infer and eliminate a troublesome quasi-periodic modulation efficiently. They concluded that the procedure along the Bayesian approach can provide us an estimation of the emission rate profile with better altitude resolution and less distortion than conventional methods. By applying the Bayesian model to the altitude profile of nitrogen monoxide (NO) density, they have clearly shown that the amplitude of the estimated quasi–periodic component is varying with height in accordance with the period of the rocket precession (Kita *et al.* (1989)).

Naturally, the Bayesian model to extract a quasi-sinusoidal component with a characteristic frequency f_c can be extended to the signal with multi-quasi-sinusoidal components. When the signal $s(n)$ consists of p quasi-periodic signals, the constraint to minimize (2.2) is replaced by that minimizing

$$\sum_n \Big(\sum_{j=0}^{2p} a(j)s(n-j)\Big)^2, \tag{2.3}$$

where $a(j)$ is a constant derived from comparing the coefficients of $s(\cdot)$ in the following equation:

$$\sum_{j=0}^{2p} a(j)s(n-j) = \prod_{j=1}^{p}\Big(1 - 2\cos(2\pi f_{c,j})B + B^2\Big)s(n). \tag{2.4}$$

Of course, $f_{c,j}$ is a characteristic frequency of j–th component and B is a backward operator: $Bs(n) = s(n-1)$. For the present there is no actual application of this model, but it may be useful for detecting a signal which consists solely of sinusoids in additive white noise.

2.2 DECAYING (OR GROWING) SINUSOIDAL WAVE MODEL

In association with the above model, we present a model which describes a signal consisting of multi-decaying (and/or growing) sinusoids. The coefficients $a(j)$ of this model in (2.3) can be obtained in the following manner. When $s(n)$ is described as a set of decaying (and/or growing) sinusoids, $s(n)$ satisfies the following linear difference equation:

$$\prod_{j=1}^{p}(z_j - B)(z_j^{\dagger} - B)s(n) = 0,$$

$$z_j = \exp\Big(2\pi(g_j + if_j)\Big), \tag{2.5}$$

where g_j and f_j are the growth rate and frequency of j–th component, respectively (Kay and Marple (1981)). Here, B is the backward operator as in (2.4), and the dagger represents the complex conjugate operation. By setting $\gamma_j = \exp(2\pi g_j)$, (2.5) can be rewritten as

$$\prod_{j=1}^{p}\Big(B^2 - 2\gamma_j\cos(2\pi f_j)B + \gamma_j^2\Big)s(n) = 0. \tag{2.6}$$

Setting $\sum_{j=0}^{2p} a(j)s(n-j) = 0$, we compare the coefficients of $s(n-j)$ with those in (2.6), for each j, and then get the coefficients of $a(j)$ as a result. In actual application, prior information for decaying (and/or growing) sinusoidal wave is to minimize the sum of the squares of $\sum_{j=0}^{2p} a(j)s(n-j)$:

This prior model implies that the signal can be expressed by a linear combination of exponentials. Using such specific, rather complicated model for an examination of the free oscillations (e.g., Earth's free oscillations (Fukao and Suda (1989)) is reasonably justified, because it precisely specifies the relationship necessary for the free oscillations.

2.3 SEASONAL COMPONENT MODEL

When a signal $s(n)$ is monthly data, it clearly shows a strong seasonal pattern with a period of $r = 12$. Here we present a seasonal component model for signal with a "cyclic" behavior with a period of r. The seasonal component model is realized by minimizing

$$\sum_n \Big(\sum_{j=0}^{r-1} s(n-j) \Big)^2 \tag{2.7}$$

(Kitagawa and Gersch (1984)). This constraint has been applied to the economic data and its practical utility has been well established (Akaike and Ishiguro (1983), Gersch and Kitagawa (1983), Kitagawa and Gersch (1984)).

For a seasonal component model with a period of r, we can consider the following simple constraint to minimize

$$\sum_n \Big(s(n) - s(n-r) \Big)^2 \tag{2.8}$$

(Akaike and Ishiguro (1983), Ishiguro (1984)). It is impossible for this seasonal component model to eliminate the direct current (DC) level component, because the solution of $s(n) = s(n-1)$ obviously satisfies this constraint (Kitagawa (1986), Higuchi (1991a)). The frequency domain property of this model clearly indicates the inability to eliminate the DC level from the estimated seasonal component (Higuchi (1991a)). Accordingly, the resulting decomposed signal by using this model for a seasonal component model intrinsically contains the DC level component. When a signal is decomposed into a trend, seasonal and other components, the constraint to minimize (2.8) should be used together with that of minimizing (2.7) because the trend component cannot be uniquely defined, for a fixed value of the hyperparameters, without accompanying the constraint of a minimization of (2.7) (Akaike and Ishiguro (1983)).

While the seasonal component model has been successfully applied to the economic data, it would be useful and efficient for the experimental data which suffers from artificial instrumental noise showing a strong cyclic pattern, due to the periodicity inherent in a experimental method such as a rotation of the rocket, or radiation of radio waves with a characteristic frequency, or the instruments using an alternating current (AC).

2.4 AR MODEL

As for a final model in this section, we show a model that imposes an auto regressive (AR) model upon a signal $s(n)$. Kitagawa and Takanami (1985) adopted this model to estimate the earthquake signal by eliminating the background and observation noises. They assume that the record of the seismograph can be decomposed into three components: earthquake signal, background noise, and observation noise. Then they use a model of $y(n) = s_1(n) + s_2(n) + e(n)$, where $s_1(n)$ and $s_2(n)$ are a background noise and an earthquake signal, respectively. Of course, $e(n)$ represents an observation noise. The prior models expressed by an AR model are imposed on both $s_1(n)$ and $s_2(n)$. In their procedure, the AR coefficient to describe a background noise, $a_1(j)$ $(j = 1, \ldots, M_1)$, is first estimated by applying the model of $y(n) = s_1(n) + e(n)$ to the seismic data which was recorded before the onset times of the earthquake signal, where M_1 is an order of the AR coefficient for the background noise. In the second procedure in an estimation of the earthquake signal, they applied the model of $y(n) = s_1(n) + s_2(n) + e(n)$ under the fixed values of $a_1(j)$. Frequency domain properties of this model can be examined by the formulation along the line of Higuchi (1991a).

3. Point-Noise Source Model

In this section, we present a couple of models in which a relative relationship between the detector and point noise source is explicitly realized as an analytic function of an angle between them. The models shown below are applicable to a case where the noise source can be sufficiently approximated by the point noise source due to its distance from the satellite. In addition, we extend those models to the multi–variate (multi–channel) time series, because it is usual that the physical quantity is simultaneously measured with several detectors in order to seek its spatial distribution.

3.1 SINGLE VARIABLE

Here we consider a single variate time series, $y(n)$, and extract an effect from the background noise due to the point noise source. In an actual data obtained by a satellite, we frequently face the situation such that the point noise source can be attributed to the sun. Accordingly, the angle between the detector and point noise source, $\theta(n)$, is known in many cases, because the angle between the detector and solar direction is precisely measured with another instrument for determination of the attitude of the satellite. If $\theta(n)$ is previously unknown, we assume that it can be described as a simple function of a time n such as

$$\theta(n) = Mod\Big(\theta_0 + \frac{2\pi n \Delta t}{T}, 2\pi\Big), \tag{3.1}$$

where θ_0 and T are an initial phase and period, respectively. Δt is, of course, a sampling time for measurements. Here, we set $\theta = 0$ in a case where a detector points just toward the point noise source. θ_0 and T are treated as hyperparameters and are determined so that **ABIC** is minimized. In an actual application to data obtained by the satellite, T should be identical with the spin period, T_s.

Equation (3.1) provides us the variation of θ in a case where the relative position of the satellite to the noise source can be satisfactorily assumed to be independent of time. However, when the satellite changes its location relative to the noise source rapidly, then (3.1) should be developed to include an effect from this rapid change. As a result, if necessary, we have to consider more complicated function for describing $\theta(n)$ as a function of n, instead of (3.1).

3.1.1 *Observational Model.* To extract the background noise from an observed time series $y(n)$ $(n = 1, \ldots, N)$, here we consider the model

$$y(n) = f(\theta(n))I(n) + t(n) + e(n), \tag{3.2}$$

where the first term in (3.2), $f(\theta(n))I(n)$, is the background noise component, $t(n)$ is a trend, and $e(n)$ is an observation noise which is assumed to be a white noise sequence with zero mean. The trend component is assumed to be free both from an effect due to the point noise source and from the observation noise. The background noise component is expressed by a product of $f(\theta(n))$ and $I(n)$, where $f(\theta)$ is an arbitrary function of θ, and $I(n)$ is a time varying intensity of the point noise source at each time n. Accordingly $f(\theta)$ is dimensionless. The representation like this for the background noise component implies that $f(\theta(n))I(n)$ shows a cyclically repeating pattern with period T and that the form of cyclic pattern changes gradually as well as the power of point noise source.

When a response of the detector to the point noise source (i.e., reception sensitivity) is known as a function of θ according to the results based on experiments at the ground, we explicitly include the relationship between them and numerically realize it as a function of θ, $f(\theta)$. For instance, in a case where we beforehand measure the sensitivities of the detector, G_i, at θ_i $(i = 1, \ldots, J)$, we simply adopt the spline smoothing approach to curve estimation in order to represent a continuous function of θ, $f(\theta)$ (Ishiguro and Arahata (1982), Silverman (1985), Wahba (1990)).

In a case where the background noise is unexpectedly observed and no information about G_i is given, we have to define the form for $f(\theta)$ by taking account of available knowledge about the response of the instrument to the point noise source. Any function can be a good candidate for $f(\theta)$ if it roughly approximates a cyclic pattern of the component which is probably attributed to the background noise. For simplicity, we sometimes select $f(\theta)$ as follows:

$$f(\theta) = \begin{cases} \cos\theta & |\theta| < 90^\circ \\ 0 & 90^\circ \leq |\theta| \leq 180^\circ. \end{cases} \tag{3.3}$$

This approximation can be justified in a situation such that the reception sensitivity of the detector is proportional to a radiation (light) flux due to the sun which is located far from the satellite and that a shape of the detector to receive the radiation is sufficiently assumed to be a plane, because the effective area for reception decreases with increasing in θ according to (3.3), as shown in Figure 1(a).

Although we can take an alternative function of $f(\theta)$, such as $\cos^2\theta$, $\exp(-\sin^2\theta)$, or quadratic curve, a choice is well justified by using the minimum **AIC** procedure that was developed as a natural extension to the maximum likelihood method (Akaike (1973), Sakamoto *et al.* (1986)). Here we remark that a computational time for estimating the background noise slightly depends on the function form for $f(\theta)$. We can therefore search for the best function form to minimizes **AIC**, without wasting more computational time.

3.1.2 *System Model.* In order to get a unique solution for decomposing the observation $y(n)$ like (3.2), we adopt the smoothness constraint for prior information of both $I(n)$ and $t(n)$. For simplicity, the smoothness constraint is for each component realized by minimizing the sum of the squares of the first difference. The smoothness constraint for $I(n)$ is based on a natural assumption such that the intensity of the point noise source changes gradually with time. In contrast, to adopt the smoothness constraint for $t(n)$ imposes restrictions on applicability of the model represented by (3.2), because in general there is no justification to assume that $t(n+1)$ takes a similar value as $t(n)$. As previously mentioned, the component due to the real signal is attributed to $t(n)$ which is the component free from both the point noise source and observation noise. Accordingly, to impose a smoothness constraint on $t(n)$ implies that the real signal at $\theta = \theta(n)$ can be considered to be almost the same as that at $\theta = \theta(n+1)$. This assumption is applicable only for the case where the phase difference in successive point, $\Delta\theta$ $(= \theta(n) - \theta(n-1)$), is small. Two situations to realize the small phase difference in (3.1) can be considered. One is that T is much longer than Δt . The other is that $|T - \Delta t|/T$ is sufficiently small. These are not particular but general situation for the measurements by the satellite, because the detector aboard the satellite is usually designed to change its view direction gradually.

3.2 EXTENSION TO MULTI–CHANNEL OBSERVATION

The satellite is usually spin–stabilized by making the direction of the spin axis pointing toward some view direction. Accordingly, the view direction of a detector aboard the spacecraft with respect to the spin axis is invariable and thus a filed of view covered by the single detector is insufficient to seek a spatial distribution of the physical quantity. To overcome this problem, several identical detectors are installed so as to cover the wider field of view. Namely, each detector has its view direction with respect to the spin axis and measures the same physical quantity simultaneously. In this case, we have to consider a model for describing a multi–variate time series, $\mathbf{y}(n) = [y_1(n), y_2(n), \ldots, y_M(n)]^T$, where $y_m(n)$ corresponds to the observation obtained by the m–th channel (detector) at time n and M is the number of detectors.

3.2.1 *Configuration.* Here we introduce a spinning spacecraft coordinate in which the spin axis is taken as the Z–axis and the X–axis is defined so as to make the XZ plan contain the point noise source. We denote the view direction from the satellite to the noise source by a vector ON. Since the point noise source is usually assumed to be located far from the satellite, its radiation flux is nearly parallel to ON.

In this coordinate, the m–th detector points toward the Om direction. The geometrical configuration of the m–th detector of the satellite in relation to the point noise source is illustrated in Figure 1(b). Along line of derivation in the Section (3.1), we define the angle between the X–axis and Om' by $\theta_m(n)$, where Om' is a projection of Om onto the XY plane. Generally, the positions of detectors are confined in the plane including the spin axis and thus $\theta_m(n)$ is common to the channels.

The angle between the direction of the point noise source (ON) and the spin axis (Z–axis) is denoted by $\phi(n)$. As mentioned above, the spin axis of the satellite is generally fixed to point toward some view direction, and then $\phi(n)$ appears to be almost constant or change very slowly unless the point noise source moves rapidly. We specify the known value of the angle between the spin axis and m–th detector by ϕ_m. Based upon these definitions, the angle between the view direction of the m–th detector (Om) and the direction of the point noise source (ON), $\psi_m(n)$, are given by

$$\psi_m(n) = \cos^{-1}\bigl(\sin\phi_m \cos\theta_m(n) \sin\phi(n) + \cos\phi_m \cos\phi(n)\bigr). \tag{3.4}$$

3.2.2 *Observational and System Models.* As in the Section (3.1), we also postulate that the background noise can be modeled as $f(\cdot)I(n)$, where $I(n)$ represents the time varying intensity of the point noise source and $f(\cdot)$ is a function which describes an effect of the angle between the view direction of the detector and direction of the point noise source on the change in the intensity of the observed background noise. While $f(\cdot)$ is a function of θ in the Section (3.1), here $f(\cdot)$ should be a function of $\psi_m(n)$: $f(\cdot) = f(\psi_m(n))$. Accordingly, the observation $y_m(n)$ is represented as

$$y_m(n) = f(\psi_m(n))I(n) + t_m(n) + e_m(n), \tag{3.5}$$

where $t_m(n)$ and $e_m(n)$ are the trend and observation noise components for the m–th channel, respectively.

As for the multi–variate time series, we also assume that $I(n)$ is gradually changing with time. In addition, a trend component of each channel is assumed to show a smooth behavior in time domain. Namely, we adopt the smoothness constraint for prior information of $I(n)$ and each $t_m(n)$ $(m = 1, \ldots, M)$. For simplicity, we adopt the simplest smoothness constraint which requires a minimization of sum of the squares of the first difference for $I(n)$ and each $t_m(n)$.

3.2.3 *Representation of State Space Model.* To formulate the model mentioned above, we represent the above Bayesian approach by using the **SSM** approach. Define the $(M+1)\times 1$ state vector at time n to be $\mathbf{z}(n) = [I(n), t_1(n), t_2(n), \ldots, t_M(n)]^T$. The state space modeling is generally given by

$$\begin{aligned} \mathbf{z}(n) &= F\mathbf{z}(n-1) + G\mathbf{u}(n) \\ \mathbf{y}(n) &= H(n)\mathbf{z}(n) + \mathbf{e}(n), \end{aligned} \tag{3.6}$$

where F and G are an $(M+1)\times(M+1)$ identical matrices. The $M\times(M+1)$ observation matrix, $H(n)$, is given by

$$H(n) = \begin{pmatrix} f(\psi_1(n)) & 1 & & O \\ f(\psi_2(n)) & & 1 & \\ \vdots & & & \ddots \\ f(\psi_M(n)) & O & & 1 \end{pmatrix} \tag{3.7}$$

The system and observation noise vectors , $\mathbf{u}(n)$ and $\mathbf{e}(n)$, are assumed to be normally distributed and independent of time, i.e.,

$$\begin{pmatrix} \mathbf{u}(n) \\ \mathbf{e}(n) \end{pmatrix} \sim N\left(\begin{pmatrix} 0 \\ \vdots \\ 0 \end{pmatrix}, \begin{pmatrix} \tau_I^2 & & & & & & \\ & \tau_1^2 & & & & & \\ & & \ddots & & & & \\ & & & \tau_M^2 & & & \\ & & & & \sigma^2 & & \\ & & & & & \ddots & \\ & & & & & & \sigma^2 \end{pmatrix} \right), \tag{3.8}$$

where τ_I^2 is the variance of the noise intensity component, $I(n)$, and τ_m^2 is also that for each trend component, $t_m(n)$. For convenience, each component in $\mathbf{e}(n)$, $e_m(n)$, is assumed to be a zero–mean Gaussian white noise sequence with common variance σ_m^2 independent of m; $\sigma_m^2 = \sigma^2$ $(m = 1, \ldots, M)$. As seen in (3.8), we assume that each observation noise component is uncorrelated with other components. Detail procedures to estimate $\mathbf{z}(n)$ are shown in Kitagawa (1981, 1983) and Kitagawa and Gersch (1984, 1985).

4. Time– and Phase– Dependent Model

In this section, we will discuss a method which can be applied to more general case where the background noise, $B(n)$, shows a more complicated dependence both on a time and on a phase θ. As mentioned in the introduction, spin synchronized signals often dominate over the naturally generated emissions. As a result, the natural signals are masked by the noise source along some view direction, and the observed data are a superposition of the background noise and natural emissions. We can therefore assume that the background noise $B(n)$ is linearly added to the real natural signal $x(n)$. Namely, we decompose the observed raw data, $y(n)$ $(n = 1, 2, \ldots, N)$ into $B(n)$ (background noise) and $x(n)$ (real signal) as follows:

$$y(n) = B(n) + x(n). \tag{4.1}$$

4.1 TIME–VARYING PHASE RESPONSE FUNCTION MODEL

Our paper addresses the problem of separating the spin–synchronized signals by using the Pioneer Venus wave data (Russell (1991), Strangeway (1991)) which is sometimes contaminated with the strong background noise. We show a typical example of the observed data, $y(n)$, and phase, $\theta(n)$, in Figures 2(a) and (b), respectively. It should be remarked that the the phase $\theta(n)$ is the angle between the detector (sensor) and the solar direction and known previously. The data is an electric field spectral density ($(V/m)^2/Hz$) for 5.4 [kHz], and then both $B(n)$ and $x(n)$ in (4.1) should be positive valued accordingly. The difference of phase of successive points in time, $|\theta(n) - \theta(n-1)|$, is set either 6.90 or 6.91 [deg.] (round off error), and there is no simple function to describe a switch from one to another. As a result, a phase $\theta(n)$ is almost uniformly distributed, but irregularly spaced in phase domain from 0° to 360° [deg.]. Since the satellite spin provides a variation of the phase $\theta(n)$ shown in Figure 2(b), the obtained data almost repeats in time domain the same pattern associated with the background noise accordingly. The extremely large values seen in Figure 2(a) are due to natural emissions ($B(n) << x(n)$).

4.1.1 *Observational Model.* As easily expected, the background noise is quite sensitive to the phase θ. We therefore assume that the background noise, $B(n)$, can be described as a function of time, n, and phase, $\theta(n)$:

$$B(n) = B(n, \theta(n)). \tag{4.2}$$

A further complication is that the phase response of $B(n)$ is time-varying. However, it can be assumed to be gradually changing with time. Accordingly we model the observed data, $y(n)$, as

$$y(n) = I(n) f(\theta(n)) + x(n), \tag{4.3}$$

where $f(\theta)$ is the phase response (stationary part of the background noise) and $I(n)$ is the envelope of the background noise intensity at each time n. Namely we assume that the background noise is modulated by the time-varying background noise intensity. This assumption is easily examined by plotting the observed data in the logarithmic scale in phase domain, because (4.3) implies that data points without a natural emission ($x(n) = 0$) should scatter around $\log f(\theta)$ according to the logarithm of the time-varying background noise intensity, $\log I(n)$:

$$\log y(n) = \log I(n) + \log f(\theta(n)) \qquad (for\ x(n) = 0). \tag{4.4}$$

Generally speaking, the mean of $I(n)$ drifts with time. Consequently, we have to take account of this long–term trend component in $I(n)$. Here we furthermore decompose $I(n)$ into two factors: $I(n) = t(n)w(n)$, where $t(n)$ and $w(n)$ are a long–term trend and stationary components, respectively. By stationary component, we mean that its distribution can be assumed to be independent of time. These assumptions presented above leads to the following model for the observations:

$$y(n) = t(n)w(n)f(\theta(n)) + x(n). \tag{4.5}$$

The data which we are now discussing are heavily contaminated with the background noise and then most of data contains quite small real signal: $x(n) \simeq 0$. Consequently, before determining $x(n)$, we will estimate the unknown form of $f(\theta)$, $t(n)$, and

$w(n)$ each, in the background noise component, $B(n,\theta)$. However, there is a significant problem which arises due to the non-linearity in the process of this decomposition. Several possible treatments to mitigate this problem can be considered. The easiest approach is to separate a procedure into two steps: one is for estimating $t(n)$ and the other is for $f(\theta)$. Here we basically adopt this procedure.

4.1.2 *Estimating of the Long–term Trend Component.* Using the decomposition of (4.4), (4.5) can be rewritten as follows:

$$\log y(n) = \log t(n) + \log w(n) + \log f(\theta(n)) \qquad (for \quad x(n) = 0). \tag{4.6}$$

Moreover a variable of $\log y(n) - \log t(n)$ gives us an opportunity to estimate $\log f(\theta)$ without being disturbed by the nonstationarity of $I(n)$. Hence, we first calculate $t(n)$ before estimating $f(\theta)$.

The process of estimating $t(n)$ begins by determining the minimum background noise level for each individual spacecraft period. This provides a rough estimate of the variation in the background noise level, $\hat{t}(n)$, which is denoted by dots in Figure 2(a). In Figure 2(a) $\hat{t}(n)$ without a dot is treated as a missing value in our procedure. We assume that the long–term trend component of the background noise intensity changes slowly with time. Hence we smooth the logarithm of $\hat{t}(n)$ by using the Bayesian smoothing technique. Here we adopt the simplest smoothness model which minimizes the sum of the squares of the first difference of $\log t(n)$. We superpose on Figure 2(a) by the broken line an initial estimate of the long–term trend component, $t^*_{(0)}(n)$, with the minimum **ABIC**, where (0) indicates the estimation based on $\hat{t}(n)$. The obtained $t^*_{(0)}(n)$ shows a straight line during this interval.

4.1.3 *Estimating of the Phase Response Function.* To estimate a phase response function of the background noise, we construct the following new time sequence

$$u(n) = \log y(n) - \log t^*_{(0)}(n) \tag{4.7}$$

which is not much influenced by the long–term trend component of the background noise intensity. If observations contain no real signal, $u(n)$ can be rewritten as

$$\begin{aligned} u(n) &= \log w(n) + \log f(\theta(n)) \\ &= e(n) + \log f(\theta(n)), \end{aligned} \qquad (for \quad x(n) = 0) \tag{4.8}$$

where $e(n) = \log w(n)$. The distribution of $u(n)$ without real signal will scatter around $\log f(\theta)$ at each θ in the phase domain according to the distribution of $e(n)$. As for the observations with significantly large real signal, the value of $u(n)$ is so far separated far enough in value from $\log f(\theta)$ at each θ that it can be considered as an outlier. We estimate the form of $\log f(\theta)$ by using $u(n)$ derived from the observation $y(n)$ which is assumed to contain no real signal.

In order to express $\log f(\theta)$ as a function of θ, for simplicity, we adopt the zero–order spline function. So that a value of $\log f(\theta)$ within a range of $(i-1)\Delta\theta \le \theta < i\Delta\theta$

(i–th bin) is represented by a single variable f_i ($i = 1, \ldots, 360/\Delta\theta$). We set $\Delta\theta = 1$ in this study. In order to detect an outlier which contains real signal, we order the data within each bin according to increasing magnitude and denote the j–th largest within i–th bin by $u_i{}^j$. Thus, $u_i^1 \leq u_i^2 \leq \cdots \leq u_i^{M_i-1} \leq u_i^{M_i}$ is the ordered set of observations within i–th bin, where the number of data points within i–th bin is specified by M_i. After rearranging the data in order of increasing magnitude, we find K_i values (out of a sample of size M_i) that are large when compared with the remaining $M_i - K_i$ values and define the rough estimate of $\log f_i$ by taking an average over the $M_i - K_i$ values of u_i^j.

The detection of unknown number of multiple outliers, has received continued attention in the statistical literature. Much of works has been done within the framework of traditional statistical hypothesis testing proposed by Grubbs (1969). An improved method based on Grubbs's (1969) test is frequently used by Tietjen and Moore (1972). Despite the simplicity of their method, there exist several difficulties (Tietjen and Moore (1972), Kitagawa (1979)). To overcome these difficulties, a Bayesian procedure has been proposed by Kitagawa and Akaike (1982), and its practical utility has been illustrated by numerical examples. In their method, observations are assumed to obey a particular Gaussian distribution with ordered means such that $M_i - K_i$ observations $u_i^1, \ldots, u_i^{M_i-K_i}$ are the realization of normally distributed variable with an unknown common mean μ_i and variance σ_i^2, and K_i observations $u_i^{M_i-K_i+1}, \ldots, u_i^{M_i}$ are obtained from Gaussian distributions with ordered means, $\mu_i^{M_i-K_i+1} \leq \ldots \leq \mu_i^{M_i}$, and common variance σ_i^2. As for the prior distribution, they assume that we have no information about the number of outliers. The readers are referred to Kitagawa and Akaike (1982) for the detailed derivations on this procedure.

For a fixed value of K_i, the logarithm of the posterior probability that each outlier u_i^j is obtained from a distribution with each μ_i^j ($j = M_i - K_i + 1, \ldots, M_i$), respectively, and that u_i^j ($j = 1, \ldots, M_i - K_i$) is drawn from a distribution with common μ_i, is given by

$$\log p(K_i|u_i) = -\frac{M_i}{2}\log\sigma_i^2 - \frac{M_i(K_i+2)}{M_i - K_i - 3} - \log M_i! + \log(M_i - K_i)!, \tag{4.9}$$

where a common additive constant is ignored and

$$\sigma_i^2 = \frac{1}{M_i}\sum_{j=1}^{M_i-K_i}(u_i^j - \mu_i)^2, \qquad \mu_i = \frac{1}{M_i - K_i}\sum_{j=1}^{M_i-K_i} u_i^j. \tag{4.10}$$

The number of outliers, K_i, is selected so as to maximize the posterior probability of u_i^j ($j = M_i - K_i + 1, \ldots, M_i$) being the outliers. It should be noticed that this posterior probability is defined by summing the posterior probability of $K_i!$ possible models, because there are $K_i!$ ways of assigning u_i^j ($j = M_i - K_i + 1, \ldots, M_i$) to the K_i distributions specified by the means μ_i^j ($j = M_i - K_i + 1, \ldots, M_i$). To calculate $K_i! - 1$ posterior probabilities except for (4.9) requires somewhat complicated numerical solution (Kitagawa and Akaike (1982)). Actually, unless there are outliers with nearly equal values, only the posterior probability of (4.9) takes a significant value among $K_i!$ models (Kitagawa and Akaike (1982)). Thus the posterior probability of u_i^j ($j = M_i - K_i + 1, \ldots, M_i$) being the outliers is simply approximated by the posterior probability of (4.9), neglecting a contribution of the posterior probabilities of $K_i! - 1$ models. We, therefore, find the value of K_i by maximizing (4.9), instead of calculating the exact posterior probabilities. After searching the optimal K_i, we define the rough estimate of $\log f_i$ by μ_i within individual bins; thus

$$\log \widehat{f}_i = \mu_i = \frac{1}{M_i - K_i} \sum_{j=1}^{M_i - K_i} u_i^j. \tag{4.11}$$

Since $\log f(\theta)$ is assumed to show a smooth behavior in the phase domain, we smooth the obtained $\log \widehat{f}_i$. Here we also adopt a Bayesian smoothing approach. The method used for smoothing here somewhat differs from that frequently used for a time series on the point that $f(\theta)$ is a periodic function. Hence f_{360} should be connected to f_1 continuously. In short, we modify the smoothness constraint from a minimization of $\sum_{i=3}^{360}(f_i - 2f_{i-1} - f_{i-2})^2$ to that of $\sum_{i=1}^{360}(f_i - 2f_{i-1} - f_{i-2})^2$, where f_0 and f_{-1} correspond to f_{360} and f_{359}, respectively. The obtained smooth $\log \widehat{f}_i$ ($\log f_i^*$) provides us the phase response function $f_{(1)}^*(\theta)$, where the subscript (1) explicitly denotes the first estimate for the phase response function $f(\theta)$.

4.1.4 *Iterative Optimization.* The defined $t_{(0)}^*(n)$ and $f_{(1)}^*(\theta)$ provide us with the initial estimate of the stationary component of the background noise intensity variation, $w_{(0.5)}^*(n)$, defined by

$$w_{(0.5)}^*(n) = \frac{y(n)}{t_{(0)}^*(n) f_{(1)}^*(\theta(n))}, \tag{4.12}$$

where a subscript (0.5) explicitly indicates an estimate when we use $t_{(0)}^*(n)$ and $f_{(1)}^*(\theta)$. Of course, the stationary component represents the stationary variation and its distribution is assumed to be characterized by a time–independent distribution function. However, it sometimes occurs that there still remains in $w_{(0.5)}^*(n)$ an effect from the long–term trend component of the background noise intensity variation. Because $t_{(0)}^*(n)$ is the initial estimate for the long–term trend component of the background noise intensity. Thus far, we adjust this initial estimate furthermore by extracting the long–term trend component from $w_{(0.5)}^*(n)$, providing that it exists.

Before extracting the long–term trend component in $w_{(0.5)}^*(n)$, we have to divide the total data set into two groups: candidates for $w_{(0.5)}^*(n)$ without the real signal, and candidates which consists of the real signal and background noise. In short, we distinguish between the candidate for a pure background noise (denoted hereafter as group–B) and that for a background noise + real signal (denoted hereafter as group–$B + x$). While the sequences of $w_{(0.5)}^*(n)$ for the group–B almost show a stationary behavior in time domain which can be characterized by a time–independent noise distribution function, $w_{(0.5)}^*(n)$ for group–$B+x$ is much larger and behaves as an abrupt positive change in the sequences of $w_{(0.5)}^*(n)$. Therefore we extract the long–term trend component in $w_{(0.5)}^*(n)$ by using only data from group–B.

The selection of the candidate for group–B can be realized in this study by examining the distribution function of

$$e_{(0.5)}^*(n) = \log w_{(0.5)}^*(n). \tag{4.13}$$

As mentioned above, $e^*_{(0.5)}(n)$ for group–$B+x$ appears as an outlier. The reason we adopt $e^*_{(0.5)}(n)$ instead of $w^*_{(0.5)}(n)$ for the selection is based on the fact that the distribution function of $e^*_{(0.5)}(n)$ for group–B can be satisfactorily approximated by the Gaussian distribution. As a short hand notation, hereafter, we denote the distribution function of $e^*_{(0.5)}(n)$ by $p(e^*_{(0.5)})$. When $e^*_{(0.5)}(n)$ is less than the threshold e_{max}, $y(n)$ is classified as a candidate for group–B. On the contrary, $y(n)$ with $e^*_{(0.5)}(n)$ larger than e_{max} is considered to be an outlier and classified as the candidate for group–$B+x$.

To divide the dataset into two groups, it still remains a problem to choose e_{max}. As previously mentioned, the real value takes only positive value in this study, and thus $y(n)$ with the smallest $e^*_{(0.5)}(n)$ (e_{min}) can be definitely attributed to group–B. Of course, when the data with the smallest value of $e^*_{(0.5)}(n)$ is attributed to as local and artificial noise, we determine e_{min} among the rest by excluding it. Furthermore, many examinations of data show that $p(e^*_{(0.5)})$ appears to be symmetric with respect to its peak. Hence, we define the threshold e_{max} by $e_{max} = e_{min} + 2(e_{peak} - e_{min})$, where e_{peak} is determined by $e^*_{(0.5)}$ at which $p(e^*_{(0.5)})$ shows a sharp peak. The value of e_{peak} is easily determined by searching the peak of the histogram of $e^*_{(0.5)}(n)$, because the number of $e^*_{(0.5)}(n)$ is in our study large enough to approximate its distribution by the histogram.

An extraction of the long–term trend component can be realized by simply smoothing $e^*_{(0.5)}(n)$ of the candidate for group–B. The slowly changing variation obtained by this smoothing is also considered to be the long–term trend component. In the smoothing procedure, $e^*_{(0.5)}(n)$ for group–$B+x$ is treated as missing value. Of course, the Bayesian approach is adopted for the smoothing and the long–term trend component is objectively and automatically estimated according to **ABIC**. In this case, we adopt the simplest smoothness model. Hereafter, this is denoted by $\widetilde{e}^*_{(0.5)}(n)$. $\widetilde{e}^*_{(0.5)}(n)$ is attributed to the long–term trend component. Hence we correct the initial estimate of the long–term trend component, $t^*_{(0)}(n)$, by

$$
\begin{aligned}
\log t^*_{(1)}(n) &= \log t^*_{(0)}(n) + \widetilde{e}^*_{(0.5)}(n) \\
&= \log t^*_{(0)}(n) + \log \widetilde{w}^*_{(0.5)}(n).
\end{aligned}
\tag{4.14}
$$

Accordingly, the corrected stationary component of the background noise intensity variation is given by

$$
\begin{aligned}
\log w^*_{(1)}(n) &= e^*_{(0.5)}(n) - \widetilde{e}^*_{(0.5)}(n) \\
&= \log w^*_{(0.5)}(n) - \log \widetilde{w}^*_{(0.5)}(n).
\end{aligned}
\tag{4.15}
$$

It should be noted that no estimate of $\log w^*_{(1)}(n)$ is given as for the candidate for group–$B+x$, because the value of $e^*_{(0.5)}(n) = \log w^*_{(0.5)}(n)$ is treated as missing in the smoothing process. Even when $\log w^*_{(0.5)}(n)$ belongs to group–$B+x$, the corrected long–term trend component, $t^*_{(1)}(n)$, can be defined by an automatic interpolation of the missing value; i.e., $\widetilde{e}^*_{(0.5)}(n)$ is defined for all n.

By using the corrected long–term trend component, $t^*_{(1)}(n)$, we can furthermore adjust the phase response function form by returning to the procedure explained in the subsection 4.1.3. In short, we replace $t^*_{(0)}(n)$ in (4.7) with $t^*_{(1)}(n)$ and successively continue the procedures shown in the subsections 4.1.3 and 4.1.4. Consequently, we can get the second estimate for the phase response function, $f^*_{(2)}(\theta)$, and the third estimate for the long-term trend component, $t^*_{(2)}(n)$. We repeat these iterative improvements for $t^*_{(k)}(n)$ and $f^*_{(k)}(\theta)$ until the distribution function of $\log w^*_{(k)}(n)$ for group-B can be satisfactorily assumed to obey the time-independent noise distribution, where k denotes the number of iterations. When the iterative process is stopped, the number of iterations is denoted by k^* We select $t^*_{(k^*)}(n)$ and $f^*_{(k^*)}(\theta)$ for the final estimators of $t(n)$ and $f(\theta)$, respectively, and denote $t^*(n)$ and $f^*(\theta)$. In the analysis of our data, we need only a few iterations and obtain the satisfactory estimation for $t^*(n)$ and $f^*(\theta)$ at one or two iterations.

4.1.5 *Estimation of Real Signal.* Based on $t^*(n)$ and $f^*(\theta)$, the estimation of the real signal begins by dividing the observations into two groups: group-B and group-$B+x$. As in the subsection 4.1.4, this classification is also realized by examining the logarithm of the stationary component in the background noise intensity variation, $\log w^*(n)$, which is defined by

$$\begin{aligned} e^*(n) &= \log w^*(n) \\ &= \log y(n) - \log t^*(n) - \log f^*(\theta(n)). \end{aligned} \tag{4.16}$$

If $e^*(n)$ is larger than e^*_{max}, $y(n)$ is assumed to belong to group-$B+x$. $y(n)$ with $e^*(n)$ less than e^*_{max} is classified into group-B.

We set e^*_{max} by simply examining the histogram of $e^*(n)$ and define its value by $e^*_{max} = e^*_{min} + 2(e^*_{peak} - e^*_{min})$ as in subsection (4.1.4), where e^*_{peak} and e^*_{min} are the peak and minimum value of the distribution of $e^*(n)$, respectively.

According to above criteria, the real signal $x(n)$ is estimated by

$$x(n) = y(n) - t^*(n)f^*(\theta(n)), \tag{4.17}$$

where $e^*(n)$ is larger than e^*_{max}. Here we define the *cleaned data* by

$$y^*(n) = t^*(n) + x(n), \tag{4.18}$$

which is free from the phase effect. Obviously, for $e^*(n)$ less than e^*_{max}, $x(n)$ should be zero. This cleaned data represents the long-term trend of the background noise intensity variation, and sometimes shows an additional real signal if we can observe it. We show the cleaned data in Figure 2(c) for the data set shown in Figure 2(a).

4.2 GENERAL MODEL

In the subsection 4.1, we assume that the background noise can be described as a function of time, n, and phase, $\theta(n)$, and took the simplest form for its representation: $B(n,\theta) = w(n)t(n)f(\theta(n))$. Namely, the form of $\log f(\theta)$ can be assumed to be independent of time after removing the long-term trend in the background noise intensity. This model directly and naturally reflects our knowledge obtained from detailed inspection of several data sets. However, occasionally a gradual change occurs in $\log f(\theta)$, which cannot be described by the model explained in (4.1).

This inability can be resolved by considering more general expression for the background noise as a function of time and phase. Namely, we divide the total data into sub-intervals and assume that the form of $f(\theta)$ is independent of time during each interval ΔP. In addition, we add a prior notion such that $B(n + \Delta P, \theta) \sim B(n, \theta)$ and $B(n, \theta) \sim B(n, \theta + \Delta\theta)$. This approach can be realized by quantizing $B(n, \theta)$ as $B(n, i\Delta\theta)$ or by representing it in terms of a spline function, in the phase domain. In this approach, the major difficulties stem from the computing time necessary for searching for $B(n, \theta)$ with the minimum **ABIC**.

The problem presented above can be interpreted to be the space-time modeling in terms of the time series analysis terminology. Instead of quantizing $B(n, \theta)$, Ogata has presented an alternative method to represent $B(n, \theta)$ by using cubic *B*-spline bases with equally spaced knots in both space- and time-axes (1988). Namely, we consider a two dimensional spline function $B(t, \theta|\mathbf{C})$ parametorized by a vector $\mathbf{C}$ on $[0, N\Delta t] \times S$, where Δt a sampling time and S is the one-dimensional torus identical with $[0, 360]$. The essential idea in his approach can be seen in Ogata and Katsura's paper (1988) for representing a surface by the combination of *B*-spline bases. As long as the dimension of $\mathbf{C}$ is not so large, his idea for the space-time modeling is possibly applicable, except for the case where we have to take into account the elimination of outliers as well as the estimation of $\mathbf{C}$. Namely, we have to develop a method to estimate $\mathbf{C}$ for representing a surface in the presence of outliers.

5. Conclusion

Although the Bayesian approach forms a general scheme for an estimation process of parameters, the solutions must be tailored to each application. In this paper we summarized several useful models based on the Bayesian approach along Akaike's (1980) line to remove the background noise closely associated with the spin of the satellite. Since data taken aboard a spinning spacecraft frequently suffers from this spin noise, these models have the potential for greatly improving the scientific return from the satellite.

Acknowledgments

I am grateful to Prof. C.T. Russell for his hospitality during my visit to University of California, Los Angeles. I thank Dr. R.J. Strangeway for providing the plasma wave data used in this study, which were obtained by the Pioneer Venus orbiter and were processed with support furnished by the National Aeronautics and Space Administration under research grants NAG2-501 and NAG2-485. Useful comments of Dr. R.J. Strangeway and G.K. Crawford are highly appreciated.

References

Akaike, H. (1973). Information theory and an extension of the maximum likelihood principle, *in proc. 2nd International Symposium in Information Theory*, (edited by B.N. Petrov and F. Csáki), Akademiai Kiado, Budapest.

Akaike, H. (1980). Likelihood and the Bayes procedure (with discussion), *Bayesian statistics*, 143–165, University press, Valencia, Spain.

Akaike, H. and Ishiguro, M. (1983). Comparative study of the X-11 and BAYSEA procedures of seasonal adjustment, *Applied Time Series Analysis of Economic Data*, (edited by A. Zellner), 17–53, U.S. Department of Commerce, Bureau of the Census, Washington, D.C.

Anderson, B.D.O. and Moore, J.B., (1979). *Optimal Filtering*, Prentice-Hall, Englewood Cliffs, N.J.

Fukao, Y. and Suda, N. (1989). Core modes of the earth's free oscillations and structure of the inner core, *Geophys. Res. Lett.*, **16**, 401–404.

Gersch, W. and Kitagawa, G. (1983). The prediction of time series with trends and seasonalities, *J. Business & Economic Statist.*, **1**, 253–264.

Gersch, W. and Kitagawa, G. (1988). Smoothness priors in time series, *Bayesian analysis of time series and dynamic models*, (edited by J.C. Spall), 431–476, Marcel Dekker, INC., New York and Basel.

Good, I.J. (1965). *The estimation of probabilities*, Cambridge, Mass.:MIT press.

Good, I.J. and Gaskins, R.A. (1971). Non–parametric roughness penalties for probability densities, *Biometrika*, **58**, 255–277.

Good, I.J. and Gaskins, R.A. (1980). Density estimation and bump–hunting by the penalized likelihood method exemplified by scattering and meteorite data (with discussion), *J. Am. Stat. Assoc.*, **75**, 42–73.

Grubbs, F. E. (1969). Procedures for detecting outlying observations in samples, *Technometrics*, **11**, 1–21.

Higuchi, T. (1991a). Frequency domain characteristics of linear operator to decompose a time series into the multi–components, *Ann. Inst. Statist. Math.*, **43**, 469–492.

Higuchi, T. (1991b). Method to Subtract an Effect of the Geocorona EUV Radiation from the Low Energy Particle (LEP) Data by the Akebono (EXOS–D) Satellite, *J. Geomag. Geoelectr.*, **43**, 957–978.

Higuchi, T., Kita, K. and Ogawa, T. (1988). Bayesian statistical inference to remove periodic noise in the optical observations aboard a spacecraft, *Appl. Opt.*, **27**, 4514–4519.

Ishiguro, M. (1984). Computationally Efficient Implementation of a Bayesian seasonal adjustment procedure, *J. Time Series Analysis*, **5**, 245–253.

Ishiguro, M. and Arahata, E. (1982). A Bayesian spline regression (in Japanese), *Proceedings of Institute of Statistical Mathematics*, **30**, 29–36.

Jackson, D. D. and Matsu'ura, M. (1985). A Bayesian approach to nonlinear inversion, *J. Geophys. Res.*, **90**, 581–591.

Kay, S.M. and Marple, Jr., S.L. (1981). Spectrum analysis –A modern perspective, *Proceedings of the IEEE*, **69**, 1380–1419.

Kita, K., Higuchi, T. and Ogawa, T. (1989). Bayesian statistical inference of airglow profiles from rocket observational data: comparison with conventional methods, *Planet Space Sci.*, **37**, 1327–1331.

Kitagawa, G. (1979). On the use of AIC for the detection of outliers, *Technometrics*, **21**, 193–199.

Kitagawa, G. (1981). A nonstationary time series model and its fitting by a recursive filter, *J. Time Series Analysis*, **2**, 103–116.

Kitagawa, G., (1983). Changing spectrum estimation, *J. Sound and Vibration*, **89**(3), 433–445.

Kitagawa, G. (1986). Decomposition of a nonstationary time series —An introduction of the program DECOMP—, (in Japanese), *Proceedings of Institute of Statistical Mathematics*, **34**, 255–271.

Kitagawa, G. and Akaike, H. (1982). A quasi Bayesian approach to outlier detection, *Ann. Inst. Statist. Math.*, **34**, Part B, 389–398.

Kitagawa, G. and Gersch, W. (1984). A smoothness priors– state space modeling of time series with trend and seasonality, *J. American Statistical Association*, **79**, 378–389.

Kitagawa, G. and Gersch W. (1985). A smoothness priors time– varying AR coefficient modeling of nonstationary covariance time series, *IEEE transactions on Automatic control*, **AC–30**, 48–56.

Kitagawa, G. and Takanami, T. (1985). Extraction of signal by a time series model and screening out micro earthquakes, *Signal Processing*, **8**, 303–314.

Leonard, T. (1978). Density estimation, stochastic process and prior information (with discussion), *J. R. Statist. Soc. B*, **40**, 113–146.

Lindley, D.V. and Smith, A.F.M. (1972). Bayes estimates of the linear model (with discussion), *J. R. Statist. Soc. B*, **34**, 1–41.

Murtagh, D.P., Greer, R.G.H., Mcdade, I.C., Llewellyn, E.J. and Bantle, M. (1984). Representative volume emission profiles from rocket photometer data, *Ann. Geophysicae*, **2**, 467–473.

Ogata. Y. (1988). Comment on Space–time modeling with long–memory dependence: Assessing Ireland's wind power resources (by John Haslett and Adrian Raftery), *Res. Mem. Inst. Statist. Math.*, No. 351, 1–5.

Ogata, Y. (1990). A monte carlo method for an objective Bayesian procedure, *Ann. Inst. Statist. Math.*, **42**, 403–433.

Ogata, Y. and Katsura, K. (1988). Likelihood analysis of spatial inhomogeneity for marked point patters, *Ann. Inst. Statist. Math.*, **40**, 29–39.

Russell, C.T. (1991). Venus lightening, *Venus Aeronomy*, 317–356, Kluwer Academic Publishers, Dordrecht, Netherlands.

Sakamoto, Y. (1991). *Categorical data analysis by AIC*, Kluwer Academic Publishers, Dordrecht, Netherlands.

Sakamoto, Y., Ishiguro, M. and Kitagawa, G. (1986). *Akaike information criterion statistics*, D. Reidel Publishing Company, Holland.

Silverman, B.W. (1985). Some aspects of the spline smoothing approach to non-parametric regression curve fitting (with discussion), *J. R. Statist. Soc. B*, **47**, 1–52.

Strangeway, R.J. (1991). Plasma Waves at Venus, *Space Sci. Rev.*, **55**, 275–316.

Tietjen, G. L. and Moore, R. H. (1972). Some Grubbs–type statistics for the detection of several outliers, *Technometrics*, **14**, 583–597.

Tikhonov, A. N. (1965). Incorrect problems of linear algebra and a stable method for their solution, *Sov. Math. Dokl.*, **6**, 988–991.

Titterington, D.M. (1985). Common structure of smoothing techniques in statistics, *International Statistical Review*, **53**, 141–170.

Wahba, G. (1990). *Spline models for observational data*, Society for industrial and applied mathematics, Philadelphia.

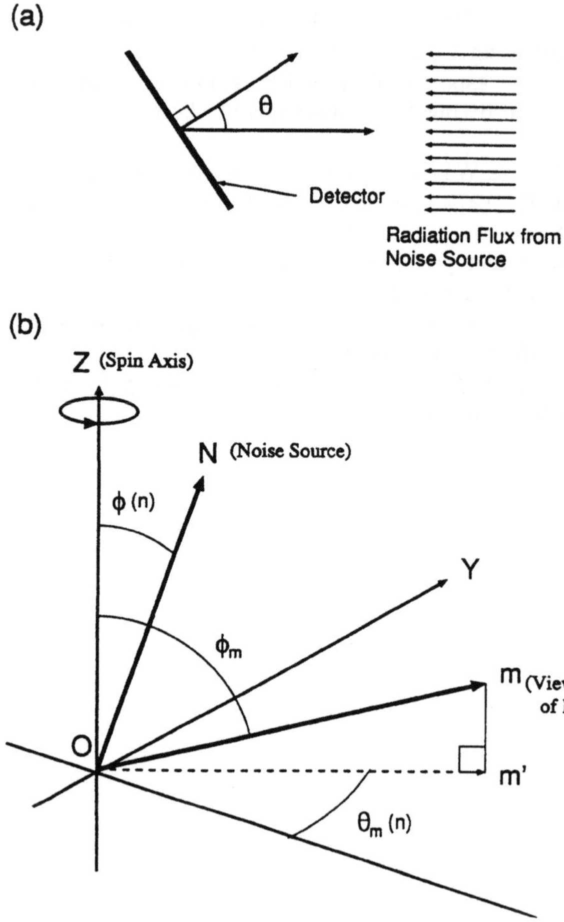

Figure 1(a) Schematic illustration of the relationship between the detector and the point noise source. In this figure, a shape of the effective area for the detector to receive the radiation is simplified to be a plane. The radiation flux due to the point noise source is approximated to be parallel to the view direction from the satellite toward the noise source. θ is the angle between the satellite–noise direction and a vector normal to the plane.

(b) Schematic configuration of a spinning satellite and noise source. In this view we take the spin axis as the Z–axis. The direction from the satellite to the point noise source, ON, lies in the XZ plane. A vector Om corresponds to the view direction of the m–th detector and Om' is its projection onto the XY plane. $\theta_m(n)$ is the angle between the X–axis and Om' as a function of time. $\phi(n)$ is defined by the angle between the Z–axis and ON. ϕ_m is the angle between the spin axis and the m–th detector.

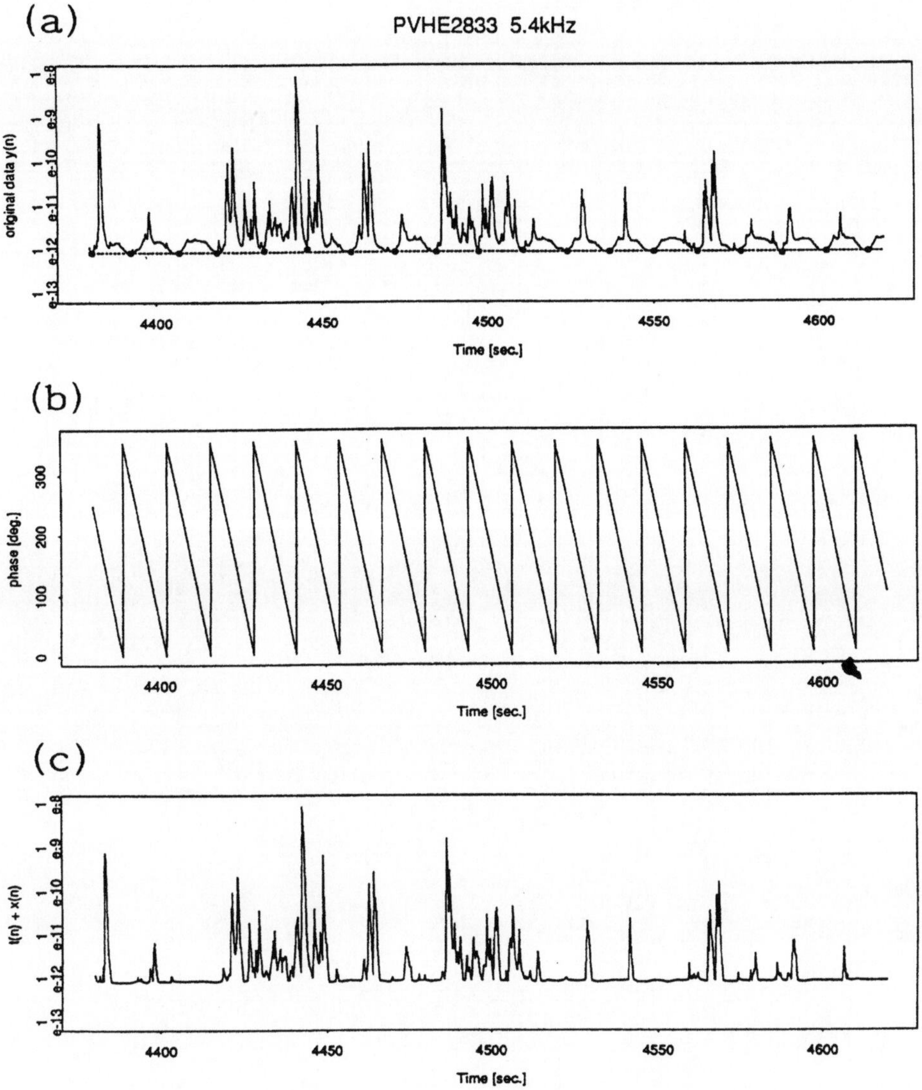

Figure. 2 (a) Raw observations of electric field spectral intensity measurements ($(V/m)^2/Hz$) for 5.4$[Hz]$. The initial estimate for the long–term trend component of the background noise intensity, $t^*_{(0)}$, is superposed by the broken line.

(b) Phase variation (degree) as a function of time.

(c) The cleaned data, $t^*(n) + x(n)$.

THE LOCAL LINEARIZATION FILTER WITH APPLICATION TO NONLINEAR SYSTEM IDENTIFICATIONS

Tohru OZAKI
Institute of Statistical Mathematics
4-6-7 Minami Azabu, Minato-ku,
Tokyo, Japan-106

Some key words: Nonlinear filter, system identification, maximum likelihood method, nonlinear dynamical systems, local linearization, Markov diffusion process, Markov chain, nonlinear random vibrations, van der Pol oscillations, innovations, nonlinear state-space representation.

1. Introduction.

Many dynamic phenomena in scientific fields are modeled by a continuous time stochastic dynamical system,

$$\dot{\underline{z}} = \underline{f}(\underline{z} \mid \underline{\theta}) + \underline{w}(t),$$

where $\underline{w}(t)$ is a Gaussian white noise. The process $\underline{z}(t)$ defined by the model is a Markov diffusion process. When we use such mathematical models in physics or engineering sciences we often have a rough idea, from some physical understanding of the process, of the size of the parameters $\underline{\theta}$ and the variance-covariance matrix Σ of the Gaussian white noise $\underline{w}(t)$ in the model. However when prediction or control of the process is concerned we sometime need more accurate estimates of the parameters. In some cases, we don't have much physical information helping us to guess the size of the parameters in the model. In such situations we often try to estimate parameters from observation data of the phenomenon as accurately as possible.

We consider, in this paper, the estimation problem for continuous time stochastic dynamical system models with discrete time observation data using a sufficiently small sampling time interval. The dimension r of the observation data $\underline{x}$ is sometimes less than the dimension k of the state $\underline{z}$. Also we sometimes cannot ignore the observation errors. Thus the estimation problem concerned is that of the following continuous time state-space model with discrete time observation model,

$$\begin{aligned} \dot{\underline{z}} &= \underline{f}(\underline{z} \mid \underline{\theta}) + \underline{w}(t) \\ \underline{x}_t &= C\,\underline{z}(t) + \underline{v}_t \end{aligned}, \qquad (1.1)$$

where C is a rxk observation matrix, e.g.

H. Bozdogan (ed.), Proceedings of the First US/Japan Conference on the Frontiers of Statistical Modeling: An Informational Approach, 217–240.

$$C = \begin{pmatrix} 0...0 & 1.....0 \\ ..... & ..... \\ 0...0 & 0.....1 \end{pmatrix}.$$

If the k-dimensional stochastic dynamical system is linear, i.e.

$$\dot{\underline{z}} = L(\underline{\theta})\ \underline{z} + \underline{w}(t), \qquad (1.2)$$

with some z-free matrix $L(\underline{\theta})$, we know that the following discrete time stochastic dynamical system model,

$$\underline{z}_{t+\Delta t} = A\,\underline{z}_t + B\,\underline{w}_{t+\Delta t}$$

is a natural and good approximation to (1.2), where

$$A = e^{L(\underline{\theta})\,\Delta t}$$

$$= \sum_{k=0}^{\infty} \frac{(\Delta t)^k}{k!} L(\underline{\theta})^k$$

and B is a matrix whose elements are functions of eigenvalues of the matrix A(see Ozaki(1986)). The process $\underline{w}_{t+\Delta t}$ is a discrete time Gaussian white noise of variance $\Delta t\Sigma$. Thus in linear cases we have the following discrete time state-space model,

$$\underline{z}_{t+\Delta t} = A\,\underline{z}_t + B\,\underline{w}_{t+\Delta t}$$

$$\underline{x}_t = C\,\underline{z}_t + \underline{v}_t \qquad (1.3)$$

When $\underline{v}_t$ is zero, it is known that $\underline{x}_t$ has ARMA(k,k-1) representation(Akaike(1974)),

$$\underline{x}_t - \phi_1\,\underline{x}_{t-1} - \phi_2\,\underline{x}_{t-2} - \ldots - \phi_K\,\underline{x}_{t-k} = CB\underline{w}_t + C(A - \phi_1\, I)B\underline{w}_{t-1} \qquad (1.4)$$

$$+ C(A^2 - \phi_1 A - \phi_2 I)B\underline{w}_{t-2} + \ldots + C(A^{k-1} - \phi_1 A^{k-2} - \ldots - \phi_{k-1} I)B\underline{w}_{t-k+1}$$

where ϕ_i's are the characteristic coefficients of the matrix A. Therefore we can estimate the parameters of the model for $\underline{z}(t)$ by the maximum likelihood method for the linear ARMA time series models without an argument of state space representation models. We note that in (1.4) the dimension of the variable $\underline{x}_t(t=1,2,...N)$ can be less than that of $\underline{z}(t)$. This is one of the reason why ARMA time series models with high orders are so powerful in many application fields where only scalar time series observation is available and the dimension of its state dynamics could be much higher. When the observation noise $\underline{v}_t$ is present we have ARMA(k,k) model instead of ARMA(k,k-1) model because $\underline{x}_i$ (i=t, t-1, ...,t-k) in (1.4) is replaced by $\underline{x}_i-\underline{v}_i$. Thus we can still estimate the parameters by the maximum likelihood method for time series models although the parametric expression of the time series model becomes more complicated.

It may be natural to expect to have a nonlinear ARMA representation for the nonlinear stochastic dynamical system

$$\dot{\underline{z}} = \underline{f}(\underline{z} \mid \underline{\theta}) + \underline{w}(t),$$

so that we can estimate the model parameters through the nonlinear ARMA time series model. However this is not possible(see Ozaki(1986)), because the introduction of the ARMA representation is based on the assumption of the linearity of the system which guarantees the Cayley-Hamilton theorem, which says

$$A^k - \phi_1 A^{k-1} - \phi_2 A^{k-2} - \ldots - \phi_k I = 0$$

for the characteristic coefficients, $\phi_1, \phi_2, ..., \phi_k$ of the matrix A. Here I is the unit matrix. If A is not constant and depends on z(t), the Caylay-Hamilton theorem does not hold.

On the other hand for the linear discrete time state space model (1.3), a parameter estimation method directly applied to the model without going through its ARMA time series model representation is also available . The direct estimation method is based on the Schweppe(1965)'s maximum likelihood method for discrete time linear state space models, where the Gaussianity of the innovations obtained by the Kalman filter (Kalman(1960), Kailath(1968)) is assumed. It will be natural to consider the extention of this direct approach from linear to nonlinear cases. To do that we need a nonlinear filtering method for the discrete time observation data which come from a continuous time dynamical system model.

The nonlinear filtering problem, although theoretically solved by Kushner(1962), has been a well-known difficult problem. This is because the exact solution of the nonlinear filtering problem leads to the infinite dimensional systems which are impossible to implement in practice. It has been said(Marcus, 1984), however, that "non-existence of finite dimensional filters does not address the issue of non exact but high-performance suboptimal filters, which is much more important from a practical point of view." Thus the problem of obtaining a good approximation to the exact solutions of nonlinear filtering problems has been studied by various approaches(see Hazewinkel and Willems(1981)). However as is mentioned in Frost and Kailath(1971) or Liang(1983), many of these results of nonlinear filtering provide a "representation" or "philosophy of approach" rather than a "solution" or a "procedure" leading to the derivation of practical estimators.

Liang(1983) and Schwartz and Stear(1968) discussed the difficulties of conventional nonlinear filters for systems with nonnegligible nonlinearities. A common difficulty seen in these empirical studies is the computational instability of nonlinear filtering. Even though the nonlinear filtering is properly defined and is supposed to have a finite valued solution, we often see a numerical explosion in mechanizing the prediction or filtering of the state estimate in discrete time steps. To overcome the computational instability problem we try to get an approximate nonlinear filter, not by approximate continuous time filter equations, but by discrete time filtering equations for an approximate discrete time model for the original continuous time dynamical system. The efficiency of this approach depends totally on how good the locally linearized discrete time model approximation is.

Before introducing our discretization scheme, we first explain, in section 2.1, why the conventional discretization schemes are unsatisfactory. In section 2.2 we show how the local linearization method preserves important characteristics such as stationarity of the original continuous time dynamical system model in the discrete time model. In section 2.3 the filtering scheme for the discretized model is presented. In section 2.4 the effectiveness of the filtering scheme was checked by applying the filter to the data generated from the van der Pol equation. In section 3.1 we give a maximum likelihood method for the discretized model using the local linearization filter. In section 3.2 we will show several applications of the present method for the identification of nonlinear oscillation models with numerical results. Asymptotic properties of the maximum likelihood estimates and the least squares estimates are briefly discussed with numerical results in section 3.3. Implication of the present method and related topics are discussed and conclusions are given in section 4.

2.The Local Linearization Filter.

For the purpose mentioned in the previous section, we try to introduce a nonlinear filter based on a "good" discrete time approximation to the continuous time stochastic dynamical system model for a nonlinear and non-Gaussian multivariate process $\underline{z}(t)$. Seeing first why conventional discretization schemes are not satisfactory helps for us to understand our new discretization scheme, .

2.1. WHY CONVENTIONAL DISCRETIZATION METHODS ARE NOT SATISFACTORY.

For simplicity of explanation we first confine ourselves to the time discretization of the scalar stochastic dynamical system model,

$$\dot{z} = f(z \mid \underline{\theta}) + w(t) \,, \tag{2.1}$$

where w(t) is a Gaussian white noise and $f(z|\underline{\theta})$ is a nonlinear function of z(t). The simplest and most common discrete time model for the scalar stochastic dynamical system (2.1) is

$$z_{t+\Delta t} = z_t + \Delta t\, f(z_t) + \sqrt{\Delta t}\, w_{t+\Delta t} \tag{2.2}$$

The discretization method is a stochastic analog of Euler's method for the discretization of deterministic dynamical system,

$$\dot{z} = f(z \mid \underline{\theta}) \,.$$

The time discretization method (2.2) is consistent, i.e.

$$\frac{z_{t+\Delta t} - z_t}{\Delta t} \rightarrow f(z_t \mid \underline{\theta}) + w(t) \quad \text{for } \Delta t \rightarrow 0 \,.$$

Actually the Ito stochastic differential equation (and the Stratonovich stochastic differential equation as well, since the variance of w(t) is independent of z(t)) is defined as a limit of this discrete time model(see Mortensen(1979)). Therefore the distribution of z_t defined by the approximate discrete model is known to converge to the distribution of z(t) in probability. The only disadvantage of this model is that the discrete time process, i.e. the Markov chain in continuous state space defined by (2.2) is not stationary for most nonlinear function $f(z|\underline{\theta})$ if the time interval Δt is fixed, no matter how small it is (Jones(1978)). We may have a non-explosive sample path if we simulate the model with a small Δt for a certain finite time interval. However the probability of computational explosion in a finite time is always 1, i.e. if we carry on the simulation it always ends with computational explosion after a finite number of steps. For example, for the process defined by (2.1) with

$$f(z \mid \underline{\theta}) = - z^3 \,,$$

the model (2.2) almost certainly explodes if z_t starts from an initial value greater than $\sqrt{2/(\Delta t)}$, even though the original continuous time process is stationary with the stationary distribution,

$$p(x) = C \exp(- \frac{x^4}{2\sigma_w^2})$$

and never explodes no matter how large its initial value. Here C is a normalizing constant and σ_w^2 is the variance of w(t).

A computationally stable time discretization scheme of deterministic dynamical system,

$$\dot{z} = f(z \mid \underline{\theta}) \tag{2.3}$$

has been used in the engineering field(p264, Sorenson(1966)). The discrete model derived from the scheme applied to the deterministic dynamical system (2.3) is

$$z_{t+\Delta t} = e^{J_t \Delta t} z_t \tag{2.4}$$

where J_t is the Jacobian of $f(z|\underline{\theta})$ at time point t. The model (2.4) is known to be computationally stable whenever the original nonlinear deterministic dynamical system is stable. When $f(z|\underline{\theta})$ is linear the trajectory of (2.4) coincides with the exact trajectory of the original system (2.3) on the discrete time points t=Δt, $2\Delta t$, $3\Delta t$, . . . If we replace the deterministic part of the model (2.2) by (2.4) we have the following discrete time stochastic dynamical system model,

$$z_{t+\Delta t} = e^{J_t \Delta t} z_t + \sqrt{\Delta t}\, w_{t+\Delta t} \tag{2.5}$$

The discrete time stochastic process z_t (2.5), which is a Markov chain with a continuous state space, is known to be computationally stable and ergodic. Unfortunately the model is not consistent unless $f(z|\underline{\theta})$ is linear. If we let $\Delta t \to 0$, the process defined by the model (2.5) converges to

$$\dot{z}(s) = J_t\, z(s) + w(s) \tag{2.6}$$

Incidentally we note that if we assume model (2.6) on the interval [t,t+Δt), the autocovariance function $\gamma_t(s)$ of the process z(s) of (2.6) is

$$\gamma_t(s) = \left(\frac{\sigma_w^2}{2J_t}\right) e^{J_t s}$$

Since

$$\frac{\gamma_t(\Delta t)}{\gamma_t(0)} = e^{J_t \Delta t},$$

it is reasonable to have $e^{J_t \Delta t}$ as a coefficient in the discretized model (2.5). This idea is used later when we introduce our local linearization scheme, which is consistent and computationally stable.

Some more sophisticated time discretization methods for the deterministic dynamical system (2.3) are known in numerical analysis, such as the Heun and Runge-Kutta methods. It may be natural to substitute the deterministic part of (2.2) by the deterministic discrete time models obtained by these more sophisticated methods. We call these discrete time stochastic dynamical system models the Heun and Runge-Kutta scheme models respectively. Actually these models are used for simulation of stochastic differential equations(Arnold(1974), McShane(1974)). Since the difference between these models and model (2.1) is of order $(\Delta t)^r$ with $r > 1$, these models preserve the consistency property of the Euler scheme model (2.2).

Unfortunately, however, the disadvantage of model (2.2), i.e. non-stationarity, is not remedied by these more sophisticated discrete time models. For example the Runge-Kutta scheme, applied to

$$\dot{z} = -z^3 + w(t),$$

gives us a stochastic discrete time dynamic model,

$$z_{t+\Delta t} = p_{81}(z_t) + \sqrt{\Delta t}\, w_{t+\Delta t} \tag{2.7}$$

where $p_{81}(z_t)$ is a 81-st order polynomial of z_t. Therefore the Markov chain defined by (2.7) is, like the one by (2.2) introduced by the Euler scheme, computationally explosive and is not stationary. We can often avoide the computational explosion within a given finite time steps, in the simulations, by taking Δt very small. However, in the inferential use of the discretized model, we *cannot* choose the scale of the white noise variance and the time interval Δt, unlike the simulation studies, since they are already fixed when the data were sampled.

In numerical analysis it is known that the error of the Euler method applied to the deterministic dynamical system (2.3) is of order $o(\Delta t)$, of the Heun method order $o(\Delta t)^2$, and of the Runge-Kutta method order $o(\Delta t)^4$. In the stochastic situation the goodness of the approximations may be evaluated by the mean square one step error,

$$E_{z_t}[\, z(t+\Delta t) - z_{t+\Delta t}\,]^2,$$

where E_{z_t} denotes the conditional expectation with respect to z_t. The Euler scheme model (2.2) is known to have the expected mean square error of order $o(\Delta t)^2$(Milshtein(1974); note that our model has, unlike those of Milshtein(1974) and Rumelin(1982), a special form, where the noise variance is constant). A higher order expected mean square error are expected from the Heun scheme and Runge-Kutta scheme models. However it is shown that the order of the expected mean square error of these models is also $o(\Delta t)^2$. Moreover it has been proved that the maximum possible speed of convergence attained by a discrete time approximate model is $o(\Delta t)^2$(Milshtein(1974), Rumelin(1982)). This result is, in a sense, natural because even though we approximate the deterministic part of the model (2.1) more accurately by the Runge-Kutta method etc., the approximation of the white noise part in these models stays as poor as the Euler scheme model, dominating the performance on the whole, and is impossible to improve in general. Thus the other discrete time models with more sophisticated time discretization schemes are no better than the Euler method (2.2).

From the above discussion we now know that we should not expect to have a new time discretization scheme with a higher order convergence speed of the approximation. What we need is a new discretization method which, like the Euler scheme method, is consistent and at the same time gives a computationally stable and stationary Markov chain like (2.5) for a fixed Δt whenever the original continuous time Markov diffusion process is stationary.

2.2. THE LOCAL LINEARIZATION SCHEME.

A key for the introduction of a new time discretization scheme lies in the scheme (2.5) which is the only one giving a computationally stable scheme among the conventional methods. This scheme is also the only one whose deterministic part gives a trajectory which exactly coincides with the true trajectory of the continuous deterministic system (2.3) at the discrete time points when $f(z|\underline{\theta})$ is linear. The scheme (2.5) can be introduced by approximating the original process by the linear process (2.6) on each short interval $[t,t+\Delta t)$, where we assume that the coefficient of the linear function of the dynamical system on the interval is given by the Jacobian J_t of $f(z|\underline{\theta})$ as $J_t z$ at time point t. A disadvantage of this scheme is that it is not consistent. This is because J_t is constant on the interval $[t,t+\Delta t)$ and the function $J_t z$ does not converge to $f(z(t)|\underline{\theta})$ for $\Delta t \to 0$. This consideration suggests that we may be able to get a computationally stable and consistent scheme if we use some different function L_t to approximate the original continuous model on each short time interval $[t,t+\Delta t)$ with, instead of (2.6), a linear stochastic dynamical system,

$$\dot{z}(s) = L_t\, z(s) + w(s), \qquad (2.8)$$

where L_t is some function of z_t and

$$L_t z_t \to f(z(t)) \quad \text{for} \quad \Delta t \to 0.$$

A scheme which satisfies this requirement is obtained by a simple assumption, i.e. "The Jacobian of the linear dynamical system,

$$\dot{z}(s) = L_t\, z(s) \qquad (2.9)$$

is equivalent to the Jacobian,

$$J_t = \left(\frac{\partial f}{\partial z}\right)_{z=z_t}$$

of the original dynamical system on each short time interval [t,t+Δt)". We call it the Local Linearization(L.L.) assumption. From this assumption we have

$$\ddot{z}(s) = J_t \dot{z}(s)\ ,$$

on [t,t+Δt). If we integrate this on the interval [t, t+t), where $0 \leq t < \Delta t$, we have

$$\begin{aligned}\dot{z}(t+\tau) &= e^{J_t \tau}\dot{z}(t)\\ &= e^{J_t \tau} f(z(t))\ .\end{aligned}$$

By integrating this again with respect to τ on the interval [0, Δt) we have,

$$z(t+\Delta t) = z(t) + J_t^{-1}(e^{J_t \Delta t} - 1) f(z(t)) \tag{2.10}$$

Since J_t is given as a function of z(t), z(t+Δt) is explicitly given as a function of z(t) at time point t. Thus the trajectory of the system (2.9) takes the value (2.10) at the discrete time point, t+Δt. On the other hand the solution z(t+Δt) of the linear system (2.9) is given explicitly as a function of L_t and z(t) by integrating $\dot{z}(s)$ of (2.9) from t to t+Δt to give,

$$z(t+\Delta t) = e^{L_t \Delta t} z(t) \tag{2.11}$$

From (2.10) and (2.11) we can write down L_t explicitly as a function of z(t),

$$L_t = \frac{1}{\Delta t}\log\{1 + J_t^{-1}(e^{J_t \Delta t} - 1) F_t\} \tag{2.12}$$

where

$$F_t = \frac{f(z(t))}{z(t)}\ .$$

L_t of (2.12) is defined on the region of z(t) where $J_t z_t \neq 0$. When $J_t z_t \neq 0$, it holds, for sufficiently small Δt, that

$$1 + J_t^{-1}(e^{J_t \Delta t} - 1) F_t > 0\ .$$

When z(t) is on the region where $J_t z_t = 0$ (which is of measure zero), the z(t+Δt) is defined separately in a manner consistent with the rest of the region(see Ozaki(1992)).

Next we try to use L_t of (2.12) to introduce a stochastic version of the discrete time approximation to the continuous time system. One simple way is, as in those conventional discretization schemes for stochastic dynamical systems, to add $\sqrt{\Delta t}\, w_{t+\Delta t}$ to the deterministic part, thus

$$z_{t+\Delta t} = e^{L_t \Delta t} z_t + \sqrt{\Delta t}\, w_{t+\Delta t}\ . \tag{2.13}$$

However if we stick to the the L.L. assumption (2.9) and extend it to stochastic dynamical systems, we have a more consistent model than (2.13). In the stochastic situation it will be natural to start from the assumption, "The stochastic dynamical system is locally linear and so it is Gaussian on each short time interval [t,t+Δt), i.e. we employ the model (2.8) on the interval to approximate the original process, where L_t is given by (2.12)". Then $\dot{z}(s)$ of (2.8) can be integrated from t to t+Δt on the interval giving

$$z_{t+\Delta t} = e^{L_t \Delta t} z_t + a_{t+\Delta t}$$

where

$$a_{t+\Delta t} = \int_t^{t+\Delta t} e^{L_t (t+\Delta t - s)} w(s) ds$$

The variance of $a_{t+\Delta t}$ is

$$\sigma_a^2 = \int_t^{t+\Delta t} e^{2L_t (t+\Delta t - s)} \sigma_w^2 ds$$

$$= \frac{(e^{2L_t \Delta t} - 1)}{2L_t} \sigma_w^2 .$$

where σ_w^2 is the variance of the continuous time white noise $w(t)$. Therefore it will be natural and more consistent if we use, instead of the model (2.13), the following discrete time model,

$$z_{t+\Delta t} = e^{L_t \Delta t} z_t + \sqrt{\frac{(e^{2L_t \Delta t} - 1)}{2L_t}} w_{t+\Delta t} \tag{2.14}$$

where $w_{t+\Delta t}$ is a discrete time Gaussian white noise of variance σ_w^2. Since L_t of (2.12) satisfies

$$L_t z(t) \rightarrow f(z(t)) \text{ for } \Delta t \rightarrow 0 ,$$

the model (2.14) is consistent. The deterministic part of the model (2.14) inherits the zero points and Jacobian of the original continuous time deterministic dynamical system (2.3). Therefore the stability property of the original continuous time stochastic dynamical system is also preserved in scheme (2.14). The ergodicity of the discrete time model is proved by direct application of Tweedie(1975)'s theorem when a proper $f(z \mid \underline{\theta})$ of the original continuous time system is given(Ozaki(1985)). Thus the Markov chain process defined by (2.14) brings us a stationary Markov chain if the original continuous time diffusion process is stationary.

The local linearization can be applied to the multi-dimensional stochastic dynamical system,

$$\dot{\underline{z}} = \underline{f}(\underline{z} \mid \underline{\theta}) + \underline{w}(t) \tag{2.15}$$

without any essential change. Since the integration of the linear multi-dimensional differential equation,

$$\dot{\underline{z}} = L \underline{z}(t)$$

is

$$\underline{z}(t) = \mathrm{Exp}(L_t) \underline{z}(0) ,$$

the discrete time model for the multi-dimensional stochastic dynamical system (2.15) is given by

$$\underline{z}_{t+\Delta t} = A(\underline{z}_t) \underline{z}_t + B(\underline{z}_t) \underline{w}_{t+\Delta t} \tag{2.16}$$

$$A(z_t) = \mathrm{Exp}(L(z_t) \Delta t)$$

$$L(z_t) = \frac{1}{\Delta t} \mathrm{Log}\{1 + J_t^{-1} (e^{J_t \Delta t} - 1) F_t\}$$

$$J_t = \left\{\frac{\partial \underline{f}(\underline{z})}{\partial \underline{z}}\right\}_{\underline{z} = \underline{z}_t}$$

F_t is such that $F_t \underline{z}_t = \underline{f}(\underline{z}_t)$. The matrix functions Exp(.) and Log(.) are defined by

$$\mathrm{Exp}(L) = \sum_{k=0}^{\infty} \frac{1}{k!} L^k$$

and

$$\mathrm{Log}(L) = \sum_{k=0}^{\infty} \frac{(-1)^k}{k!} (L - I)^k$$

respectively. The elements of the matrix B_t are explicitly given as a function of eigenvalues of the matrix L_t(Ozaki,1986,1989).

2.3. THE LOCAL LINEARIZATION FILTER.

It is well known that nonlinear filtering is a difficult problem even in the scalar case. Its main difficulty is the computational instability which comes from the nonlinearity of the state dynamics. Our idea of obtaining a computationally stable filtering scheme for,

$$\dot{\underline{z}} = \underline{f}(\underline{z} \mid \underline{\theta}) + \underline{w}(t) \qquad (2.17)$$
$$\underline{x}_t = C\,\underline{z}(t) + \underline{v}_t$$
$$C = \begin{pmatrix} 0 \ldots 0 & 1 \ldots\ldots 0 \\ \ldots\ldots & \ldots\ldots \\ 0 \ldots 0 & 0 \ldots\ldots 1 \end{pmatrix}$$

is to introduce a stable approximate discrete time L.L. model from (2.17), instead of introducing an approximate continuous time filtering scheme, and to give a discrete time filtering scheme for the L.L. model. In (2.17), $\underline{w}(t)$ is a continuous time k-dimensional Gaussian white noise with variance-covariance matrix $\Sigma_{\underline{w}}$ and $\underline{v}_t$ is a discrete time r-dimensional Gaussian white noise with variance-covariance matrix $\Sigma_{\underline{v}}$ and $\underline{w}(t)$ and $\underline{v}_t$ are independent. Thus from (2.17), using the L.L. method in the previous section, we have,

$$\underline{z}_{t+\Delta t} = A(\underline{z}_t)\,\underline{z}_t + B(\underline{z}_t)\,\underline{w}_{t+\Delta t} \qquad (2.18)$$
$$\underline{x}_t = C\,\underline{z}_t + \underline{v}_t$$

where $A(z_t)$ and $B(z_t)$ are given in (2.16). $\underline{w}_{t+\Delta t}$ is a k-dimensional discrete time Gaussian white noise with variance-covariance matrix $\Sigma_{\underline{w}}$. Also between the i-th component $w^{(i)}$ of $\underline{w}_{t+\Delta t}$ and j-th component $v^{(j)}$ of $\underline{v}_t$ it holds that $E[w^{(i)}v^{(j)}]=0$ for 1£ i ≤k and 1≤j≤r. Based on the minimum variance principle(Sorenson(1966)), which is to minimize

$$E[(\underline{z}(t) - \underline{z}_{t|t})'(\underline{z}(t) - \underline{z}_{t|t})] \,,$$

the filtering equation for (2.18) is obtained as follows,

$$\underline{z}_{t+\Delta t \mid t} = A(\underline{z}_t)\,\underline{z}_{t \mid t} \qquad (2.19)$$

$$\underline{z}_{t \mid t} = \underline{z}_{t \mid t-\Delta t} + K_t\,\underline{\nu}_t$$

$$\underline{\nu}_t = \underline{x}_t - C\,\underline{z}_{t \mid t-\Delta t}$$

$$K_t = P_t C' (C P_t C' + \Sigma_{\underline{v}})^{-1}$$
$$P_t = E[(\underline{z}(t) - \underline{z}_{t|t-\Delta t})(\underline{z}(t) - \underline{z}_{t|t-\Delta t})']$$
$$V_t = P_t - K_t C P_t$$

$$= P_t - P_t C' (C P_t C' + \Sigma_{\underline{v}})^{-1} C P_t$$

where v_t is the innovation of the filtering model, K_t is the filter gain and P_t is the variance-covariance matrix of the one step ahead prediction error of the state z_t. The evolution of P_t is obtained from

$$P_{t+\Delta t} = E[(z_{t+\Delta t} - z_{t+\Delta t|t})(z_{t+\Delta t} - z_{t+\Delta t|t})']$$
$$= E[\{A(\underline{z}_t)(\underline{z}_t - \underline{z}_{t|t}) + B(\underline{z}_t)\underline{w}_{t+\Delta t}\}\{A(\underline{z}_t)(\underline{z}_t - \underline{z}_{t|t}) + B(\underline{z}_t)\underline{w}_{t+\Delta t}\}'] .$$

Since we are assuming $A(\underline{z}_t)$ and $B(\underline{z}_t)$ are constant on $[t,t+\Delta t)$, we have

$$P_{t+\Delta t} = A(\underline{z}_t)V_t A(\underline{z}_t)' + B(\underline{z}_t)\Sigma_{\underline{w}} B(\underline{z}_t)' \tag{2.20}$$

For small Δt, the function B(.) is almost constant, whereas A(.) is not. Since we assume that the system (2.18) is linear on $[t,t+\Delta t)$ and the system transition is characterized by $A(z_t)$ as in (2.19), it will be reasonable to replace $A(z_t)$ and $B(z_t)$ in (2.19) and (4.4) by $A(z_{t|t})$ and $B(z_{t|t})$ respectively. Then we have

$$\underline{z}_{t+\Delta t|t} = A(\underline{z}_{t|t})\underline{z}_{t|t} , \tag{2.21}$$

and

$$P_{t+\Delta t} = A(\underline{z}_{t|t})V_t A(\underline{z}_{t|t})' + B(\underline{z}_{t|t})\Sigma_{\underline{w}} B(\underline{z}_{t|t})' \tag{2.22}$$

The above filtering equations were first introduced in Ozaki(1990a), where its effectiveness was confirmed by comparing the results by the conventional Extended Kalman filtering method using simulation data..

2.4. APPLICATION TO THE VAN DER POL OSCILLATORS.

We applied the local linearization filter to the data(see Fig.1) generated from

$$\ddot{x}(t) - 3\{1 - x(t)^2\}\dot{x}(t) + x(t) = 0 , \tag{2.23}$$

which is called the van der Pol equation. Liang(1983) says about the example, "This example was selected not only because of its third order nonlinearities but also because Kushner(1967) had previously shown that for this state-estimation problem, even with a linear measurement model the linearized filter was extremely unstable and was completely useless within a fraction of a time unit, and even the implementation of a Gaussian second order filter proved to be unstable." In Liang(1983)'s simulation studies the observation noise σ_v^2 is chosen to be $\sigma_v^2=0.1$.

From (2.23) we have the local linearization filter (2.19) whose A(.) and B(.) matrices are given by the following $\underline{f}(\underline{z}_t)$, J_t and F_t ,

$$\underline{f}(\underline{z}_t) = [3\{1 - x_t^2\}\dot{x}_t - x_t, \ \dot{x}_t]'$$

$$J_t = \begin{pmatrix} 3(1 - x_t^2) & -6x_t\dot{x}_t - 1 \\ 1 & 0 \end{pmatrix}$$

$$F_t = \begin{pmatrix} 3(1 - x_t^2) & -1 \\ 1 & 0 \end{pmatrix} .$$

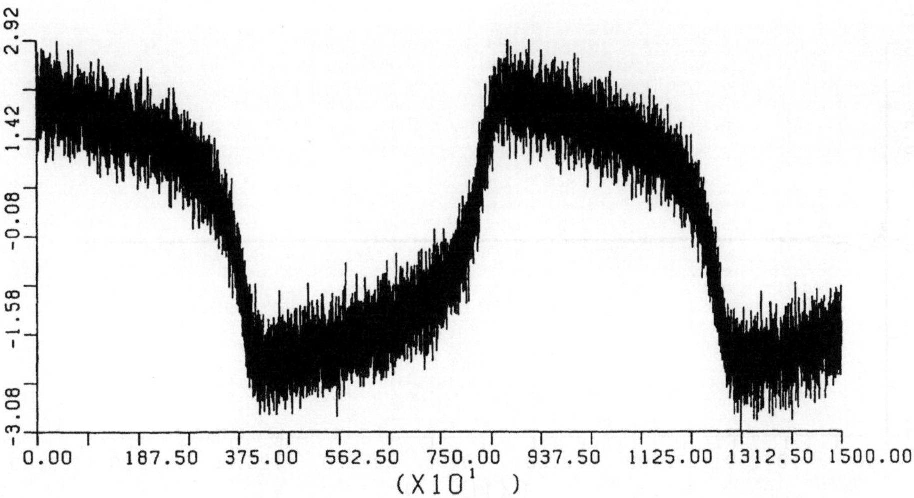

Figure 1:van der Pol osillation data generated from (2.23) with σ_v^2=0.1.

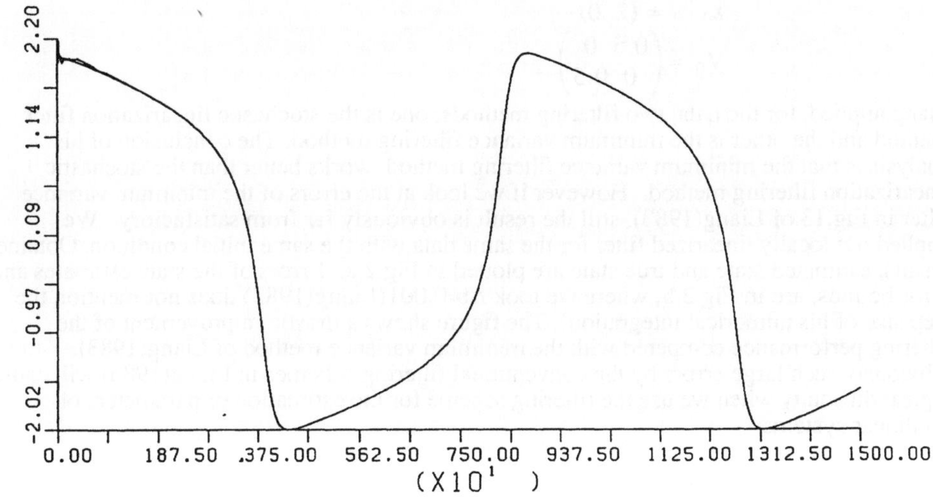

Figure 2.a : Estimated states and true states.

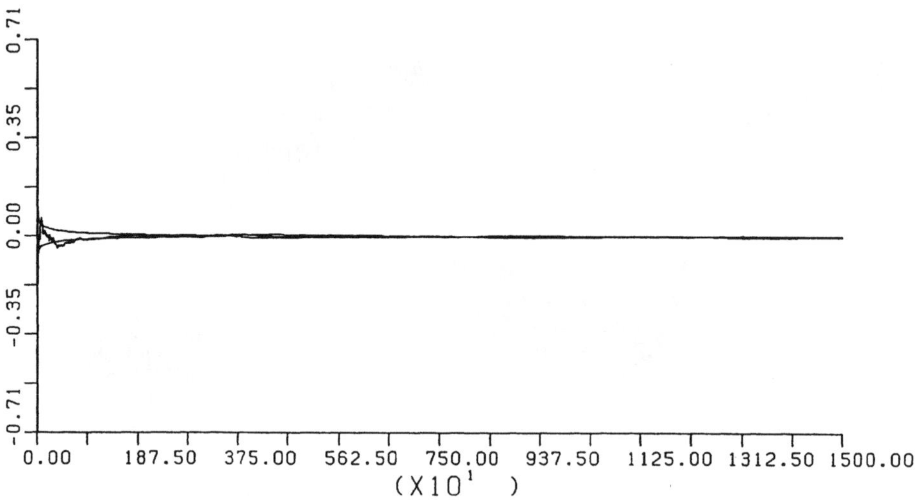

Figure 2.b : Errors of the estimates and error bounds.

Fig.1 shows the simulation data generated from the model (2:23). In the paper the initial estimate of the state and the variance-covariance matrix of the initial estimate were taken as follows,

$$z_{0\,|\,0} = (2, 0)'$$
$$V_0 = \begin{pmatrix} 0.5 & 0 \\ 0 & 0.5 \end{pmatrix}.$$

Liang applied, for the data, two filtering methods; one is the stochastic linearization filter method and the other is the minimum variance filtering method. The conclusion of his analysis is that the minimum variance filtering method works better than the stochastic linearization filtering method. However if we look at the errors of the minimum variance filter in Fig.13 of Liang(1983), still the result is obviously far from satisfactory. We applied our locally linearized filter for the same data with the same initial condition. Obtained results, estimated state and true state are plotted in Fig.2.a. Errors of the state estimates and error bounds, are in Fig.2.b, where we took Δt=0.001(Liang(1983) does not mention the step size of his numerical integration). The figure shows a drastic improvement of the filtering performance compared with the minimum variance method of Liang(1983). Obviously such large errors by the conventional filtering schemes in Liang(1983) will cause a great difficulty when we use the filtering scheme for the estimation of parameters of nonlinear systems.

3. System identification.

With (2.16) we can transform the time series data $\underline{z}_1, \underline{z}_2, ..., \underline{z}_N$ into Gaussian white noise $\underline{w}_1, \underline{w}_2, ..., \underline{w}_N$, and we can estimate the model parameters by maximizing the likelihood in terms of the parameter $\underline{\theta}$. This is the way how parameters are estimated in ordinary linear or nonlinear time series models. However when we observe an r-dimensional(r< k) vector data $\underline{x}_i$, instead of the k-dimensional state vector $\underline{z}_i$, and when observation errors are present, we cannot obtain the residuals directly from the model structure as for (2.16). We need a nonlinear filtering method to whiten the r-dimensional data, $\underline{x}_1, \underline{x}_2, ..., \underline{x}_N$, into r-dimensional innovations, $\underline{v}_1, \underline{v}_2, ..., \underline{v}_N$.

3.1. THE MAXIMUM LIKELIHOOD METHOD

Before we go into the maximum likelihood method for the nonlinear state space models, let us see how the linear state space model parameters are estimated by the maximum likelihood method. The parameters of a discrete time linear state space model,

$$\begin{aligned} \dot{\underline{z}} &= L(\underline{\theta})\ \underline{z} + \underline{w}(t) \\ \underline{x}_t &= C\ \underline{z}(t) + \underline{v}_t \end{aligned} \tag{3.1}$$

$$C = \begin{pmatrix} 0...0 & 1.....0 \\ ..... & ..... \\ 0...0 & 0.....1 \end{pmatrix}$$

can be estimated by maximizing the likelihood,

$$p(\underline{x}_1, \underline{x}_2, \ldots, \underline{x}_N | \underline{\theta})$$

of the model for the given data $\underline{x}_1, \underline{x}_2, ..., \underline{x}_N$. If we use the Kalman filter, the data $\underline{x}_1, \underline{x}_2, ..., \underline{x}_N$ are transformed to innovations $\underline{v}_1, \underline{v}_2, ..., \underline{v}_N$ as

$$\underline{v}_i = \underline{x}_i - \hat{\underline{x}}_{i|i-1} \tag{3.2}$$

where $\hat{\underline{x}}_{i|i-1}$ is the one-step ahead prediction value of $\underline{x}_i$ at time point i-1 and is a function of $\underline{z}_{i-1}$, V_{i-1} and $\underline{\theta}$. If the initial values $\underline{z}_o$ and V_o are known, then the Jacobian of the transformation from $\underline{x}_1, \underline{x}_2, ..., \underline{x}_N$ to the innovations $\underline{v}_1, \underline{v}_2, ..., \underline{v}_N$ is 1. Usually they are unknown and may be specified by a probability density function $p(\underline{z}_0, V_0 | \underline{\theta})$. Then the likelihood is written down in terms of the innovations, obtained by applying the Kalman filter, as

$$\begin{aligned} p(\underline{x}_1, \underline{x}_2, \ldots, \underline{x}_N | \underline{\theta}) &= \int\int p(\underline{x}_1, \underline{x}_2, \ldots, \underline{x}_N | \underline{z}_0, V_0, \underline{\theta})\, p(\underline{z}_0, V_0 | \underline{\theta})\, d\underline{z}_0\, dV_0 \\ &= \int\int p(\underline{v}_1, \underline{v}_2, \ldots, \underline{v}_N | \underline{z}_0, V_0, \underline{\theta})\, p(\underline{z}_0, V_0 | \underline{\theta})\, d\underline{z}_0\, dV_0 \end{aligned}$$

There are many ways of specifying $p(\underline{z}_0, V_0 | \underline{\theta})$. One simple and practical way of specification is to use the delta function. If we assume a delta function for $\underline{z}_o$ and V_o , the likelihood becomes

$$p(\underline{x}_1, \underline{x}_2, \ldots, \underline{x}_N | \underline{z}_0, V_0, \underline{\theta}) = p(\underline{v}_1, \underline{v}_2, \ldots, \underline{v}_N | \underline{z}_0, V_0, \underline{\theta}),$$

where we have two more parameter sets $\underline{z}_0$ and V_0. Since the innovations are Gaussian white noise, we have

$$p(\underline{v}_1, \underline{v}_2, \ldots, \underline{v}_N | \underline{z}_0, V_0, \underline{\theta}) = \prod_{i=1}^{N} \frac{1}{\sqrt{2\pi |\Sigma_i|}} \exp\left(-\frac{1}{2} \underline{v}_i' \Sigma_i^{-1} \underline{v}_i\right) \tag{3.3}$$

Then we have

$$(-2)\log p(\underline{x}_1, \underline{x}_2, \ldots, \underline{x}_N | \underline{z}_0, V_0, \underline{\theta}) = \sum_{i=1}^{N} (\log |\Sigma_i| + \underline{v}_i' \Sigma_i^{-1} \underline{v}_i) + N \log 2\pi \quad . \tag{3.4}$$

Then the maximum likelihood estimates of the parameters $\underline{z}_0$, V_0 and $\underline{\theta}$ are obtained by minimizing (-2)log-likelihood (3.4). The maximum likelihood method based on this innovation likelihood has been used for the estimation of parameters of linear state space models(Schweppe(1965), Harvey(1984) and Kitagawa(1981)).

In the nonlinear case,

$$\dot{\underline{z}} = \underline{f}(\underline{z} | \underline{\theta}) + \underline{w}(t) \tag{3.5}$$
$$\underline{x}_t = C\,\underline{z}(t) + \underline{v}_t$$

we expect to use the L.L. filter to transform the data $\underline{x}_1, \underline{x}_2, \ldots, \underline{x}_N$ into Gaussian white noise innovations $\underline{v}_1, \underline{v}_2, \ldots, \underline{v}_N$ as we did in the linear case. When $f(\underline{z}|\underline{\theta})$ is nonlinear the state $\underline{z}(t)$ is not Gaussian. However if the sampling interval Δt is sufficiently small, we can assume that the data $\underline{x}_1, \underline{x}_2, \ldots, \underline{x}_N$ are generated from the discrete time state space model,

$$\underline{z}_{t+\Delta t} = A(\underline{z}_t)\,\underline{z}_t + B(\underline{z}_t)\,\underline{w}_{t+\Delta t} \tag{3.6}$$
$$\underline{x}_t = C\,\underline{z}(t) + \underline{v}_t \,,$$

whose state space process $\underline{z}_t$ is driven by $\underline{w}_t$. Then we can get discrete time Gaussian white noise innovations, from observations $\underline{x}_t$ by applying the L.L. filter, since $\underline{w}_{t+\Delta t}$ is a discrete time Gaussian white noise. It is also proved by Frost and Kailath(1971) that the continuous time innovation process for a non-Gaussian diffusion process is a Gaussian white noise. Since the continuous time innovations are defined as a limit of discrete time innovations as

$$\underline{v}(t) = \lim_{\Delta t \to 0} \underline{v}_t$$
$$= \lim_{\Delta t \to 0} (\underline{x}_t - C\,\underline{z}_{t|t-\Delta t})$$

their result implies that the innovation $\underline{v}_t$ obtained by the L.L. filter is very close to a Gaussian white noise. This consideration justifies and guarantees the use of the following representation for the approximation of the (-2)log-likelihood,

$$(-2)\log p(\underline{x}_1, \underline{x}_2, \ldots, \underline{x}_N | \underline{z}_0, V_0, \underline{\theta}) = \sum_{i=1}^{N} (\log |\Sigma_i| + \underline{v}_i' \Sigma_i^{-1} \underline{v}_i) + N \log 2\pi \tag{3.7}$$

where Σ_i is the variance-covariance matrix of the innovation v_i and is given by the L.L. filter (4.3) as

$$\Sigma_i = CP_i C' + \Sigma_{\underline{v}} \tag{3.8}$$

Thus the maximum likelihood estimates of the parameters $\underline{z}_0$, V_0 and $\underline{\theta}$ of (3.1) are obtained by minimizing (3.7).

3.2. PARAMETER ESTIMATION IN PRACTICE.

It is a common view of applied time series analysts that most nonlinear models carry a lot of theoretical material without much useful applications. Unfortunately this view cannot be denied when we look through the articles and books on system identifications where almost all numerical results treated are about linear models. This is mainly because, for most typical nonlinear models such as van der Pol oscillators, the numerical difficulties of nonlinear models' computational instabilities have been unavoidable. As we saw in the previous sections, the L.L. filter provide us with a computationally stable scheme. With this L.L. filter combined with the maximum likelihood method, we try to see if the identification of nonlinear random vibration systems is possible and usable in practice. For this purpose we use the following two examples of typical nonlinear random vibrations.

Example 1. $\ddot{x}(t) - 3\{1 - x(t)^2\}\dot{x}(t) + x(t) = n(t)$ (3.9)

Example 2. $\ddot{x}(t) + 0.25\,\dot{x}(t) + \{14.8 + 4\,x(t)^2\}x(t) = n(t)$ (3.10)

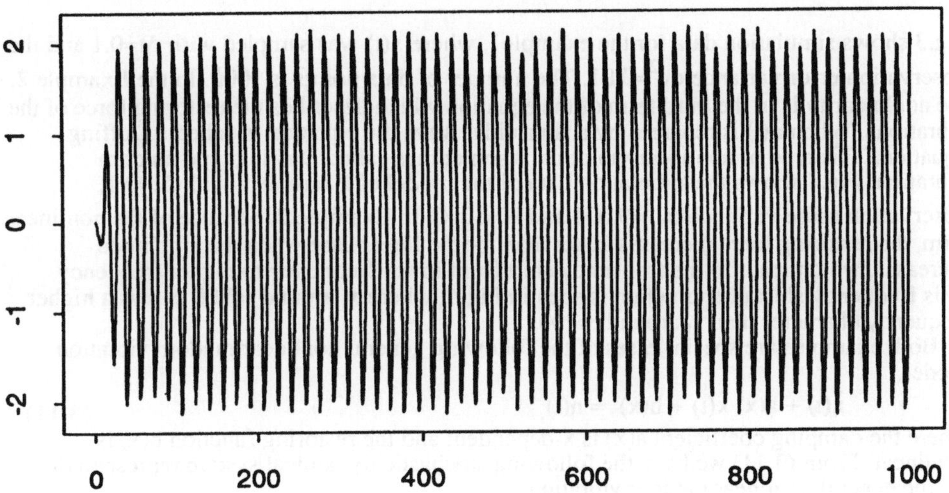

Figure 3. : Simulated data of model (3.9) with $\sigma_n^2=1$ and $\sigma_v^2=0.01$. (N=1000, Δt=0.1)

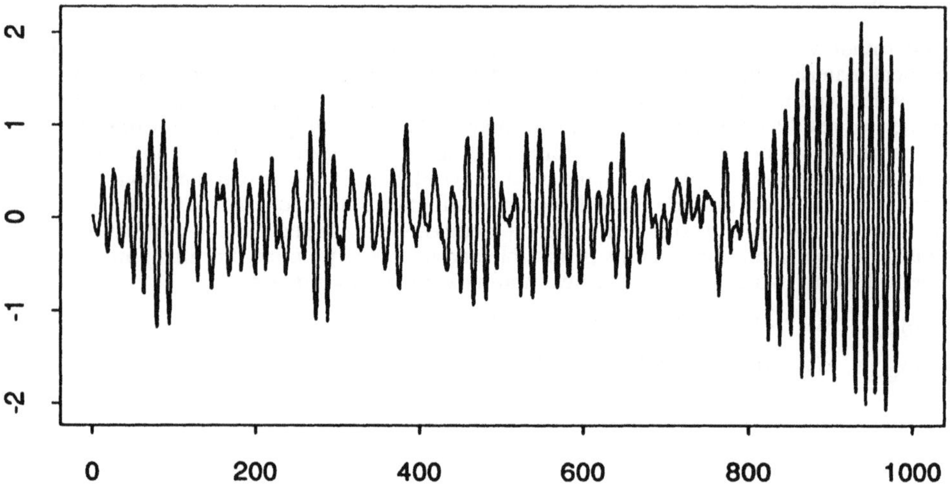

Figure 4. : Simulated data of model (3.10) with $\sigma_n^2=4$ and $\sigma_v^2=0.001$. (N=1000, $\Delta t=0.1$)

Fig.3 shows simulation data for the example1, where x(t) was sampled with $\Delta t=0.1$ and the observation error variance $\sigma_v^2=0.01$. The number of data points is 1000. In the Example 2, the nonlinearity is in the third term of the equation which describes the restoring force of the vibrations. When n(t) is replaced by a sinusoidal wave the equation is called a Duffing equation. This type of nonlinear vibration is sometimes called Duffing type random vibration. Fig.4 shows the simulation data of the model of Example 2 with $\Delta t=0.1$, observation noise variance $\sigma_v^2=0.001$ and N=1000. In Example 2, because of the nonlinear term, the restoring force increases faster than for the linear case when the amplitude increases, which leads to the frequency shift from lower frequency to higher frequency. This frequency shift phenomenon is seen in Fig.4, where x_t around t=900 shows a higher frequency character than x_t around t=700.

Both examples are special cases of the following general nonlinear random vibration model,

$$\ddot{x}(t) + a(x)\,\dot{x}(t) + b(x) = n(t), \tag{3.11}$$

where the damping coefficient a(x) is x-dependent and the restoring function b(x) is nonlinear. From (3.11) we have the following stochastic dynamical system representation for the general nonlinear random vibration,

$$\dot{\underline{z}} = \underline{f}(\underline{z}) + \underline{w}(t) \tag{3.12}$$

$$\underline{z}(t) = \{\dot{x}(t), x(t)\}'$$

$$\underline{w}(t) = \{n(t), 0\}'$$
$$\underline{f}(\underline{z}) = \{a(x)\dot{x} + b(x),\ \dot{x}\}'.$$

Its locally linearized model is

$$\underline{z}_{t+\Delta t} = A(\underline{z}_t)\,\underline{z}_t + \underline{n}_{t+\Delta t},$$

where

$$A(\underline{z}_t) = \mathrm{Exp}(L_t \Delta t) \tag{3.13}$$
$$\underline{n}_{t+\Delta t} = \int_t^{t+\Delta t} \exp\{L_t(t + \Delta t - u)\}\underline{w}(u)du$$
$$L_t = \frac{1}{\Delta t}\mathrm{Log}\{I + J_t^{-1}(\exp^{J_t \Delta t} - I)\, F_t\}$$
$$J_t = \begin{pmatrix} a(x) & a'(x)\dot{x} + b'(x) \\ 1 & 0 \end{pmatrix},$$
$$F_t = \begin{pmatrix} a(x) & b(x)/x \\ 1 & 0 \end{pmatrix}.$$

We note that the variance-covariance matrix S_t of $\underline{n}_{t+\Delta t}$ has full rank 2, even though the variance-covariance matrix of the continuous time white noise $\underline{w}(t)=(n(t),0)'$ has rank one. This is because of the integration of $\underline{w}(t)$ over $[t,t+\Delta t)$. S_t is decomposed as follows,

$$\Sigma_t = \sigma^2 U_t \begin{bmatrix} \lambda_1 & 0 \\ 0 & \lambda_2 \end{bmatrix} U_t',$$

using the unitary matrix U_t. The elements of U_t and eigenvalues λ_1 and λ_2 are given as functions of eigenvalues of L_t and are explicitly given as functions of $\dot{x}_t$ and x_t(see Ozaki(1986)). Therefore we have, from (3.12), the following discrete time state space representation,

$$\underline{z}_{t+\Delta t} = A(\underline{z}_t)\,\underline{z}_t + B(\underline{z}_t)\,\underline{w}_{t+\Delta t}$$
$$x_t = C\,\underline{z}_t + v_t$$
$$B(\underline{z}_t) = U_t\begin{pmatrix} \sqrt{\lambda_1} & 0 \\ 0 & \sqrt{\lambda_2} \end{pmatrix} \tag{3.14}$$
$$C = (0, 1),$$
$$\underline{z}_t = (\dot{x}_t, x_t)',$$

where $\underline{w}_{t+\Delta t}$ is a discrete time Gaussian white noise of variance-covariance matrix $\sigma^2 I$ and $A(\underline{z}_t)$ is given in (3.13). Using the L.L.filter equations (2.19) and (2.22) for (3.14) we can write down the likelihood of the model for the data $\underline{x}_1, \underline{x}_2, \ldots, \underline{x}_N$.

We applied the maximum likelihood method based on the L.L.filter and obtained the results in Table 1. In the computation of the maximum likelihood estimation we fix the initial values $\underline{z}_0$ and V_0 as $\underline{z}_0=(0.1, 0.01)'$ and

$$V_0 = \begin{pmatrix} 0.1 & 0 \\ 0 & 0.01 \end{pmatrix}.$$

The estimated parameters are quite satisfactory and close to the true values in both cases.

Table 1 Maximum likelihood estimates of the model (3.9) and the model (3.10).

	Model (3.9)			Model (3.10)	
	true	estimated		true	estimated
a_1	-3	-2.9178	a	0.25	0.1709
a_2	0	-0.1599×10^{-1}	b_1	14.8	15.5977
a_3	3	2.9436	b_2	0	0.4649
b	14.8	14.7528	b_3	4	3.8065
σ_n^2	1	1.7432	σ_n^2	4	3.5093
σ_v^2	0.01	0.9132×10^{-2}	σ_v^2	0.001	0.9990×10^{-3}

3.3. MAXIMUM LIKELIHOOD ESTIMATES AND LEAST SQUARES ESTIMATES.

For the estimation of parameters of state-space representation models, apart from the maximum likelihood method, there are some other methods such as the least squares estimation method, which suggests to choose the parameter minimizing the sum of squares of the prediction errors,

$v_t(\underline{\theta})=x_t-x_{t|t-1}(\underline{\theta})$, thus

$$\hat{\underline{\theta}}_N = \arg\min_{\underline{\theta}} \{\frac{1}{N}\sum_{t=1}^{N}\frac{1}{2}v_t^2(\underline{\theta})\}$$

Ljung(1976), extending the least squares method, introduced the prediction error estimation method which suggests to choose the parameter,

$$\hat{\underline{\theta}}_N = \arg\min_{\underline{\theta}} \{\frac{1}{N}\sum_{t=1}^{N} l(v_t(\underline{\theta}))\}$$

where $l(.)$ is a scaler-valued(typically positive) function. Asymptotic properties, such as consistency and normality, were proved, and conditions for the convergence have been refined(Caines(1976), Ljung(1978), Caines(1978) and Ljung and Caines(1979)) for the prediction error estimation method. Incidently, we note that the importance of the subject of the present paper was pointed out, in relation to the prediction error estimation method, by Caines(1978), saying "As a result, we have 'reduced' the identification of (stationary) nonlinear systems to the set of difficult open problems known as nonlinear filtering, and nonlinear system identification constitutes, in the author's view, an exciting area for future research."

Since our maximum likelihood method is a special case of the prediction error estimation method($l(.) = -\log p(.)$), the maximum likelihood estimates we have seen in the previous section share the same asymptotic convergence properties as the prediction error estimation method. However what interests applied scientists more is not whether the consistency is weak or strong, but whether the estimates are computationally obtainable in practice and whether the obtained estimates are reliable for ordinary data analysis where number of data points are limited. If the estimates are obtainable we can easily check the asymptotic convergence properties of the estimation method by repeatedly applying the method to simulation data using a modern fast computer.

Using our nonlinear filtering method, i.e. the L.L.filtering scheme, we applied the maximum likelihood method and the least squares method to the twenty sets of data generated from the nonlinear model (3.10). The number of data points of the first each ten data sets is 100 and the number of data points for the latter each ten data sets is 1000. In both methods we used the same numerical optimization procedure(Ishiguro and Akaike(1989)). The estimation results are in Table 2 and Table 3 respectively.

Table 2. MLE and LSE (N=100) for (3.10).

	a	b_1	b_2	b_3	σ_n^2	σ_v^2
TRUE	0.25	14.8	0	4	4	0.001
MLE1	0.16	18.38	-0.91	-1.03	2.68	0.00120
2	0.32	16.27	2.24	3.17	3.14	0.00091
3	0.78	14.44	0.59	3.01	5.31	0.00083
4	0.42	14.95	-2.47	3.89	3.39	0.00106
5	0.28	16.37	-0.75	4.06	3.02	0.00092
6	0.07	12.22	1.81	8.16	4.34	0.00157
7	0.28	17.05	1.19	3.05	3.62	0.00099
8	0.52	13.46	1.41	13.69	3.92	0.00101
9	0.08	15.87	1.13	2.99	4.35	0.00112
10	0.11	13.92	0.43	4.04	4.60	0.00084
LSE1	0.16	18.35	-0.95	-0.99	1.93	0.00086
2	0.32	16.20	2.16	3.33	1.3×10^7	3800
3	0.78	14.95	0.74	1.84	1.94	0.00031
4	0.42	14.94	-2.68	3.95	6.7×10^4	21
5	0.28	16.54	-0.70	3.75	3.00	0.00091
6	0.07	12.17	1.93	8.21	3.52	0.00126
7	0.28	17.23	1.28	2.76	5.06	0.00138
8	0.52	13.19	1.36	13.30	0.02	0.4×10^{-5}
9	0.07	15.85	1.16	3.00	2.1×10^5	53
10	0.10	13.68	0.41	4.20	0.10	0.00002

Table 3. MLE and LSE (N=1000) for (3.10).

	a	b_1	b_2	b_3	σ_n^2	σ_v^2
TRUE	0.25	14.8	0	4	4	0.001
MLE1	0.17	15.60	0.46	3.81	3.51	0.00100
2	0.21	14.49	-0.23	4.25	3.86	0.00099
3	0.33	14.39	-0.35	3.01	3.78	0.00110
4	0.41	14.70	0.05	4.29	3.86	0.00116
5	0.37	14.13	-0.36	4.76	3.31	0.00090
6	0.23	14.91	-0.15	4.11	4.40	0.00105
7	0.32	13.89	-0.03	4.80	4.32	0.00099
8	0.29	14.09	0.29	4.38	4.00	0.00109
9	0.14	14.53	-0.04	4.18	3.62	0.00108
10	0.14	14.82	-0.40	3.74	4.78	0.00107
LSE1	0.18	15.62	0.48	3.79	5.42	0.00153
2	0.21	14.49	-0.23	4.24	5.03	0.00129
3	0.32	14.38	-0.38	4.69	5.7×10^4	16
4	0.41	14.73	0.02	4.24	4.12	0.00124
5	0.37	14.25	-0.33	4.61	1.7×10^4	37
6	0.23	14.78	-0.05	4.14	0.36	0.00009
7	0.27	13.78	-0.11	4.77	0.49	0.00012
8	0.29	14.02	0.33	4.45	1.7×10^3	0.47
9	0.14	14.55	-0.03	4.17	9.9×10^4	29.54
10	0.14	14.82	-0.39	3.74	3.2×10^3	0.94

From the tables we can see that the maximum likelihood method works reasonably well even for fairly short data. The least squares method, on the other hand, sometimes gives very unreliable estimates even for long data. The extraordinary large values of the least squares estimates for σ_n^2 and σ_v^2 are not caused by the employed nonlinear optimization method. The gradient of these optimized points were all reasonably small(less than 10^{-3}). The original data of these abnormal estimates do not show any special different feature(see Fig.5 and Fig.6). The present numerical results imply that the maximum likelihood method is much more reliable than the least squares method, although both methods share the same asymptotic convergence properties in theory.

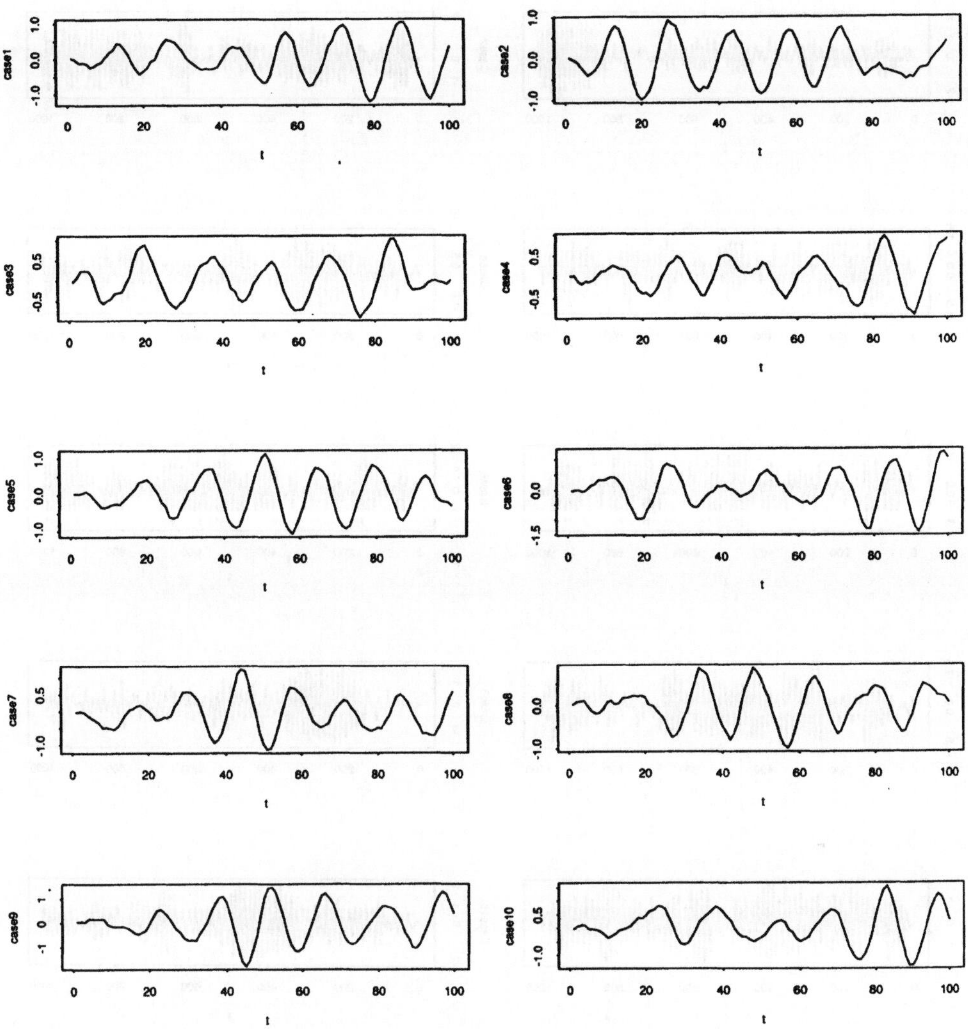

Figure 5. Ten sets of simulated data of model (3.10) with $\sigma_n^2=4$ and $\sigma_v^2=0.001$. (N=100, Δt=0.1)

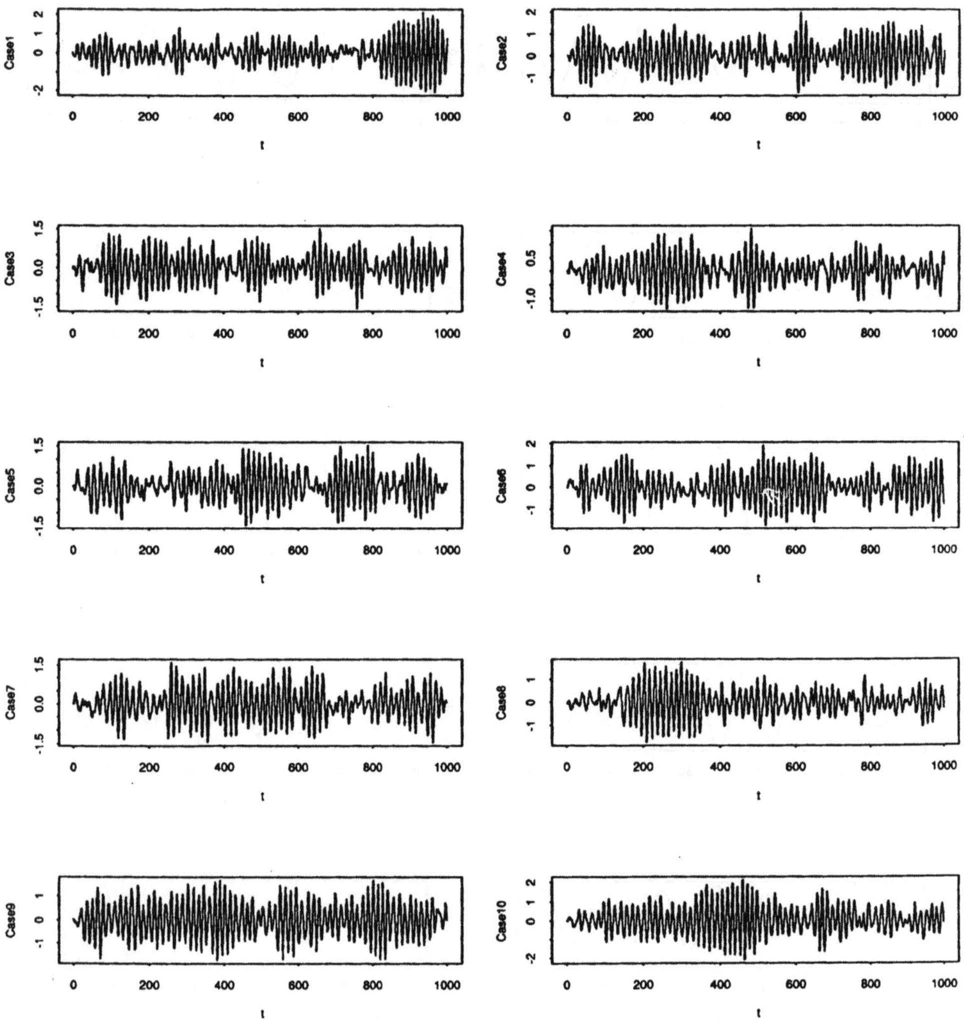

Figure 6. Ten sets of simulated data of model (3.10) with $\sigma_n^2=4$ and $\sigma_v^2=0.001$ (N=1000, $\Delta t=0.1$)

4. Conclusions.

Many interesting models, such as deterministic nonlinear dynamical systems, stochastic nonlinear dynamical systems and diffusion processes, used in sciences, were unfortunately not estimable from observation data.

We have seen that the local linearization scheme provides us with a method to treat these continuous time models, by automatically converting them into discrete time models, on the same basis as nonlinear time series models. Further we have seen that the local linearization filter, applied to the state space representation models of these discrete time models, made it possible to estimate the parameters of these models from time series data.

Acknowledgement

The present work owes a great deal to Akaike's introduction of Boltzman entropy in statistics, which directed the interest of many statisticians away from proving asymptotic theory of unobtainable estimates to devising concrete methods of estimating useful models in natural and social sciences. On the occasion of Prof.Akaike's 65th birthday, I would like to express hearty thanks to him. Sharing an office with him when I was a junior young researcher is one of my most valuable experiences.

References

Akaike, H.(1974) "Markovian representation of stochastic processes and its application to the analysis of autoregressive moving average processes", *Ann. Inst. Statist. Math. 26*, 363-387.

Arnold, L.(1974) *Stochastic Differential Theory and Applications*, John Wiley & Sons.

Caines,P.E.(1976) "Prediction error identification methods for stationary stochastic processes", *IEEE Trans. Automatic Control, AC-21*, 500-505.

Caines,P.E.(1978) "Stationary linear and nonlinear system identification and predictor set completeness", *IEEE Trans. Automatic Control, AC-23*, 583-594.

Frost,P.A. and Kailath, T.(1971) "An innovation approach to least squares estimation-part III: nonlinear estimation in white Gaussian noise", *IEEE Trans. Automatic Control, Vol.AC-16*, 217-226.

Harvey, A. C.(1984)"A unified view of statistical forecasting procedures", *Journal of Forecasting, vol.3* , 245-275

Hazewinkel, M. and Willems, J. (1981) *Stochastic Systems: The Mathematics of Filtering and Identification and Applications*, Proceeding of the NATO Advanced Study Institute held at Les Arcs, Savoie, France, June 22 - July 5, 1980., D.Reidel Publishing Company

Ishiguro,M. and Akaike,H.(1989) "DALL: Davidon's algorithm for log likelihood maximization - A Fortran subroutine for statistical model builders -, *Computer Science Monographs, No.25*, Institute of Statistical Mathematics, Tokyo.

Jones, D.A.(1978)"Nonlinear autoregressive processes, *Proc. Roy. Soc. London Ser. A 360*, 71-95.

Kailath, T.(1968)"An innovations approach to least-squares estimation-Part I:Linear filtering in additive white noise", *IEEE Trans. Automatic Control, Vol.AC-13*, 646-654.

Kalman, R.E.(1960)"A new approach to linear filtering and prediction problems", *Trans. ASME, Series D, Journal of Basic Engineering, vol 82*, 35-45.

Kitagawa, G.(1981)"A nonstationary time series model and its fitting by a recursive filter", *J. Time Ser. Anal. 2*, 103-116.

Kushner, H.J.(1962) "On the differential equations satisfied by conditional probability densities of Markov processes, with applications" *J.SIAM Control, SerA, Vol 2*, 106-119.

Kushner, H.J.(1967) "Approximations to optimal nonlinear filters", *IEEE trans. Automatic Control, Vol.AC-12*,546-556.
Liang,D.F.(1983) "Comparisons of nonlinear recursive filters for systems with nonnegligible nonlinearities", in *Control and Dynamical Systems, Advances in Theory and Applications, Vol 20* : Nonlinear and Kalman Filtering Techniques C.T.Leondis Eds. Academic Press, New York, 341-401.
Ljung, L.(1976) On the consistency of prediction error identification methods. In R.K.Mehra and D.G.Lainiotis, eds. : *System Identification - Advanced Studies*. Academic Press, New York.
Ljung,L.(1978) "Convergence analysis of parametric identification methods", *IEEE Trans. Automatic Control, AC-23*, 770-783.
Ljung,L. and Caines,P.E.(19789) "Asymptotic normality of prediction error estimation for approximate system models", *Stochastics, 3*, 29-46.
Markus, S.I.(1984) "Algebraic and geometric methods in nonlinear filtering", *SIAM J. Control and Optimization, Vol.22*, 817-844.
McShane, E.J.(1974) *Stochastic Culculus and Stochastic Models*, Academic Press.
Milshtein(1974)"Approximate integration of stochastic differential equations, *Theory Prob. Appl., 23*, 396-401.
Mortensen(1979)"Mathematical problems of modeling stochastic dynamic systems, *J. Statist. Phys. 1*, 271-296.
Ozaki, T.(1985) "Nonlinear time series models and dynamical systems", in *Handbook of Statistics, Vol 5*, Hannan, E.J. et al Ed., North-Holland, 25-83.
Ozaki, T.(1986) "Local Gaussian modelling of stochastic dynamical systems in the analysis of nonlinear random vibrations", in *Essays on Time Series and Allied Processes, Festschrift in honour of Prof.E.J.Hannan*, Applied Probability Trust, 241-255.
Ozaki, T.(1989) "Statistical identification of nonlinear random vibration systems", *J. of Appl. Mechanics, vol 56*, 186-191.
Ozaki, T.(1990a) A local linearization approach to nonlinear filtering , Research Memo. No.378, Institute of Statistical Mathematics, Tokyo, to appear in *International J. on Control*.
Ozaki, T.(1990b) Identification of nonlinearities and non-Gaussianities in time series , to appear in the proceeding of 1990 Summer International Interdisciplinary Workshop on Time Series Analysis at the IMA, July 2-27, 1990, Springer-Verlag.
Ozaki, T.(1992) "A bridge between nonlinear time series models and nonlinear stochastic dynamical systems: a local linearization approach", *Statistica Sinica*, 2, 113-135.
Rumelin, W.(1982)"Numerical treatment of stochastic differential equations", *SIAM J. Numer. anal. Vol.19* 604-613.
Schwartz,L. and Stear,E.B.(1968) "A computational comparison of several nonlinear filters", *IEEE Trans.Automatic Control, vol.AC-13*, 83-86.
Schweppe, H.(1965)"Evaluation of likelihood functions for Gaussian signals", *IEEE Trans. Automatic Control, Vol.AC-11*, 61-70.
Sorenson, H.W.(1966) "Kalman filtering technique", in *Advances in Control Systems, Theory and Applications, Vol 3*, Academic Press, 219-292.
Tweedie, R.L.(1975) "Sufficient conditions for ergodicity and stationarity of Markov chains on a general state space. *Stoch. Proc.Their Appl. 3*, 385-403.

INFERENCE OF EVOLUTIONARY TREES FROM DNA AND PROTEIN SEQUENCE DATA

MASAMI HASEGAWA
The Institute of Statistical Mathematics
4-6-7 Minami-Azabu, Minato-ku, Tokyo 106 Japan

1. Introduction

The evolutionary history of life is inscribed in the DNA and protein sequences of living organisms. Owing to the rapid accumulation of sequence data of "DNA" and its product "protein" from diverse organisms, there is an increasing demand for methods of data analysis to infer evolutionary trees (molecular phylogenies) based on statistical models for base or amino acid substitutions.

DNAs and proteins are sequences of four types of bases and of twenty types of amino acids, respectively. At each site of a sequence, one type of base or amino acid may change to another one during evolution, and the substitutions occur independently between lineages after they separate in an evolutionary tree. These substitutions can be best regarded as stochastic (Kimura, 1983). Members of a set of DNA or protein sequences descended from a common ancestral sequence are called homologous to each other, and homologous sequences in contemporary organisms are the results of the stochastic evolutionary process. We infer the evolutionary tree from these sequences.

The widely used methods for inferring molecular phylogenies, such as the maximum parsimony and distance matrix methods (e.g., Nei, 1987), were fairly heuristic. Since they are not based on clearly described models of substitution process, the underlying assumptions of the methods are not clear. Neyman (1971) was the first to use the maximum likelihood (ML) to infer evolutionary trees from DNA sequences. Later, Felsenstein (1981) developed a practical method of the ML for analyzing DNA sequence data under the assumption of i.i.d. among sites without assuming an equal rate among lineages. To apply the method, he regards an evolutionary tree as a statistical model containing several unknown parameters, such as branch lengths and evolutionary rates. The parameters are estimated by the ML, and the tree with the highest likelihood is chosen as the most likely candidate for the true tree.

In this paper, I shall review recent developments of the ML method and its applications in molecular phylogenetics that have been made in this institute.

2. Protein Phylogeny

The phylogenetic tracing can be extended back to the early evolution of life on earth, such as the branching between humans and bacteria. All the extant life on earth

H. Bozdogan (ed.), Proceedings of the First US/Japan Conference on the Frontiers of Statistical Modeling: An Informational Approach, 241–248.

have common basic machineries of heredity and biochemistry, and therefore they all must have descended from a single common ancestor. Phylogenetic study of amino acid sequences of proteins, that are universally found in all the organisms and are conservative in amino acid substitutions, is suitable for tracing the early history of life (Hasegawa et al., 1993a, 1993b), while ribosmal RNA sequences widely used in this field have several problems (Hasegawa and Hashimoto, 1993).

For DNA and RNA evolution, a transition matrix among the four species of bases is 4×4, while, for protein evolution, it is 20×20 (among twenty species of amino acids) and is more intractable in ML analyses. Kishino et al. (1990) and Adachi and Hasegawa (1992a) developed a ML method for inferring protein trees that takes into account unequal transition probabilities among pairs of amino acids by using an empirical transition matrix compiled by Dayhoff et al. (1978) (the Dayhoff model). It does not assume rate constancy among lineages, and hence it gives a robust estimate of a tree topology even if the rate differs among lineages (Hasegawa and Fujiwara, 1993). Furthermore, Adachi and Hasegawa (1992b), Hasegawa et al. (1993a, 1993b), Adachi et al. (1993) and Hashimoto et al. (1993) compared several alternative models for amino acid substitutions with different numbers of parameters by using AIC (Akaike, 1974). It turned out that the Dayhoff model is by far the best among existing models of amino acid substitutions for various sets of real protein data.

We have applied the ML method for protein phylogeny in the study of the early evolution of life (Hasegawa et al., 1990b, 1993a, 1993b; Hashimoto et al., 1992), and of several phylogenetic problems (Kishino et al., 1990; Hasegawa et al., 1992; Hashimoto et al., 1993). Even if the Dayhoff model is the best among existing models, we must admit that it is merely an approximation of the actual process of amino acid substitutions during evolution. If the assumed underlying model is not close to the actual evolutional process, then the inference of the tree could be incorrect. Therefore, we must continue to improve the model in the future, and AIC can be used as a good criterion for the model selection. Nevertheless, simulation studies have shown that the ML method is robust to some extent against the violation of the assumed Markov model in inferring tree topologies (Hasegawa et al., 1991; Fukami-Kobayashi and Tateno, 1991; Hasegawa and Fujiwara, 1993), and therefore we think that the Dayhoff model is useful enough in inferring tree topologies.

3. Confidence Limit of the ML Tree

Hasegawa and Kishino (1989a) noted that, since the likelihood is dependent on a particular realization of a random variable, it is important to estimate the variance of the difference of log-likelihood between alternative tree topologies. They estimated the variance by using the bootstrap resampling (Felsenstein, 1985). Kishino and Hasegawa (1989) developed a simple method to estimate approximately the variance of the difference of log-likelihood between trees without repeating bootstrap resampling and ML estimation for each sample. Later, Kishino et

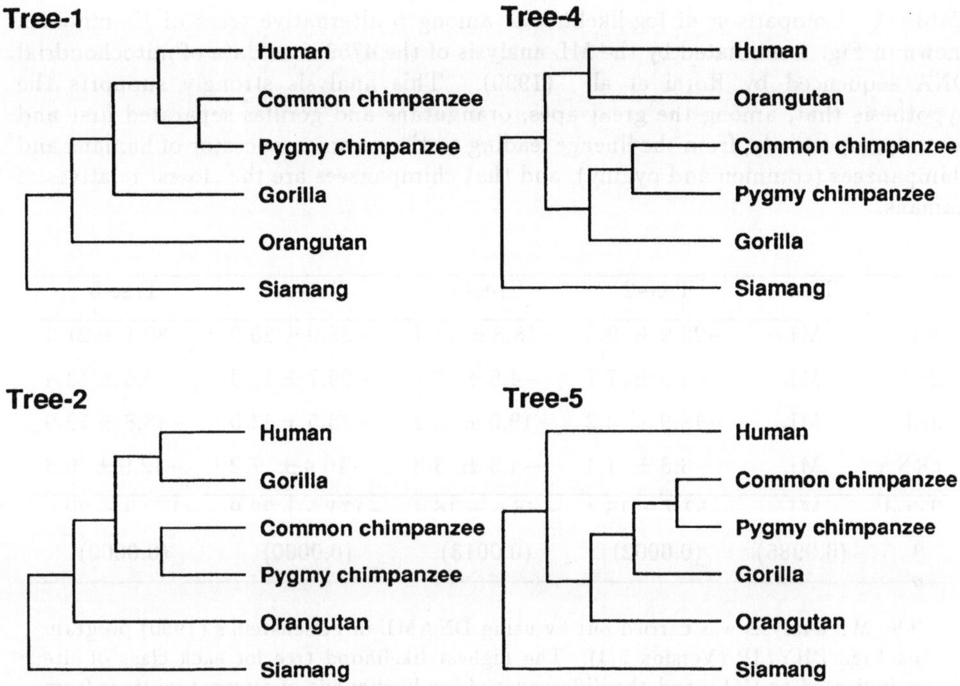

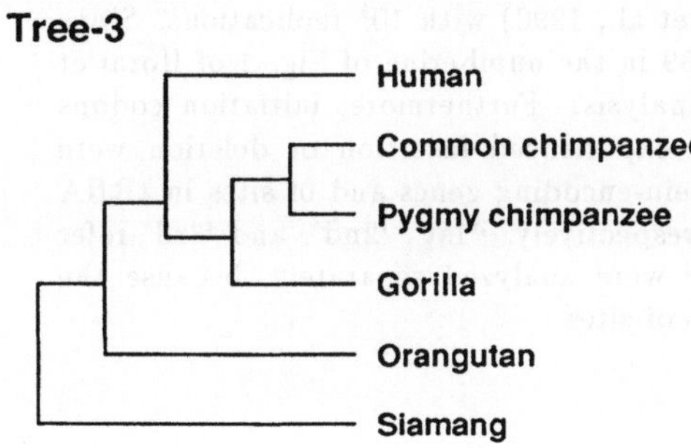

Figure 1. Candidate tree topologies among humans, chimpanzees (common and pygmy chimpanzees), gorillas, and orangutans with siamangs as an outgroup.

Table 1. Comparison of log-likelihoods among 5 alternative trees of Hominoidea shown in Fig. 1 estimated by the ML analysis of the 4759 sites data of mitochondrial DNA sequenced by Horai et al. (1990). This analysis strongly supports the hypothesis that, among the great apes, orangutans and gorillas separated first and second, respectively, from the lineage leading to the common ancestor of humans and chimpanzees (common and pygmy), and that chimpanzees are the closest relatives of humans.

	Tree-1	Tree-2	Tree-3	Tree-4	Tree-5
1st	ML	-25.9 ± 9.6	-18.8 ± 11.4	-88.0 ± 20.9	-86.1 ± 21.3
2nd	ML	-4.6 ± 7.7	-4.5 ± 7.7	-29.7 ± 13.3	-29.6 ± 13.4
3rd	ML	-18.9 ± 8.2	-19.0 ± 8.2	-43.5 ± 13.0	-43.8 ± 12.9
tRNA	ML	-3.3 ± 4.4	-4.3 ± 3.8	-10.4 ± 7.2	-12.9 ± 6.3
Total	ML	-52.7 ± 15.4	-46.6 ± 16.4	-171.5 ± 28.9	-172.3 ± 29.0
	(0.9985)	(0.0002)	(0.0013)	(0.0000)	(0.0000)

The ML analysis was carried out by using DNAML in Felsenstein's (1990) program package PHYLIP (Version 3.3). The highest likelihood tree for each class of site is indicated as 'ML', and the differences of log-likelihoods of alternative trees from that of the ML tree are shown with their SE following ± estimated by Kishino and Hasegawa's formula (1990). Bootstrap probabilities given in parentheses were estimated by the RELL method (Kishino et al., 1990) with 10^4 replications. Since the alignment is ambiguous at sites 869–889 in the numbering of Fig. 1 of Horai et al., these sites were excluded from the analysis. Furthermore, initiation codons of protein-encoding genes and sites that experienced insertion or deletion were excluded. The numbers of codons in protein-encoding genes and of sites in tRNA genes under analyses are 1337 and 755, respectively. '1st', '2nd', and '3rd' refer to codon positions of a codon, and they were analyzed separately, because the evolutionary rate differs among the classes of sites.

al. (1990) developed two simple methods to estimate approximately a bootstrap probability; that is, a frequency of a particular tree being the highest likelihood tree among alternatives during bootstrap resampling. These methods estimate bootstrap probabilities of alternative trees from a ML analysis without performing ML estimation for each resampled data.

One of these methods, called the RELL (resampling estimated log-likelihood), estimates bootstrap probabilities by bootstrap resampling the estimated log-likelihoods of sites rather than the data followed by the ML estimation of the resampled data. By comparing bootstrap probabilities estimated by the RELL method from real data with those estimated by repeating bootstrap resampling and ML estimation, Hasegawa and Kishino (1993) have shown that the simple method is a good approximation to the computationally intensive method. The methods have been applied to evaluate phylogenetic placing of humans within primates (Kishino and Hasegawa, 1989; Hasegawa et al., 1990a; Hasegawa, 1990, 1991).

Fig. 1 represents five alternative trees among humans, chimpanzees (common and pygmy chimpanzees), gorillas, and orangutans with siamangs as an outgroup. Any one of the trees has ever been advocated by some morphologists. Table 1 gives the results of ML analysis for the 4759 sites data of mitochondrial DNA sequenced by Horai et al. (1992). Consistently with Horai and his coworkers, this analysis supports Tree-1 and excludes other possibilities for the mitochondrial gene tree.

4. Estimation of Branching Dates in a Tree

It has been suggested that the rate of substitutions of DNA or protein during evolution is nearly uniform among different lineages (Kimura, 1983). This characteristic of molecular evolution, called the "molecular clock", is in sharp contrast with morphological evolution in which constancy of the rate can seldom be observed.

Although the approximate rate constancy is generally a rule in molecular evolution, several instances of rate nonconstancy have been suggested (e.g., Hasegawa et al., 1987; Hasegawa and Kishino, 1989b; Adachi and Hasegawa, 1992b; Adachi et al., 1993), and therefore in Table 1 we inferred the tree topology without assuming the rate constancy . Once the tree topology or the branching order is known, we can evaluate the extent to which the rate of base or amino acid substitutions is equal among different lineages, and can estimate the dates of branchings in the tree. Kishino and Hasegawa (1990) developed a statistical method for estimating branching dates from DNA sequence data by taking account of possible rate variations among lineages.

We assume that each site evolves according to a Markov process in which a nucleotide i (T, C, A, or G) is replaced by another nucleotide j in an infinitesimally short interval of time, dt, with a probability of $P_{ij}(dt)$, as follows,

$$P_{ij}(dt) = \begin{cases} \alpha\pi_j \, dt & \text{for transition} \quad (T \leftrightarrow C,\ A \leftrightarrow G) \\ \beta\pi_j \, dt & \text{for transversion} \quad (T, C \leftrightarrow A, G). \end{cases} \tag{1}$$

where π_j is the composition of nucleotide j (Hasegawa et al., 1985). The rate parameters α for transition and β for transversion can be different among different branches in a tree. We assume that all of the nucleotide sites are independently and homogeneously variable with the substitution probability represented by Eq. (1).

The α's, β's, and branching dates are parameters of the model. In a model where evolutionary rates are not equal among branches, different sets of rate parameters,

α_k and β_k, are assigned to branches along which rate differs. An extreme case is that every branch has its own parameters, α_k and β_k ($k = 1, 2, \ldots$, number of branches) (this was just the situation when we inferred the branching order), and the other extreme is rate constancy throughout the tree. Between these two extremes, there are many intermediate models in which several, but not all, branches share a common set of rate parameters. Among these models, we select a model that gives the minimum AIC. From the minimum AIC model, we estimate the rate parameters α's and β's, as well as the branching dates.

Application of this method to DNA sequence data of several genes has indicated that humans diverged from chimpanzees about 4–5 million years ago (Hasegawa, 1990, 1991, 1992; Hasegawa et al., 1990a; Hasegawa and Kishino, 1991). Furthermore, the most recent common mitochondrial DNA ancestor of modern humans is estimated to be 211,000 $\pm$ 111,000 years ago ($\pm$: 1SE) (Hasegawa and Horai, 1991; Hasegawa et al., 1993c).

5. Future Directions

Although the ML method is robust to some extent against the violation of the assumed model in inferring tree topologies (Hasegawa and Fujiwara, 1993), the estimates of branch lengths depends on the assumed underlying model for base or amino acid substitutions. Therefore, for the estimation of branching dates, we must use a realistic model as far as possible. In analyzing base sequences of protein encoding genes, the i.i.d. assumption does not hold, because a substitution of a base depends on the other bases in the codon. In order to overcome this difficulty, it might be desirable to deal with a transition matrix of 64×64 among codons, but it is intractable. Therefore, an analysis of amino acid sequence level is more appropriate for these genes, provided that a realistic model for amino acid substitution is attainable. For this purpose, the Dayhoff model may not be sufficiently realistic, and the model must be improved in the future. And AIC should remain to be useful in improving the model.

Molecular phylogenetics is seen as making an estimate of an unknown quantity, in the presence of uncertainty, and using a probabilistic model of the evolutionary process (Felsenstein, 1988). In the field of molecular phylogenetics, however, it has not necessarily been recognized widely that a statistical framework is appropriate in analyzing sequence data, and not so many satatisticians are working there in spite of the importance of the problem. There remain many problems untouched in this field waiting for statisticians' contribution.

References

Adachi, J. and Hasegawa, M. (1992a) MOLPHY: Programs for Molecular Phylogenetics, I — PROTML: Maximum Likelihood Inference of Protein Phylogeny, *Computer Science*

Monographs, No.27, Institute of Statistical Mathematics, Tokyo.

Adachi, J. and Hasegawa, M. (1992b) Amino acid substitution of proteins coded for in mitochondrial DNA during mammalian evolution. *Japanese Journal Genetics*, Vol. 67, No. 3, 187–197.

Adachi, J., Cao, Y. and Hasegawa, M. (1993) Tempo and mode of mitochondrial DNA evolution in vertebrates at the amino acid sequence level: rapid evolution in warm-blooded vertebrates. *Journal of Molecular Evolution* (in press).

Akaike, H. (1974) A new look at the statistical model identification. *IEEE Transactions on Automatic Control*, Vol. AC-19, No. 6, 716–723.

Dayhoff, M.O., Schwartz, R.M., and Orcutt, B.C. (1978) A model of evolutionary change in proteins, *Atlas of Protein Sequence and Structure* (ed. M.O. Dayhoff), Vol. 5, Suppl. 3, 345–352, National Biomedical Research Foundation, Washington, D.C.

Felsenstein, J. (1981) Evolutionary trees from DNA sequences, a maximum likelihood approach. *Journal of Molecular Evolution*, Vol. 17, 368–376.

Felsenstein, J. (1985) Confidence limits on phylogenies, an approach using the bootstrap. *Evolution*, Vol. 39, No. 4, 783–791.

Felsenstein, J. (1988) Phylogenies from molecular sequences, inference and reliability. *Annual Review of Genetics*, Vol. 22, 521–565.

Felsenstein, J. (1990) PHYLIP, Version 3.3, University of Washington, Seattle.

Fukami-Kobayashi, K. and Tateno, Y. (1991) Robustness of maximum likelihood tree estimation against different patterns of base substitutions. *Journal of Molecular Evolution*, Vol. 32, 79–91.

Hasegawa, M. (1990) Phylogeny and molecular evolution in primates. *Japanese Journal of Genetics*, Vol. 65, No. 4, 243–265.

Hasegawa, M. (1991) Molecular phylogeny and man's place in Hominoidea. *Journal of Anthropological Society of Nippon*, Vol. 99, No. 1, 49–61.

Hasegawa, M. (1992) Evolution of hominoids as inferred from DNA sequences, *Topics in Primatology, Vol. 1. Human Origins*, (eds. T. Nishida, W.C. McGrew, P. Marler, M. Pickford, and F.B.M. de Waal), 347–357, Univ. Tokyo Press, Tokyo.

Hasegawa, M. and Fujiwara, M. (1993) Relative efficiencies of the maximum likelihood, maximum parsimony, and neighbor joining methods for estimating protein phylogeny. *Molecular Phylogenetics and Evolution* (in press).

Hasegawa, M. and Hashimoto, T. (1993) Ribosomal RNA trees misleading? *Nature*, Vol. 361, No. 6407, 23–23.

Hasegawa, M. and Horai, S. (1991) Time of the deepest root for polymorphism in human mitochondrial DNA. *Journal of Molecular Evolution*, Vol. 32, No. 1, 37–42.

Hasegawa, M. and Kishino, H. (1989a) Confidence limits on the maximum-likelihood estimate of the hominoid tree from mitochondrial-DNA sequences. *Evolution*, Vol. 43, No. 3, 672–677.

Hasegawa, M. and Kishino, H. (1989b) Heterogeneity of tempo and mode of mitochondrial DNA evolution among mammalian orders. *Japanese Journal of Genetics*, Vol. 64, No. 4, 243–258.

Hasegawa, M. and Kishino, H. (1991) DNA sequence analysis and evolution of Hominoidea. *New Aspects of the Genetics of Molecular Evolution*, (eds. M. Kimura and N. Takahata), 303–317, Springer-Verlag, Berlin.

Hasegawa, M. and Kishino, H. (1993) Simple methods for estimating bootstrap probability of a maximum likelihood tree. *Molecular Biology and Evolution* (in press).

Hasegawa, M., Kishino, H. and Yano, T. (1985) Dating of the human-ape splitting by a molecular clock of mitochondrial DNA. *Journal of Molecular Evolution*, Vol. 22, No. 2, 160–174.

Hasegawa, M., Kishino, H. and Yano, T. (1987) Man's place in Hominoidea as inferred from molecular clocks of DNA. *Journal of Molecular Evolution*, Vol. 26, No. 1-2, 132–147.

Hasegawa, M., Kishino, H., Hayasaka, K., and Horai, S. (1990a) Mitochondrial DNA evolution in primates, transition rate has been extremely low in lemur. *Journal of Molecular Evolution*, Vol. 31, No. 2, 113–121.

Hasegawa, M., Iwabe, N., Mukohata, Y., and Miyata, T. (1990b) Close evolutionary relatedness of archaebacteria, Methanococcus and Halobacterium, to eukaryotes demonstrated by composite phylogenetic trees of elongation factors EF-Tu and EF-G, eocyte tree is unlikely. *Japanese Journal of Genetics*, Vol. 65, 109–114.

Hasegawa, M., Kishino, H. and Saitou, N. (1991) On the maximum likelihood method in molecular phylogenetics. *Journal of Molecular Evolution*, Vol. 32, 443–445.

Hasegawa, M., Cao, Y., Adachi, J. and Yano, T. (1992) Rodent polyphyly? *Nature*, Vol. 355, No. 6361, 595–595.

Hasegawa, M., Hashimoto, T., Adachi, J., Iwabe, N., and Miyata, T. (1993a) Early divergences in the evolution of eukaryotes, ancient divergence of *Entamoeba* that lacks mitochondria revealed by protein sequence data. *Journal of Molecular Evolution* (in press).

Hasegawa, M., Hashimoto, T. and Adachi, J. (1993b) Origin and evolution of eukaryotes as inferred from protein sequence data, *The Origin and Evolution of Prokaryotic and Eukaryotic Cells*, (eds. H. Hartman and K. Matsuno), World Sci. Publ. (in press).

Hasegawa, M., Di Rienzo, A., Kocher, T.D., and Wilson, A.C. (1993c) Toward a more accurate time scale for the human mitochondrial DNA tree. *Journal of Molecular Evolution* (in press).

Hashimoto, T., Adachi, J. and Hasegawa, M. (1992) Phylogenetic place of *Giardia lamblia*, a protozoan that lacks mitochondria. *Endocytobiosis and Cell Research*, Vol. 9, 59–69.

Hashimoto, T., Otaka, E., Adachi, J., Mizuta, K., and Hasegawa, M. (1993b) The giant panda is most close to a bear, judged by α- and β-hemoglobin sequences. *Journal of Molecular Evolution* (in press).

Horai, S., Satta, Y., Hayasaka, K., Kondo, R., Inoue, T., Ishida, T., Hayashi, S., and Takahata, N. (1992) Man's place in Hominoidea revealed by mitochondrial DNA genealogy. *Journal of Molecular Evolution*, Vol 35, No. 1, 32–43.

Kimura, M. (1983) *The Neutral Theory of Molecular Evolution.* Cambridge University Press, Cambridge.

Kishino, H., and Hasegawa, M. (1989) Evaluation of the maximum likelihood estimate of the evolutionary tree topologies from DNA sequence data, and the branching order in Hominoidea. *Journal of Molecular Evolution*, Vol. 29, No. 2, 170–179.

Kishino, H., and Hasegawa, M. (1990) Converting distance to time, an application to human evolution. *Methods in Enzymology*, Vol. 183, 550–570.

Kishino, H., Miyata, T., and Hasegawa, M. (1990) Maximum likelihood inference of protein phylogeny and the origin of chloroplasts. *Journal of Molecular Evolution*, Vol. 30, 151–160.

Nei, M. (1987) *Molecular Evolutionary Genetics.* Columbia University Press, New York.

Neyman, J. (1971) Molecular studies of evolution, a source of novel statistical problems, *Statistical Decision Theory and Related Topics*, (eds. S.S. Gupta and J. Yackel), 1–27, Academic Press, New York.

NEW STRUCTURE CRITERIA IN GROUP METHOD OF DATA HANDLING

T. LANGE
Department of Electrical Engineering
College of Technology and Economics
Berlin Lichtenberg / Germany

1. Introduction

The task of statistical modeling can be expressed as shown in Fig. 1. It can be divided into two parts:

1. parameter estimation ($\mathbf{a}^*$),
2. selection of the model structure ($\mathbf{M}^*$, $\mathbf{k}^*$).

The task of modeling, especially the automatic selection of the model structure based on the **Group Method of Data Handling GMDH** of IVAKHNENKO (1975) will be discussed in connection with the condition that the amount of experimental data is quite limited.

The availability of a limited amount only of experimental data points often leads to a considerable **uncertainty** in the task of complex system modeling because the basic assumptions for regression analysis in this case are questionable. Under these conditions it is especially difficult to choose the structure criteria for the selection of a model.

2. Special problems in case of uncertainty

Several investigations, carried out by the author, showed that the origin of uncertainty is a very diverse problem which depends not only on the absolute amount of primary data but also on a lot of correlated factors:

a) amount of measuring points
b) length and position of interpolation region
c) value of coefficients of the true model
d) proportion of correlation rank of columns of matrix **X** and the factors a,b,c,
e) nature of the output disturbance
f) completeness of the input vector $\underline{\mathbf{x}}$

H. Bozdogan (ed.), Proceedings of the First US/Japan Conference on the Frontiers of Statistical Modeling: An Informational Approach, 249–266.

Three components of statistical model:

1. a system

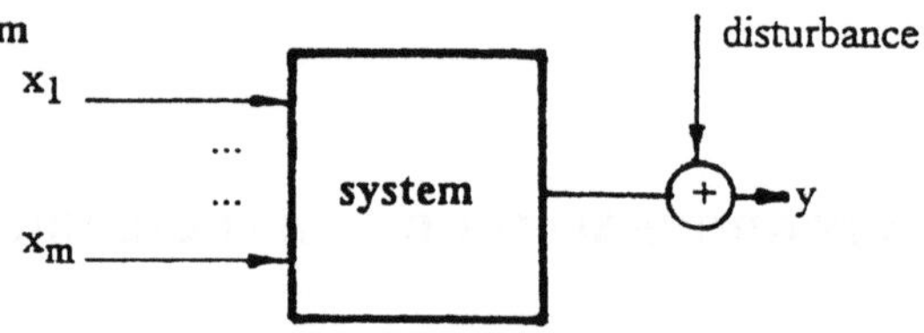

2. a data table (data set):

	x_1	x_2	...	x_m	y
1			...		
2			...		
.			.		
.			.		
.			.		
l			...		

3. a model set-up (Kolmogorov-Gabor-polynomial):

$$y = \sum_{i=0}^{m} a_0 + \sum_{i=0}^{m} \sum_{j=0}^{m} a_i x_i + \sum_{i=0}^{m} \ldots \sum_{q=0}^{m} a_{ij} x_i x_j + \ldots + a_{i \ldots q} x_i \ldots x_q \quad (1)$$

The following

- **optimal value of the coefficient vector a^***
- **amount of relevant inputs M^***
- **polynom's optimal power k^***

have to be found by means of a suitable coefficient criterion Q_c and an appropriate structure criterion Q_s.

Fig.1: The common task of structure modeling

3. Typical effects during object simulation with Non-Gaussian disturbances as symptoms of condition of uncertainty

Often it is impossible to exclude the influence of the following three kinds of disturbances with extreme Non-Gaussian distribution.
It has to be considered that the mathematical expectation of Gaussian disturbances must be strong $M(\delta)=0$ or, in a broader sense $M(\delta)$ must be invariant with respect to the argument. From this fact the independence of the expectation on the argument may be deduced: $M(\delta)=$ const.
The following three practically relevant cases violate, besides other Gauss-Markov-conditions, the above mentioned elementary conditions of Gaussian disturbances.

a) **Multiplicative-additive disturbance** may be characterized by

$$y_{\delta i} = y_{gi} * (1 + \Theta_i * p) = y_{gi} + y_{gi} * \Theta_i * p \quad (2)$$

where
Θ is an uniformly distributed random number between -1 and +1,
p s the disturbance-signal-ratio in percent and
$i = 1,\dots,N$ is the index of the data point.

This kind of disturbance implies $M(\delta)$=const. and is a function of y_{gi}

$$\delta = f(y_g) = f(a_0+a_1x_1+\dots+a_Mx_m{}^k) = \Theta * p * (a_0+a_1x_1+\dots+a_Mx_m{}^k) \quad (3)$$

This linear mapping of y_g onto δ leads to the violation of a further Gauss-Markov-Condition. In case of $M(\delta_i{}^T\delta_i) = 0$ we get $\mathrm{cov}(\sigma) <> \sigma^2 * I$, where σ^2 is the dispersion of the disturbance. Practically this causes the algorithm of structure finding to accumulate additional "unwelcome" variables x_{m+1} to x_{m+u} into the function $\delta = f(y_g)$; this leads to additional numerical problems due to their linear dependence on $x_1 - x_m$.

b) **Outlier-disturbances** are determined by

$$y_{\delta i} = y_g + \delta_i \quad (4)$$

where $\delta i = \varepsilon_i + \gamma_i$

with ε_i follows a Gaussian disturbance,
$\gamma_i = 0$ for all $i <> j$ and
j is the index of data points which do not match this Gaussian disturbance (outlier).

Under such conditions the fact that N $\not\to \infty$ plays an important role because Least Squares estimations looses their optimality (TUKEY 1975).

c) Disturbance due to "**nonobservable variables**":

$$y_{\delta i} = y_g + \delta_i \qquad (5)$$

where $\delta_i = a_{M+1}x_{m+1} + ... + a_{M+v}x_{m+v} + \varepsilon_i$

and x_{m+1} to x_{m+v} are nonobservable or unknown but relevant variables of influence; this occurs typically in the investigation of large-scale systems.

Apart from these cases uncertainty can appear if the above mathematical means are used for the **modeling of system dynamics**, what is described mathematically by equation (2). As follows from equation (2), here the case is considered that the disturbance impacts the system's input.

Since the system's output signal y is disturbed, the input values $x_{q+1}, ..., x_{m-1}, x_m$ of the **dynamic model** are also disturbed. This disturbance of model inputs leads to the violation of the Gauss-Markov-conditions.
Even if the disturbance at the **system output** is a Gaussian one and even if the amount of data points is sufficient, the Gauss-Markov-conditions will be violated.

Obviously, in all these cases the RSS (**Residual Sum of Squares**) as structure criterion may not be the best.

During the investigation of non-standard disturbances on the basis of simulation experiments the following effects could be detected several times (LANGE 1989):

1. If non-standard disturbances influenced the output of the system, or if in case of Gaussian disturbances the amount of measured data was quite limited, a shift of the RSS functional and its optimum away from the origin was detected (see Fig. 2). This RSS shift corresponds to the growth of the vector norm of the estimated parameters.

   In connection with this phenomenon the investigation of disturbances of the kind "nonobservable variable" was the most important one because this kind of disturbance corresponds to the structure of partial models which are used during the automatic selection of model structure. These partial models are strongly simplified and biased. That concerns, for instance, the model's set-up on every layer of the multilayered GMDH algorithm, which has the form:

$$z_i = A_1 v_i + A_2 v_i v_j \qquad (6)$$

where

vi and v_j are preliminary variables and

z_i is a preliminary output of the multilayered GMDH algorithm.

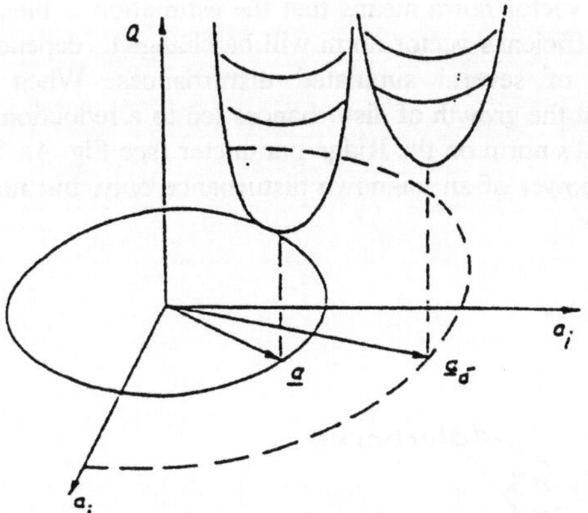

$Q\_ = Q\_(\underline{a})$ - is the LSM-functional of the simulated system without disturbances
$\underline{a}$ - vector of coefficients
$Q_\delta = Qd(\underline{a}_\delta)$ - LSM-functional of the simulated system under disturbance

Fig. 2: Shift of the RSS functional

2. In case of the above effect (see item 1) the tendency could be detected that the growth's curvature of the vector norm of estimated parameters depends on the complexity of the model (the higher the complexity, the steeper the curvature of the growth of the vector norm, see Fig. 3).

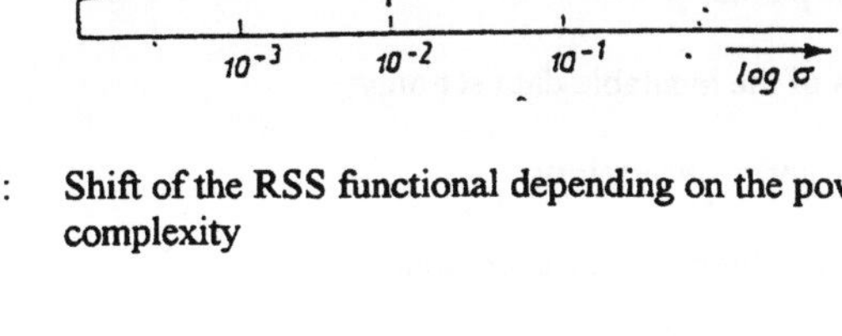

Fig. 3: Shift of the RSS functional depending on the power of disturbance and on model complexity

3. The growth of the coefficient's vector norm means that the estimation is biased. This fact led to the idea to observe how the coefficient's vector norm will be changed depending on the value of the Ridge-parameter in case of several simulated disturbances. When carrying out this investigation it was detected that the growth of disturbances led to a reduction of the curvature of the dependence of the coefficient's norm on the Ridge-parameter (see Fig. 4). This effect could be used for the estimation of the power of an unknown disturbance only, but not for the automatic selection of the model's structure.

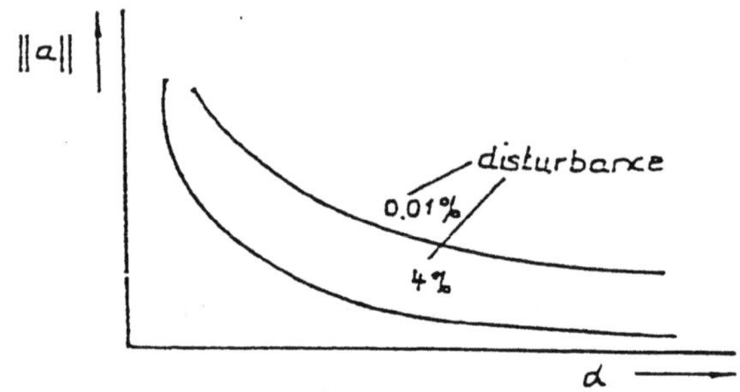

Fig. 4.: Dependence of the coefficient norm $\| \underline{\mathbf{a}} \|$ on the Ridge-parameter α

These detected effects were the starting point for asking the following questions:

1. How is it possible to use the **data set** to decide whether Least Squares is applicable for a given data set ?"

2. If the answer is "No", then what should be used instead of Least Squares ?

3. How should the quality of a model be assessed numerically in the above case ?

To solve the above problems several new criteria were proposed and investigated by the author. In addition, some known criteria were modified and investigated.

These investigations can be divided into three points:

1. the identification of uncertainty by means of the available data set only;

2. the choice of a suitable structure criterion under uncertainty;

3. the evaluation of the quality of the structure criteria under uncertainty.

4. The identification of the existence of uncertainty by means of the available data set

To identify the existence of uncertainty it is proposed to use the common Bootstrap method (EFRON 1983). Several simulation experiments showed that uncertainty leads to a bad convergence of parameters (growth of the coefficient norm). The Bootstrap characteristics can therefore be used for the visual recognition of uncertainty (see Fig. 5).

The Bootstrap characteristics, shown in Fig. 5, have been built for coefficients of models with uniform structure:

$$y = a_0 + a_1x + a_2x^2 + a_3x^3$$

Information about uncertainty can be obtained from the Bootstrap characteristics by the following rule: The more shallow the Bootstrap figure, the higher the rank of uncertainty.
So, it can be recognized from the shown Bootstrap characteristics that in opposite to the first example in case of the second example the identification has been executed under uncertainty, because the Bootstrap figure is too flat.

The visual recognition by means of Bootstrap characteristics is usable for preliminary investigations by dialogue (for example, in form of Bootstrap characteristics for the estimated coefficients on the first layer of GMDH algorithms), but this cannot be recommended for the automatic selection of model structure.

However, the Bootstrap estimation of the mean square deviation of the coefficients of models on the first layer of GMDH algorithms is a degree of uncertainty and can be used for the automatic selection of the model structure:

$$M[(\underline{a}_B - \underline{a})^T(\underline{a}_B - \underline{a})] = \frac{\sum_{j=1}^{NB} [(\underline{a}_{Bj} - \underline{a})^T(\underline{a}_{Bj} - \underline{a})]}{NB} \tag{7}$$

where

- $\underline{a}$ is the coefficient vector calculated on the original data set;
- $\underline{a}_B$ is the Bootstrap coefficient;
- $\underline{a}_{Bj}$ is the coefficient vector calculated on the j-th Bootstrap data set;
- M is the mathematical expectation;
- NB is the amount of Bootstrap's data sets.

This criterion was named **"CIU - Criterion for Indication of Uncertainty"** (LANGE 1987).

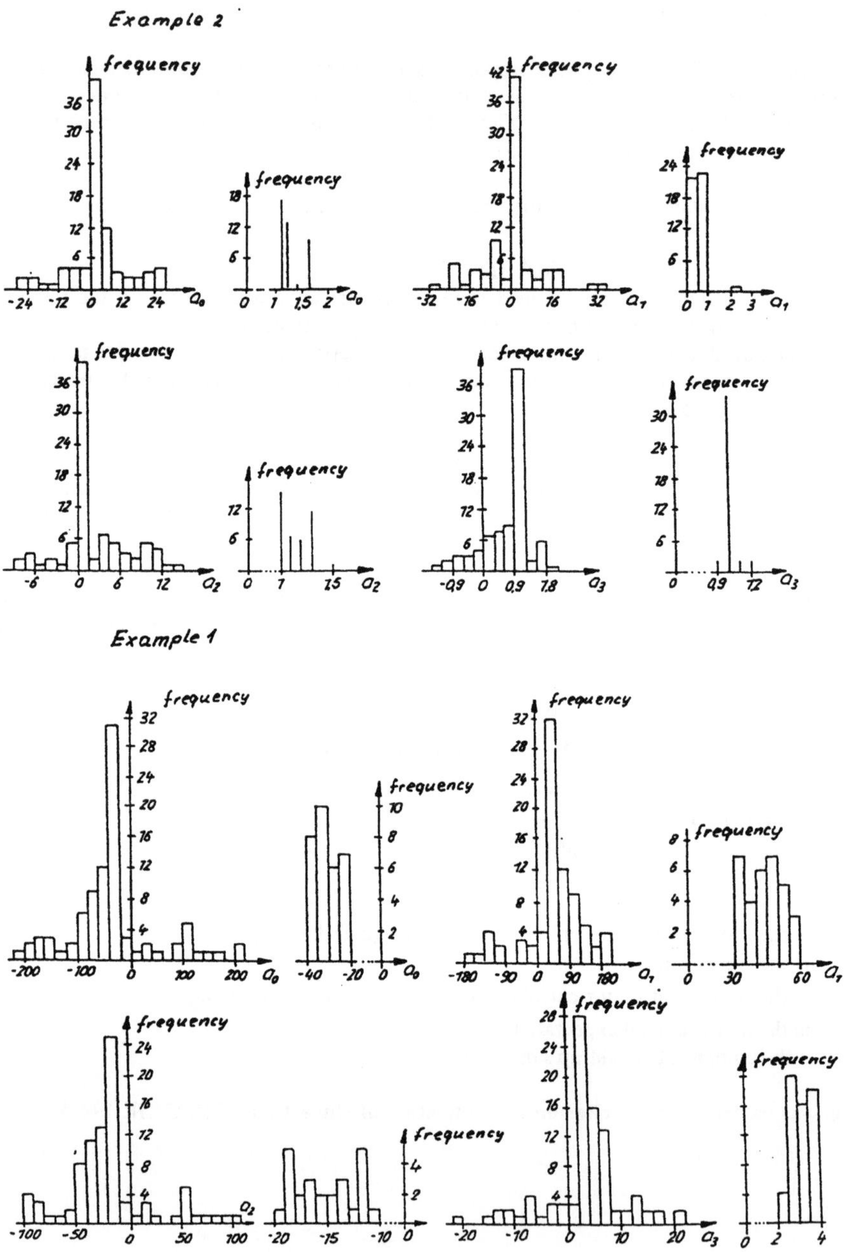

Fig. 5: Bootstrap characteristics for the visual recognition of uncertainty

5. The choice of the structure criterion

As an answer to the question, which structure criterion has to be chosen if non-standard disturbances influence the system output, two alternatives are possible. These alternatives are based on two kinds of model comparison (LANGE 1990):

i) the analytic comparison of the models,

ii) the heuristic comparison of the models.

To define these kinds of model comparison it is necessary to introduce the term "nearness of models":

"Nearness of models" M_I, M_{II} **is characterized by**

$$\rho_{R1}(M_I, M_{II}) =< \varepsilon\ (l, \delta) \quad (8)$$

in which

ρ_{R1}	is the metric distance,
l	is the amount of data in the data set,
δ	is the disturbance at the output.

If $\rho_{R1}(M_I, M_{II})$ **is small, the models are near to each other. The term "nearness of models" is a substitute for the convergence of the models when the amount of data is limited:**

$$\lim_{l \to \infty} \rho_{R1}(M_I, M_{II}) = 0 \quad (9)$$

i) The analytic approach to compare models:

One of the models will be named the **true** one and the other the **real** one. The true model should provide output values which are identical with the output values of the investigated system if there are no disturbances. In consequence, the true model must be an unbiased one. Naturally, in this case the aim of modeling is to find a real model which is as close to the true model as possible. The real model is searched under the condition that the disturbance was considered as $\delta = 0$ concerning the true model. Thus the quality of modeling depends on the power of the real disturbance. In addition, this quality also depends on the amount of available data.

The distance between models can be expressed mathematically, if certain properties of the disturbance and of the model are known.

ii) The heuristic approach to compare models:

In this case, models M_I and M_{II} will be considered as models with the same structure but estimated on different data sets.

These two approaches to compare the model quality correspond to the two different possibilities of comparing everyday models in reality (as a matter of principle), which are often met in life.

The first version (that means the analytic comparison of models) provides the comparison with an ideal model; this requires an a_priori knowledge of properties of the ideal model under real conditions.

The second version (that means the heuristic comparison of models) corresponds to verifying the stability of the real model in case of certain changes in the conditions. As well, this resistance will be the criterion of quality. This version demands a minimum of information only.

It is possible to show that the structure criteria, based on the analytic kind of model comparison (like most of structure criteria), consist of two contradictory functionals or of one main functional,in which the compromise between these contradictory functionals is already taken into consideration. This statement is explained as follows:

The distance between the ideal and the real model (that means the criterion itself) can be written as mean square error:

$$M[(\underline{\hat{a}} - \underline{a})^T(\underline{\hat{a}} - \underline{a})] = [D(\underline{\hat{a}}) + R^2(\underline{\hat{a}})] \longrightarrow \min \qquad (10)$$

where

- $\underline{\hat{a}}$ is the coefficient vector of the ideal model,
- $\underline{\hat{a}}$ is the coefficient vector of the real model,
- M[.] is the mathematical expectation,
- $D(\underline{\hat{a}})$ is the dispersion of the vector,
- $R(\underline{\hat{a}})$ is the bias (shift) of the vector.

In this case the dispersion $D(\underline{\hat{a}})$ and the shift $R(\underline{\hat{a}})$ are two contradictory criteria and the minimization of the mean square error will be a multi-criteria task (a task of Pareto-optimization (ESTER, 1987)) to be solved.

The compromise between two criteria can be expressed by different ways depending on the concrete reasons of uncertainty:

- amount of data points **l**,
- relation between the amount of data **l** and the amount of parameters of the model **h**,
- regularity of the input matrix, etc.

Depending on the amount of data points **l**, on the amount of model parameters **h** and on the significance level η, this compromise is often hidden in a correction term or a correction coefficient for the RSS (which is connected with the dispersion), for instance, in the criteria, proposed by AKAIKE (1981) and VAPNIK (1984).

In many criteria the main functional is deformed, for example, in functionals for robust estimation or in **gnostic functionals**, proposed by KOVANIC (1984). In these cases the multi-criteria task is often not visible.

To make this multi-criteria problem clear, on the one hand, and to explain, on the other hand, such terms like robustness and sensitivity of the criterion with the aim to classify the great variety of criteria by fields of application, the terms

- "robustness (or sensitivity) in sector I (central sector)",
- "robustness (or sensitivity) in sector II (marginal sector)"

should be introduced (see Fig. 6 and 7).

In Fig. 6b those criteria $Q(\hat{a})$ are demonstrated which are robust (curve (a)) or sensitive (curve (b)) in **the marginal sector II**. The **flat functional (a)** belongs to a **robust criterion** and the **steep functional (b)** to a **sensitive criterion**. In Fig. 6a the empirical distributions $P_{em}(\hat{a})$ of estimated coefficients are shown. Here, in the marginal sector II the area under distribution (a) is smaller than the area under distribution (b).

In Fig. 7 an example for the robustness (curve (a)) and the sensitivity (curve (b)) of criteria **in the central sector I** is shown. Here, the **steep curves (a)** correspond to a **robust criterion** and the **flat curves (b)** - to a **sensitive criterion**.
Robustness in the marginal sector II means that the functional (criterion) **must be flat** in this sector. On the other hand, robustness in the central sector I means that the functional **must not be flat** in the central sector.

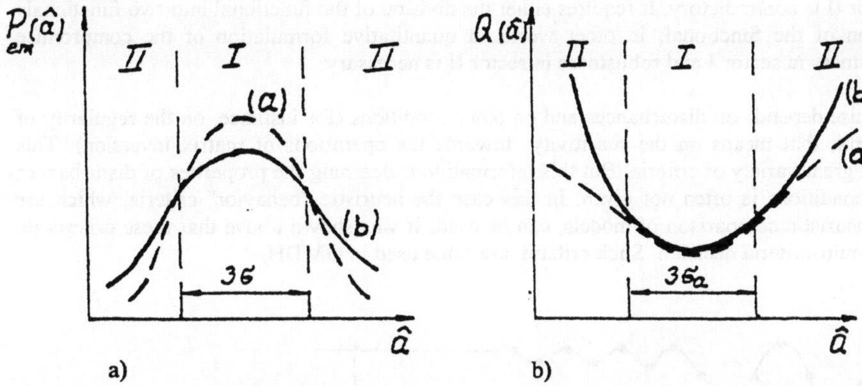

Fig. 6: Example for the robustness (a) and sensitivity (b) of criteria in the marginal sector II

It seems to be better to discuss the robustness in the marginal sector II as illustrated in Fig. 6a:

"Robustness in the marginal sector II" corresponds to the normal and widespread term robustness" (robust criterion, robust estimation (HAMPEL, 1986), which means robustness against disturbances shifting the mean value of the estimation, such as outlier disturbances.

The "robustness and the sensitivity in the central sector I" is illustrated in Fig. 7b:

If the functional in the central sector I is flat, every numerical coincidence can move the searched coefficients. In Fig. 7a this effect is expressed by a flat distribution in the central sector I; this corresponds to a great dispersion.

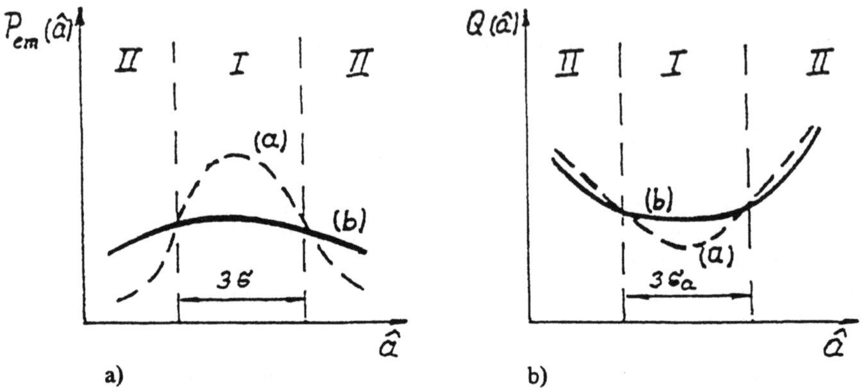

Fig. 7: Example for the robustness (a) and sensitivity (b) of criteria in the central sector I

The demand of having robustness of a square functional both in the central sector I and in the marginal sector II is contradictory. It requires either the division of the functional into two functionals or a correction of the functional. In other words, a quantitative formulation of the compromise between robustness in sector I and robustness in sector II is necessary.

This compromise depends on disturbances and on other conditions (for instance, on the regularity of the input matrix, that means on the sensitivity towards the operations of matrix inversion). This brings about a great variety of criteria. But this information concerning the properties of disturbances and of other conditions is often not given. In this case the heuristic "behavior" criteria, which are based on the heuristic comparison of models, can be used. It was shown above that these criteria do not require a multi-criteria decision. Such criteria are often used in GMDH.

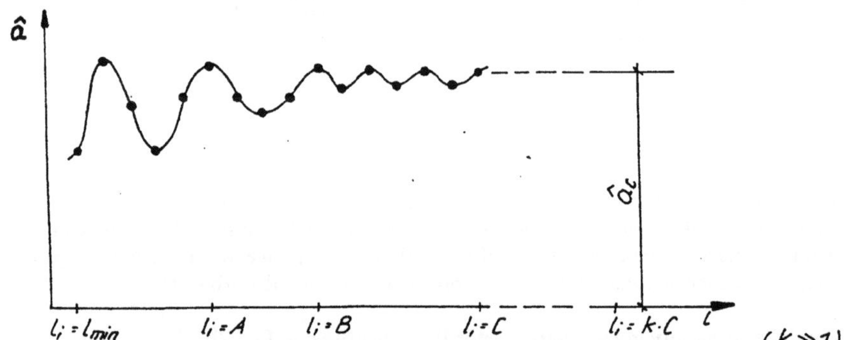

$\hat{a}_c$ is the regression coefficient (estimated by $l_i = kC$ observations),

l_i is the amount of data points in the i-th sample,

l_{min} is the minimum amount of data points in the sample ($l_{min} = > M + 1$; $k >> 1$).

Fig.8: Behavior of estimated coefficient values in dependence on the amount of observations

Simulations showed that in most of observed cases the dependence of the estimated coefficient values and also of the PSS-values on the amount of experimental information is similar to a transient process (see Fig. 8 and 9).

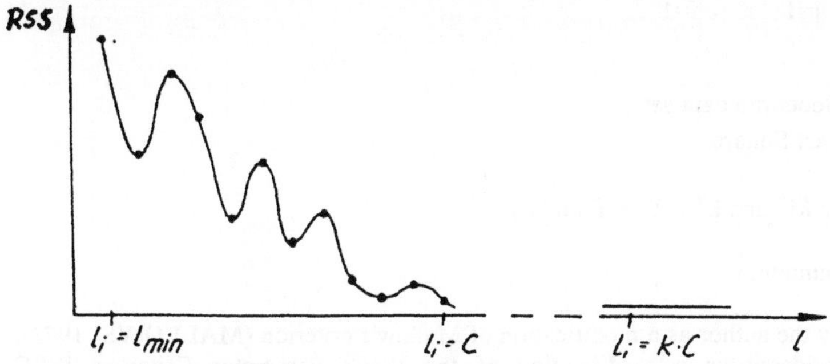

Fig.9: Behavior of RSS-values in dependence on the amount of data points

If a constant amount of observations from different samples are used, different curves of the same kind as shown in Fig. 8 and 9 will be obtained.

This effect allows to interpret the heuristic self-organizing "behavior"-criteria as follows:

In principle, the "behavior"-criteria use the information from different data samples, which have, however, the same size in most cases. That means that for every model structure under consideration a quantity will be estimated which provides a rough information about the empirical distribution of the values of **a**, RSS etc., which were estimated on the basis of different samples of the same size, built from the original data. The particular structure will be chosen which possesses the highest resistance against a data change for the given sample size.

The following new structure criteria were proposed or modified and investigated:

- the **Gnostic Structure Criteria GSC**:

$$GSC = \left\{ \left[\sum_{i=1}^{N} (y_i - f(x,a^*))^2 \right] / \left[\sum_{i=1}^{N} (y_i - f(x,a^*))^{-2} \right] \right\}^{1/4} \tag{11}$$

Here, the optimal value a^* will be estimated by Least Square, but the optimal values of structure M^* and k^* by

GSC —> minimum.

This criterion was derived from the Gnostic Theory (KOVANIC 1984) for structure selection.

- the **Criterion of Nonlinear Sum**:

$$(1/M) * \sum_{j=1}^{m} a_j^2 < (2/M) \sum_{j=1}^{M} (1/NB) \sum_{i=1}^{NB} (a_i^* - a_{ij})^2 = CNS \tag{12}$$

where

NB is the amount of Bootstrap data sets,

a^* is estimated by Least Square.

The optimal structure values M^* and k^* will be found by

CNS —> minimum.

This criterion is proposed by the author as a modification of Mallow's criterion (MALLOWS, 1973), but it could also be considered as a modification of the Weak Parameter Criterion WPC (BROERSEN 1985).

- the **Disturbance Criterion DC2**:

The models will be influenced by continual synthetic disturbance and their behavior will be observed as follows:

First, the discrete values of a^* will be estimated by

$$Q_j = \sum_{i=1}^{N} \frac{[y_i - f(x,a_j^*)]^2}{\{[1 - x_i^T(X^TX + \alpha I)^{-1} * x_i]^2\}^j} \longrightarrow \min \tag{13}$$

where

α is the Ridge-parameter.

Then, the values of

$$n_j = \sum_{i=1}^{N} \frac{[y_i - f(x,a_j^*)]^2}{[1 - x_i^T(X^TX + \alpha I)^{-1} * x_i]^2} \tag{14}$$

will be calculated for $j = 0$ and $j = 1$.

At last, the optimal structure values M^* and k^* will be found by

$$DC2 = \frac{n_0}{n_1} \longrightarrow \text{minimum.} \tag{15}$$

All these criteria were investigated by simulation experiments. For comparison, RSS and PSSMOD, as modified by Tamura and Kondo, were used. The results of these investigations are described in LANGE (1992).

The latest criterion to be proposed and being investigated now is

- the **Local Data Uncertainty Criterion LDUC**:

$$LDUC = tr \left| \frac{\sum_{i}^{B} (\hat{y}_{Bi} - y_{Bi})^2}{B - h} (X_B^T X_B)^{-1} \right| + \sum_{j=1}^{h} (\hat{a}_{Aj} - \hat{a}_{Bj})^2 \tag{16}$$

where

$A+B=l$ is the sum of two data sets of different size (the primary sample of l data points has been subdivided into two parts A and B);

$\hat{y}_B$ is the calculated output of model A (which was found by the data sample A) but the input matrix X_B is used for the model output;

y_B is the output measured for the data sample B, if the input matrix X_B is given;

h is the amount of coefficients of the model;

$\hat{a}_A$, $\hat{a}_B$ are coefficients of the same structure but estimated by different data samples A and B.

This criterion appears similar to mean square error, but in fact it is not connected with the analytic comparison of models. The criterion describes both the resistance of the estimated output and the resistance of the estimated coefficients concerning data changes for every model structure under consideration and for a fixed limited data set. The term "limitedó means that the size of the data sample corresponds to the part in the Fig. 8 and 9, where the transient process is not finished yet ($l_i < C$).

On the other hand, as opposite to the criterion of regularity and to the minimum-of-bias criterion (IVAKHNENKO 1987), the LDUC provides the opportunity to have both of the above criteria in the form of one criterion only, but without the disadvantages of the combined criteria (IVAKHNENKO 1971, 1976).

6. The evaluation of quality of the structure criterion used in case of uncertainty

First, to evaluate the quality of decision rules, the Tchebyshev inequality (VAPNIK, 1984) can be used. But since the true dispersion of the criterion is unknown and the amount of available data is quite limited, the values of the criterion can be considered as a random variable. Thus the Tchebyshev inequality can be transformed into the following:

$$M(\hat{Q}_l) =< \hat{Q}_l + \{[D(Q)]/\eta\}^{1/2} \qquad (17)$$

where

$\hat{Q}_l$ is the criterion value, calculated by the given data table.

Here, every criterion value, calculated by the i-th Bootstrap-table (i = 1, 2, ... , NB) consisting of l data, has to be considered as a realization of the random variable Q_l on the basis of the i-th data sample (refer to EFRON, 1983). By the help of NB realizations of Q_l the dispersion of the estimation was calculated.

Second, the following criterion can be used both for the evaluation of the quality of decision and the estimation of uncertainty:

$$\mathbf{CDR} = \left| \frac{Q_{ub} - Q_l}{Q_l} \right| \qquad (18)$$

where

Q_l is the value of the structure criterion, calculated **on the original data set;**

Qub is the numerical limiting value of all values of the structure criterion, calculated **on NB Bootstrap tables.**

This criterion is named **Criterion of Data Reliability**.

On the basis of the results of simulation investigations it can be assumed for structure modeling that uncertainty exists for a given data set, if the following inequality is met:

$$\mathbf{CDR} = \left| \frac{Q_{ub} - Q_l}{Q_l} \right| => 3$$

If the CDR takes a double figure value the uncertainty is so great that no structure criterion leading to useful results can be proposed at the current state of investigations.

7. Short list of results

The results of investigations an be summarized as follows:

1. **The Criterion for Indication of Uncertainty CIU** is a highly reliable indicator of uncertainty.

2. The **Criterion of Nonlinear Sum CNS** is well suited for finding wrong regressors but only in the presence of outlier-disturbances.

3. The **Gnostic Structure Criterion GSC** shows a good behavior if the disturbance influencing the original data set is only a multiplicative-additive or of the outlier type.

4. The nature of disturbance is rarely known in advance; therefore, the properties of the structure criterion should not depend on the kind of disturbance. The **Disturbance Criterion DC2** is a criterion that satisfies this request.

5. The **Criterion of Data Reliability CDR** can be used both for the estimation of uncertainty of data and the evaluation of the quality of structure criteria provided a data sample is given.

References:

Akaike, H. (1981). Raswitije statistitscheskich metodow, (in Russian), *Trends and Progress in System Identification*, (eds. P. Eykhoff), Pergamon Press, Oxford-New York-Toronto-Sydney-Paris-Frankfurt.

Broersen, P.M.T. (1985). Selecting the Order of Aotoregressive Models from Small Samples, *IEEE Transaction on Acoustics, Speech, and Signal Processing*, Vol. ASSP-33, No.4, 874-879.

Efron, B., Diaconis, P. (1983). Statistitscheskije metody s inversnym ispol'zovanijem EVM, (russian translation from the original), *V Mire Nauki*, No. 7, 60-73.

Ester, J. (1987). *Systemanalyse and mehrkriterielle Entscheidung*, (in German), Verlag Technik, Berlin.

Hampel, F.R., et. al. (1986). *Robust Statistics. The Approach Based on Influence Functions*, John Wiley & Sons, New York.

Ivakhnenko, A.G. (1971). *Sisteme evristitscheskoj samoorganizacii v technitscheskoj kibernetiki*, (in Russian), Technika, Kiev.

Ivakhnenko, A.G. (1975). *Dolgosrotchnoje prognosirovanije i upravlenije sloshnymi sistemami*, (in Russian) Technika, Kiev.

Ivakhnenko, A.G. et. al. (1976). *Prinjatie reschenii na osnove samoorganizacii*, (in Russian), Sov.Radio, Moscow.

Ivakhnenko, A.G. and Juratschkovski, J.P. (1987). *Modelirovanije sloshnych sistem po eksperimental'nym dannym*, (in Russian), Radio i Svaz, Moscow.

Kovanic, P. (1984). Gnostical theory of individual data, *Control and Information Theory*, Vol. 13, No.4, 259 - 274, Publishing house of the hungarian academy of science, Budapest.

Lange,T., Gnatowski, K., Umbreit, S. (1987). Zur Wahl geeigneter Kriterien fuer die Struktursuche bei Modellierungsaufgaben, *Wiss.Z. TH Ilmenau*, Vol. 33, No.2, 67 -78.

Lange, T. (1989). Multi-criteria Decision Approach to Structural Modelling with Uncertainty Conditions, Syst.Anal.Model.Simul., Vol.6, No.2, 147 - 154.

Lange T. (1990). Strukturkriterien für kurze Stichproben, *Wiss. Z. TH Ilmenau*, Vol.36, No.3, 111-122.

Lange, T. (1992). *Strukturselektionskriterien für die Modellbildung bei Existenz von Unsicherheit*, (in German), Habilitationsschrift (theses / in preparation), Berlin.

Tukey, J.W. (1975). Instead of Gauss-Markov least squares, what? *Appl.Statist.*, 351 - 372

Mallows, C.L. (1973). Some comments on C_p. *Technometrics*, No.15, 661-675.

Vapnik, V.N. (1984). *Algoritmy i programmy vosstanovlenija zavisimosti*, (in Russian), Nauka, Moscow.

THE USE OF THE KULLBACK-LEIBLER DISTANCE FOR LEARNING IN NEURAL BINARY CLASSIFIERS

D. PADOS, P. PAPANTONI-KAZAKOS,
D. KAZAKOS, and A. KOYIANTIS
Department of Electrical Engineering
Thornton Hall
University of Virginia
Charlottesville, VA 22903

1. Introduction

Recent discoveries by cognitive scientists have resulted in renewed interest in neural network structures, which perform elementary functions and "learn" via interaction with their environment. Given a neural architecture and a function per element, this latter learning corresponds to *supervised learning,* in neural network and pattern recognition terminology, and corresponds to the sequential adaptation or estimation of pertinent parameters via training data. When neural networks operate in stochastically described environments, supervised learning becomes a statistical sequential estimation problem, dealt with by stochastic approximation methods. There is rich literature in such methods represented by the works of Adhelhamid (1973), Beran (1976), Blum (1954), Fabian (1968), Fisher (1922), Gerencser (1986), Kashyap et al (1966, 1970), Kiefer et al (1952), Kushner (1987), Kushner et al (1978), Ljung (1977), Ljung et al (1983), Robbins et al (1951), and Young et al (1972). In parallel, robust parameter estimation techniques which can be extended to the stochastic approximation domain can be found in the works by Beran (1977, 1978), Bickel (1975), Huber (1973, 1981), and Kazakos et al (1989).

In the neural networks literature, supervised learning has been basically limited to techniques arising from the Robbins/Monro (1951) method and its extensions, with performance criterion the least squares error. The representative works on the subject are those by Barron et al (1977, 1988), Elman et al (1988), Gorman et al (1988), Minsky et al (1969), Rosenblatt (1957), Rosenblatt et al (1986), Werbos (1974), White (1989), Widrow (1962), and Widrow et al (1960, 1987). We note that Werbos' (1974) extension is known as the back propagation learning algorithm.

Our main argument in this paper is that a supervised learning algorithm should arise from a performance criterion which is pertinent to the network objective. When the latter objective is hypothesis testing, the least squares error is an irrelevant criterion. The appropriate performance criteria are then either the Chernoff bound (Chernoff (1952)) or the Kullback-Leibler distance, also known as the relative information measure. The Chernoff bound has been used successfully in the evaluation of multihypothesis performance by Kazakos (1990, 1991, 1992), while the value of the Kullback-Leibler distance in model selection (Akaike (1974) and Rissianen (1978)), in recognition of change in stochastic processes (Bansal et al (1986, 1989)), and generally in

H. Bozdogan (ed.), Proceedings of the First US/Japan Conference on the Frontiers of Statistical Modeling: An Informational Approach, 267–292.

hypothesis testing (Blahut (1974), Fisher (1934), Kazakos et al (1989), Wald (1943)) is well established.

Pursuing our argument, we consider here network structures whose objective is binary hypothesis testing. We assume then that the performance measure at the operational stage, for the recognition of the acting hypothesis, is the Neyman-Pearson criterion (see Halford et al (1991), Batalama et al (1991), and Papantoni-Kazakos et al (1991)), where the latter maximizes the information provided by the network regarding the acting hypothesis (see Papantoni-Kazakos et al (1991)). In this case, we use the Kullback-Leibler distance between the observed and the desired Type-1 or Type-2 error probabilities, as the performance criterion, to derive adaptive learning algorithms for pertinent network parameters. Our algorithm takes advantage of the analytical and convex characteristics of the Kullback-Leibler distance and of the reduction of free parameters that results from constructive and objective-oriented neural network designs (see Halford et al (1991), Batalama et al (1991), and Papantoni-Kazakos et al (1991)), to accelerate convergence to a global optimum, under some general conditions.

2. Preliminaries

Neural networks are structures designed to perform parallel computation operations, for the satisfaction of a prespecified objective. While it is desirable that a neural network be "robust" in terms of satisfying a number of objectives via appropriate adaptation of some of its operational parameters, the main character of the neural operations is actually dictated by the objective and the pertinent performance criteria. Alternatively, objective-oriented network designs can lead to simple but powerful structures with "robust" per element operations and to a dramatic reduction of the number of "free parameters" involved in the *supervised learning* stage. For example, let us consider a Hopfield network structure (Hopfield (1982)) dedicated to binary hypothesis testing (see Figure 1), where $\{x_{in}\}$ signify external data and $\{u_{in}\}$ signify element outputs, where elements receive a single scalar datum per time instant, and where x_{in} and u_{in} represent the incoming datum and the output for element i at the nth time iteration, respectively. The outputs $\{u_{in}\}$ are then binary, 0 or 1, numbers, and each element can provide the hypothesis testing decision (0 for hypothesis H_0 and 1 for hypothesis H_1) at any time instant. In Halford et al (1991), objective-oriented design of the Hopfield network in Figure 1 was performed, for binary hypothesis testing and for stationary and memoryless hypotheses. As a result, optimal neural operations for some general classes of hypotheses and strong relationships between network parameters and performance characteristics were discovered. In particular, for $U(x) \overset{\Delta}{=} \begin{cases} 1 ; x \geq 0 \\ 0 ; x < 0 \end{cases}$, it was then found that:

$$u_{i1} = U(g(x_{i1}) - \lambda_1) \; ; \; \forall i \tag{2.1}$$

$$u_{i2} = U(w \sum_i u_{i1} + g(x_{i2}) - \lambda_2) \; ; \; \forall i$$

When the log likelihood ratio of the two hypotheses is a strictly monotone function, then the operation g(x) in (2.1) can be any strictly monotone and bounded function, which

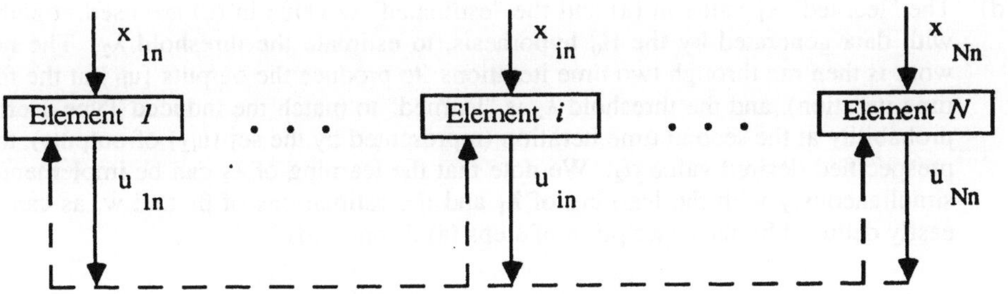

Figure 1

corresponds to a "robust" network design. The "weight" w in (2.1) is not an arbitrary constant, as conventionally assumed in the neural networks literature: instead, it is a function of the Type-1 and Type-2 error probabilities attained by each element at the first time iteration. Specifically, if α_n and $1-\beta_n$ denote the Type-1 and the Type-2 error probabilties attained by each element at the nth time iteration, respectively, then,

$$w = \log \frac{\beta_1(1-\alpha_1)}{\alpha_1(1-\beta_1)} \tag{2.2}$$

Thus, for the first two time iterations of a Hopfield network deployed for binary hypothesis testing, only the two thresholds, λ_1 and λ_2, need to be estimated via ***supervised learning*** techniques, while, (as found in Halford et al (1991), two time iterations suffice for superior performance, for relatively small number of network elements). In a Neyman-Pearson framework, this learning process is implemented in the following steps:

(a) Training data from the H_0 hypothesis are generated. The threshold λ_1 is "learned" to match the induced Type-1 error probability at the first time iteration (represented by the set $\{u_{i1}\}$ of outputs), to a prespecified desired value α_1.

(b) The "learned" λ_1 value in (a) is used and data from the H_1 hypothesis are generated. The outputs $\{u_{i1}\}$ are then used to "estimate" the Type-2 error probability $1-\beta_1$ (at the first time iteration), where $\hat{\beta}_1 = (kN)^{-1}\sum_{i=1}^{kN} u_{i1}$. We note that step (b) can be incorporated in step (a), if we alternate between feeding the network with data from hypotheses H_0 versus H_1. Then, at each algorithmic iteration, the value of the threshold λ_1 is first adjusted (using data from H_0), and then the value of β_1 is subsequently updated (using the adjusted λ_1 value and data from H_1).

(c) The "learned" in (b) value β_1 is used (together with the prespecified desired value α_1), to compute the estimated w value in (2).

(d) The "learned" λ_1 value in (a) and the "estimated" w value in (c) are used, together with data generated by the H_0 hypothesis, to estimate the threshold λ_2. The network is then ran through two time iterations (to produce the outputs $\{u_{i1}\}$ at the first time iteration), and the threshold λ_2 is "learned" to match the induced Type-1 error probability at the second time iteration (represented by the set $\{u_{i2}\}$ of outputs), to a prespecified desired value α_2. We note that the learning of λ_2 can be implemented simultaneously with the learning of λ_1 and the estimations of β_1 and w, as can be easily deduced by the description of steps (a) through (d).

From the steps of the learning process described above, it is clear that the only crucial issue is the development of a learning algorithm for either one of the two thresholds, λ_1 and λ_2. Indeed, identical algorithms should be clearly used for the evaluation of both threholds. Thus, without lack in generality and avoiding unnecessary complexities, it suffices to concentrate on the design of a recursive learning algorithm for the threshold λ_1. Selecting the Neyman-Pearson operational model, the initial important problem is the criterion by which the "matching" of the induced Type-1 error probability to a desired value α will be measured, where this criterion should be pertinent to the network objective, namely binary hypothesis testing.

In the statistics literature, it is well established that the appropriate performance criteria in hypothesis testing are either the Chernoff distance or the Kullback-Leibler distance. The former is a strong distance representing the moment generating function of log likelihood ratios, while the latter can be viewed as the expected value of log likelihood ratios that are generated by stationary and memoryless hypotheses, it is weaker than the Chernoff distance, and represents the asymptotic performance of binary hypothesis testing schemes when the hypotheses generate i.i.d. data. In this paper, we will use the Kullback-Leibler distance as the "matching" criterion in the development of learning algorithms for parameters such as the threshold λ_1. Our approach, which is explained in Section 3 below, is a combination of the Newton's numerical method for deterministic convex functions and stochastic approximation techniques.

3. Approach To Learning - Heuristics

Consider a neural network, such as the Hopfield structure in the previous section, whose objective is binary hypothesis testing in a Neyman-Pearson sense. Let then the desired value α of the Type-1 error probability p be given, where p is controlled by a scalar parameter λ. Feeding the network with data from the H_0 hypothesis, whose statistics are unknown, and observing binary network outputs only, the objective is to recursively adapt the parameter λ, to match the induced Type-1 error probability $p=p(\lambda)$ with the desired value α. The "matching" criterion used will be the Kullback-Leibler distance, $K(\alpha,p)$, where,

$$K(\alpha,p) \stackrel{\Delta}{=} \alpha \log \frac{\alpha}{p} + (1-\alpha) \log \frac{1-\alpha}{1-p} \tag{3.1}$$

The Kullback-Leibler distance whose analytical and strict convexity properties are well-established, attains its global minimum (equal to zero) at $p=\alpha$. The idea here is to use these properties, in conjunction with Newton's numerical method (see Hestens (1980)) and the recursive estimation techniques for probability masses induced by ergodic processes, to recursively and simultaneously update (or estimate) the λ and p values. Applied on the Kullback-Leibler function in (3.1), Newton's method is expected to accelerate the convergence of p to α. We present the logic in our approach, below.

Consider the Kullback-Leibler function in (3.1), where p is a function of λ, say $p = h(\lambda)$. Let us assume that $h(\lambda)$ is an analytic function, and let then $\dot{h}(\lambda)$ and $\ddot{h}(\lambda)$ denote first and second order derivatives at λ, respectively. Let also $\dot{K}(\alpha,p)$ and $\ddot{K}(\alpha,p)$ denote first and second order derivatives of the Kullback-Leibler function in (3.1) with respect to p, respectively. The idea is to use the Kullback-Leibler function, to recursively estimate the parameter λ. Viewing the latter function as a function of λ, we then obtain,

$$\frac{\partial}{\partial\lambda} K(\alpha,p) = \dot{h}(\lambda)\dot{K}(\alpha,p)$$

$$\frac{\partial^2}{\partial\lambda^2} K(\alpha,p) = \ddot{h}(\lambda)\dot{K}(\alpha,p) + [\dot{h}(\lambda)]^2\,\ddot{K}(\alpha,p)$$

$$\left[\frac{\partial}{\partial\lambda} K(\alpha,p)\right] \Big/ \left[\frac{\partial^2}{\partial\lambda^2} K(\alpha,p)\right] = \frac{1}{\dot{h}(\lambda)}\left\{\frac{\ddot{h}(\lambda)}{[\dot{h}(\lambda)]^2} + \frac{\ddot{K}(\alpha,p)}{\dot{K}(\alpha,p)}\right]^{-1} \tag{3.2}$$

$$= \frac{1}{\dot{h}(\lambda)}\left[\frac{\ddot{h}(\lambda)}{[\dot{h}(\lambda)]^2} + \frac{(p-\alpha)^2 + \alpha(1-\alpha)}{p(1-p)(p-\alpha)}\right]^{-1}$$

where,

$$\dot{K}(\alpha,p) = \frac{p-\alpha}{p(1-p)} \tag{3.3}$$

$$\ddot{K}(\alpha,p) = \frac{(p-\alpha)^2 + \alpha(1-\alpha)}{p^2(1-p)^2}$$

Then, $K(\alpha,p)$ possesses a unique extremum with respect to λ iff $\dot{h}(\lambda) \neq 0\ \forall\lambda$. This extremum, at $\lambda = \lambda_o$ such that $p(\lambda_o) = \alpha$, is a minimum as shown by $\frac{\partial^2}{\partial\lambda^2} K(\alpha,p)$ in (3.2). Therefore, if p were a known analytic function $h(\lambda)$ with $\dot{h}(\lambda) \neq 0\ \forall\lambda$, a numerical method, such as the one by Newton, could be applied for the computation of the parameter value (referring to λ) which attains the global minimum of the Kullback-Leibler function. Denoting then by $\hat{\lambda}_n$ the value of λ found at the n-th computational step, and by $\hat{p}_n$ the value $h(\hat{\lambda}_n)$, Newton's method gives the following recursion:

$$\hat{\lambda}_{n+1} = \hat{\lambda}_n - \frac{1}{\dot{h}(\hat{\lambda}_n)}\left[\frac{\ddot{h}(\hat{\lambda}_n)}{[\dot{h}(\hat{\lambda}_n)]^2} + \frac{(\hat{p}_n-\alpha)^2 + \alpha(1-\alpha)}{\hat{p}_n(1-\hat{p}_n)(\hat{p}_n-\alpha)}\right]^{-1} \tag{3.4}$$

In the problem at hand, the function $h(\lambda)$ is controlled by both the operations performed per neural element and the density function that generates the data. Since the latter is unknown, so is $h(\lambda)$. Thus, $\hat{p}_n$ is here actually estimated from binary observations $\{u_i\}$, and the function $h(\lambda)$ is subject to selection. First, let us either adopt an initial

$h(\lambda)$ choice such that $|\ddot{h}(\lambda)| << 1$ everywhere, or simply ignore the term $\ddot{h}[\dot{h}]^{-2}$ in (3.4). Then, recall that if the binary observations $\{u_i\}$ are generated by an ergodic process, then $\hat{p}_n$ is given by $n^{-1}\sum_{i=1}^{n} u_i$, which can be written recursively, as follows:

$$\hat{p}_{n+1} = \hat{p}_n + \frac{u_{n+1} - \hat{p}_n}{n+1} \tag{3.5}$$

The recursion in (3.5) corresponds to the case where the parameter λ remains unchanged, while the recursion in (3.4), (with or without the inclusion of the term $\ddot{h}[\dot{h}]^{-2}$), implies "confidence" in the estimate $\hat{p}_n$. Thus, to attain relative confidence in the $\{\hat{p}_n\}$ estimates, the recursive adaptations of λ should not be too frequent, while monotonically increasing data window sizes for each new adaptation of the λ value result in increasing accuracy of the $\{\hat{p}_n\}$ values. In parallel, in view of the changing λ values, the induced then bias in the estimate in (3.5) may be reduced, via the addition of an analogous to $|\hat{\lambda}_{n+1} - \hat{\lambda}_n|\dot{h}(\hat{\lambda}_n)$ correction term, at the time instants when the λ values are updated. From the above arguments, we are led to simultaneous $\hat{\lambda}_n$ and $\hat{p}_n$ adaptations, described by the algorithm below.

ALGORITHM

Initial values: Select an initial pair $(\hat{\lambda}_1, \hat{p}_1)$ of values, two positive integers N and m, and a function $h(\lambda)$. The selection of the latter is discussed in the next section.

Computational Steps:

(a) Given computed values $(\hat{\lambda}_n, \hat{p}_n)$, compute $\hat{\lambda}_{n+1}$ as follows:

$$\hat{\lambda}_{n+1} = \hat{\lambda}_n - \frac{1}{\dot{h}(\hat{\lambda}_n)} \frac{\hat{p}_n(1-\hat{p}_n)(\hat{p}_n-\alpha)}{(\hat{p}_n-\alpha)^2+\alpha(1-\alpha)} \;;\text{ if } \begin{array}{c} n=kN+\frac{m}{2}k(k-1) \\ k=1,2,\ldots \end{array} \tag{3.6}$$

$$\hat{\lambda}_{n+1} = \hat{\lambda}_n \;;\text{ otherwise}$$

Let $g(x)$ be such that $b_l \le g(x) \le b_u$. Then, for some small positive value δ, the $\hat{\lambda}_{n+1}$ value in (3.6) is "corrected" to $b_l + \delta$; if $\hat{\lambda}_{n+1} < b_l + \delta$, and to $b_u - \delta$; if $\hat{\lambda}_{n+1} > b_u - \delta$.

(b) Use $\hat{\lambda}_{n+1}$ and a newly generated random input from H_0 to produce the binary output u_{n+1}. Then, update $\hat{p}_n$ as follows:

$$\hat{p}_{n+1} = \hat{p}_n + \frac{u_{n+1}-\hat{p}_n}{n+1} - \frac{2kN+mk(k-1)}{2(k+1)N+m(k+1)k} \frac{\hat{p}_n(1-\hat{p}_n)(\hat{p}_n-\alpha)}{(\hat{p}_n-\alpha)^2+\alpha(1-\alpha)} \;;\text{ if } \begin{array}{c} n=kN+\frac{m}{2}k(k-1) \\ k=1,2,\ldots \end{array} \tag{3.7}$$

$$\hat{p}_{n+1} = \hat{p}_n + \frac{u_{n+1} - \hat{p}_n}{n+1} \text{ ; otherwise.}$$

For some small positive value ζ, the $\hat{p}_{n+1}$ value in (3.7) is "corrected" to ζ; if $\hat{p}_{n+1} < \zeta$, and to $1-\zeta$; if $\hat{p}_{n+1} > 1-\zeta$.

Remarks: (1) From the description of the simultaneous adaptations in the algorithm, it is not hard to see that, when random data from the hypothesis H_0 are generated at each step, then, the sequence $\{\hat{\lambda}_n, \hat{p}_{n-1}\}_n$ forms a two-dimensional vector Markov process. (2) In the algorithm, N data are generated by the initial value $\hat{\lambda}_1$, while the data window size increases by m, for each newly generated λ estimate: for the kth such estimate, N+m (k-1) data are generated. The last term in the first part of (3.7) corresponds to the correction term for the reduction of the bias, discussed earlier. (3) An important issue in the design of the algorithm is the selection of the function $h(\lambda)$ in (3.6), which will be examined in Section 4. (4) It is perhaps appropriate to mention here that the Kullback-Leibler performance criterion has been used for the estimation of the weights in Boltzman machines, (see Ackley et al (1985), and Amari et al (1992)), where in such machines the weights characterize completely the steady-state statistics of the machine-states. In fact, Boltzman machines are used to generate statistics of binary sequences. In contrast, in the problem considered in this paper, which refers to a neural network with deterministic binary per neuron mappings, rather than stochastic such mappings as in Boltzman machines, the statistics of the output binary sequences are determined by both the threshold parameter of interest and the unknown statistics of the input data. Thus, while the use of the Kullback-Leibler distance results in a simple recursive estimation algorithm for Boltzman machines, the same distance induces a coupled system of two simultaneous stochastic approximation estimation techniques in our case. In addition, in Ackley et al (1985) and Amari et al (1992), only the first order derivative of the Kullback-Leibler distance is used in the derivation of the corresponding learning algorithm, while here, we use first and second order derivatives as in Newton's method, which is expected to result in faster algorithmic convergence. □

4. Some Critical Design Issues

In the design of the algorithm presented in Section 3, the first critical issue is the selection of the function $h(\lambda)$. This function is controlled by the nature of the parameter λ, the neural operations, and the unknown statistics of the hypothesis H_0. In this section, we study the $h(\lambda)$ selection for the case where λ is the threshold λ_1 (see expression (2.1)) in the Hopfield structure discussed in Section 2: the selection for the threshold λ_2 in (2.1) is similar. In this case, assuming stationary and memoryless H_0 hypothesis generated by some density function $f_0(x)$, we conclude that if $f_0(x)$ were known, then $h(\lambda)$ would be selected as follows.

$$h(\lambda) = P_0(g(X) \geq \lambda) = \int_{x:g(x)\geq\lambda} f_0(x)dx \tag{4.1}$$

If the function g(x) is in addition analytic and strictly monotonically increasing (such as the sigmoid function) and if the density $f_0(x)$ is continuous everywhere, then, for $F_0(x)$ denoting the cumulative distribution function corresponding to the density $f_0(x)$, the function $h(\lambda)$ would be as follows,

$$h(\lambda) = 1 - F_0(g^{-1}(\lambda))$$

with

$$[\dot{h}(\lambda)]^{-1} = -\frac{\dot{g}(g^{-1}(\lambda))}{f_0(g^{-1}(\lambda))} \tag{4.2}$$

where $y = g^{-1}(x)$ implies $x = g(y)$.

Since the density $f_0(x)$ is unknown, the issue arising is: given the function g(x), given that it is analytic and strictly monotonically increasing, given possibly a class of densities that may generate the H_0 hypothesis, how should the function $[\dot{h}(\lambda)]^{-1}$ be designed. Before we proceed with the examination of this issue, we will present some simple properties of the so selected function g(x), in a proposition.

Proposition 1

Let g(x) be analytic and strictly monotonically increasing. Let the two hypotheses, H_0 and H_1, be generated by memoryless and stationary processes, represented by the density functions $f_0(x)$ and $f_1(x)$, respectively, whose only known property is that $\log [f_1(x)/f_0(x)]$ is a strictly monotonically increasing function. Then, the use of the function g(x) is equivalent to the optimal Neyman-Pearson test at the pair (f_0, f_1).

Proof

Both g(x) and $\log [f_1(x)/f_0(x)]$ lead to the sufficient statistic x. Thus, they represent identical Neyman-Pearson tests. □

Remarks: (1) As a simple consequence of the proposition, when the function g(x) is used, and when the true densities induce a strictly monotonically increasing log likelihood ratio function, then the optimal threshold λ_1 selection in (2.1), at a given Type 1 error probability level, also induces the minimum Neyman-Pearson Type-2 error probability. (2) If $f_0(x)$ is analytic and such that $\dot{f}_0(x)/f_0(x)$ is strictly monotonically decreasing, and if $f_1(x) = f_0(x-m)$ for some positive constant m, then the condition in the proposition is satisfied. (3) When, in addition to being analytic and strictly monotonically increasing, the function g(x) is also bounded (from both above and below), and its derivative $\dot{g}(x)$ is also such that $\dot{g}(\infty) = \dot{g}(-\infty) = 0$, then $\dot{g}(x)$ behaves as a density function. Indeed, $\dot{g}(x)$ can be then expressed as follows:

$$\dot{g}(x) = c\, q(x) \tag{4.3}$$

where $c = [g(\infty) - g(-\infty)]$, and where q (x) is positive everywhere, it integrates to one, and $q(\infty) = q(-\infty) = 0$. Functions g(x) that possess the above property are, for example, the sigmoid function, or the functions,

$$g(x) = 2d\,\Phi(x)\,;\ \dot{g}(x) = 2d\,\phi(x) \qquad (4.4)$$

$$g(x) = 2d\,[\Phi(x) - 1/2]\,;\ \dot{g}(x) = 2d\phi(x) \qquad (4.5)$$

where d is some positive finite constant and where $\phi(x)$ and $\Phi(x)$ represent the density and the cumulative distribution functions of the zero mean and unit variance Gaussian random variable at the point x, respectively. The function g(x) in (4.4) is positive everywhere, while that in (4.5) is odd. □

Let us now return to the design of the function $[\dot{h}(\lambda)]^{-1}$, in the case where the function g(x) is such that (4.3) holds for some density function q(x), and the H_0 hypothesis is generated by some density function in a given class C. Due to (4.2), in conjunction with (4.3), the issue is then the selection of some density function f^* in C, which will provide $[\dot{h}(\lambda)]^{-1}$ as follows:

$$[\dot{h}(\lambda)]^{-1} = -c\,\frac{q(g^{-1}(\lambda))}{f^*(g^{-1}(\lambda))}\,;\ f^* \varepsilon C \qquad (4.6)$$

Towards selecting f^*, we first express some known results from Markov theory, in the form of a propositon.

<u>Proposition 2</u>

Let $\{X_n\}$ be a discrete-time Markov process. The process is asymptotically ergodic if and only if it converges to a stationary and ergodic process. In the latter case, the rate with which this convergence occurs is determined by the characteristics of the drift process $\{X_{n+1} - X_n\}$. If $\{X_n\}$ converges, in particular, to a Gaussian stationary and ergodic process, then, the rate of the convergence is determined by the ratio, $E\{X_{n+1} - X_n\}[E\{[X_{n+1} - X_n]^2\} - E^2\{X_{n+1} - X_n\}]^{-1/2}$. □

At this point, we are interested in the selection of $[\dot{h}(\hat{\lambda}_n)]^{-1}$ in (3.6). Thus, we are using Proposition 2 with conditioning on $\hat{p}_n$. That is, in the determination of $h(\hat{\lambda}_n)$, we use as criterion the ratio,

$$R(h,f) \triangleq E_f\{\hat{\lambda}_{n+1} - \hat{\lambda}_n \mid \hat{p}_n\}[E_f\{[\hat{\lambda}_{n+1} - \hat{\lambda}_n]^2 \mid \hat{p}_n\} - E_f^2\{\hat{\lambda}_{n+1} - \hat{\lambda}_n \mid \hat{p}_n\}]^{-1/2},$$

where h is the function in (3.6) and where f denotes the density function that generates the actual data. For g(x) as in (4.3), for some given class C of densities, and for $h(\lambda)$ as in (4.6) for some $f^* \varepsilon C$, we conclude that R (h, f) takes the form in (4.7), if a random datum x is generated by f and if $\hat{\lambda}_n$ is selected equal to g(x). We note that such $\hat{\lambda}_n$ selection corresponds to a worst case, where the history of the $\{\hat{\lambda}_n\}$ recursions is ignored and $\hat{\lambda}_n$ is generated randomly via the unknown acting density f.

$$R(h,f) = R(f^*,f) \triangleq \left\{ \frac{\int_{-\infty}^{\infty} \left[\frac{q(x)}{f^*(x)}\right]^2 f(x)dx}{\left[\int_{-\infty}^{\infty} \frac{q(x)}{f^*(x)} f(x)dx\right]^2} - 1 \right\}^{-1/2} \quad ; f, f^* \varepsilon C \tag{4.7}$$

We will thus use a minimax approach on the functional in (4.7), to design the function $[\dot{h}(\lambda)]^{-1}$ in (4.6). In particular, we will search for the f^* in C that satisfies the following condition:

$$R(f^*,f^*) = \sup_{f\varepsilon C} R(f, f) = \sup_{f\varepsilon C} R(f^*, f) \tag{4.8}$$

The problem posed by the condition in (4.8) has a solution if the class C is convex and compact, and f^* represents then the *least favorable* density in C, with respect to the $R(f_1, f_2)$ criterion in (4.7). We now proceed with a lemma.

Lemma 1

Let C be a convex and compact class of density functions. Then, the density f^* in C that satisfies the condition in (4.8) is that which satisfies the condition:

$$\int_{-\infty}^{\infty} \frac{q^2(x)}{f^*(x)} dx = \inf_{f\varepsilon C} \int_{-\infty}^{\infty} \frac{q^2(x)}{f(x)} dx \tag{4.9}$$

Proof

(a) From (4.7), we observe that $R(f, f) = \left\{ \int_{-\infty}^{\infty} \frac{q^2(x)}{f(x)} dx - 1 \right\}^{-1/2}$. Thus, the supremum of $R(f, f)$ corresponds to the infimum in (4.9).

(b) $R(f^*,f)$ possesses a supremum with respect to f^* in C, attained at $f^* = f$. Thus, $R(f^*,f) \leq R(f,f)$; $\forall f^* \varepsilon C$. But, since $R(f^*,f^*) = \sup_{f\varepsilon C} R(f,f)$, we conclude that $R(f^*,f) \leq R(f^*,f^*)$; $\forall f \varepsilon C$. Thus, $R(f^*,f^*) = \sup_{f\varepsilon C} R(f^*,f)$. □

Drawing from Lemma 1, we can now present choices for the function in (4.6), for various given classes of densities. Our results are due to calculus of variations applied on (4.9), and can be found, for example, in Kazakos and Papantoni-Kazakos (1989).

Classes of Densities and the h(x) Function

1. Let C represent the outlier model around some known density f_o, at contamination level ε. That is, for given ε in (0,1), we have:

$$C = \{f : f(x) - (1-\varepsilon)f_o(x) \geq 0; \forall x\}$$

Then, for $f_o(x) \neq q(x)$,

$$f^*(x) = \max((1-\varepsilon)\, f_o(x),\ \mu\, q(x))$$
$$\mu : \int f^*(x)\, dx = 1 \tag{4.10}$$
$$\left[\dot{h}(\lambda)\right]^{-1} = -c \min\left\{\frac{q(g^{-1}(\lambda))}{(1-\varepsilon) f_o(g^{-1}(\lambda))},\ \frac{1}{\mu}\right\}$$

2. Let C be the class of all densities with given finite support [a, b]. Then,

$$f^*(x) = [b-a]^{-1}\ ;\ \forall\, x \in [a,b]$$
$$[\dot{h}(\lambda)]^{-1} = -c(b-a)\, q(g^{-1}(\lambda)) \tag{4.11}$$

3. Let C include all possible densities. Then,

$$f^*(x) = q(x)$$
$$[\dot{h}(\lambda)]^{-1} = -c \tag{4.12}$$

This concludes Section 4. In the next section, we will study the convergence properties of the algorithm in Section 3.

5. Convergence Studies

We study the convergence properties of the algorithm in Section 3 for the same problem as that considered in Section 4: that is, for the case where the parameter λ is the threshold λ_1 in the Hopfield structure discussed in Section 2. Let then H_0 be generated by a stationary and memoryless process that possesses a first order density function denoted f(x), and let the function g(x) be strictly monotonically increasing. We start with a theorem.

Theorem 1

Consider the sequence $\{\hat{p}_n\}$ of probability estimates induced by the algorithm, when data are generated by a memoryless and stationary process whose first order density is f(x). Then, the subsequence $\{\hat{p}_{kN+\frac{m}{2}k(k-1)}\}_{k\geq 1}$ converges to the value α, almost surely, for any $N\geq 1$ and $m\geq 0$. Furthermore, the variables $\{Y_k \triangleq \hat{p}_{kN+\frac{m}{2}k(k-1)} - \hat{p}_{(k-1)N+\frac{m}{2}(k-1)(k-2)}\}_{k\geq 1}$ are then asymptotically zero mean Gaussian: the variance of Y_k is of order $[N+m(k-1)][2N+m(k-1)]^{-2}k^{-2}$, where the latter determines the order of convergence of the subsequence $\{\hat{p}_{kN+\frac{m}{2}k(k-1)}\}_{k\geq 1}$ to the value α.

Proof

To avoid complexities in terminology, we first denote $\hat{p}_{kN+\frac{m}{2}k(k-1)}$ by $\hat{q}_k$, and $\hat{\lambda}_{kN+\frac{m}{2}k(k-1)}$ by $\hat{\mu}_k$. Then, by repetitions of the recursions in (3.7), we find:

$$\hat{q}_{n+1} = \hat{q}_n + \frac{2[N+mn]}{(n+1)[2N+mn]}\left\{\frac{1}{N+mn}\sum_{i=1}^{N+mn} u_i(\hat{\mu}_{n+1}) - \hat{q}_n\right\}$$

$$-\frac{n^2[2N+m(n-1)][2N+m(n-1)+2n^{-1}]}{(n+1)^2[2N+mn]^2}\,\frac{\hat{q}_n(1-\hat{q}_n)(\hat{q}_n-\alpha)}{(\hat{q}_n-\alpha)^2+\alpha(1-\alpha)} \tag{5.1}$$

From (5.1), we easily conclude that,

$$\lim_{n\to\infty} E\{\hat{q}_{n+1} - \hat{q}_n \mid \hat{\mu}_{n+1}, \hat{q}_n = \alpha\} = 0;\ \forall N\geq 1,\ \forall m\geq 0,\ \forall \hat{\mu}_{n+1} \tag{5.2}$$

[a] Since the expected drift conditioned on $\hat{q}_n=\alpha$ converges to zero, (see (5.2)), to show almost sure convergence, (see Markov Theory or Blum's (1954) conditions as reduced here), it suffices to show that, there exists $n_o<\infty$, such that:

$$\sup_{|\hat{q}_n-\alpha|>\delta} (\hat{q}_n-\alpha)E\{\hat{q}_{n+1}-\hat{q}_n \mid \hat{\mu}_{n+1},\hat{q}_n\}<0;\forall\delta>0;\forall n>n_o \tag{5.3}$$

Via (5.1), the condition in (5.3) can be written as follows, where we denote $F(g^{-1}(\hat{\mu}_{n-1}))$ by F_n, for simplicity in notation, and where we assume $n >> N$.

$$F_n > l_n(x_n) \stackrel{\Delta}{=} [1-\alpha-x_n]\left[1 - \frac{nx_n(\alpha+x_n)}{x_n^2+\alpha(1-\alpha)}\right];\ \forall\ x_n \stackrel{\Delta}{=} \hat{q}_n-\alpha>\delta>0;\forall n>n_o \tag{5.4}$$

$$F_n < \mu_n(y_n)\stackrel{\Delta}{=} 1+(\alpha-y_n)\left[n-1 - n(1-\alpha)\frac{\alpha-y_n}{y_n^2+\alpha(1-\alpha)}\right];\ \forall\ y_n \stackrel{\Delta}{=} \alpha-\hat{q}_n > \delta>0;\forall n>n_o \tag{5.5}$$

Since F_n is such that $0 < F_n < 1$, then, to guarantee the satisfaction of both the inequalities in (5.4) and (5.5), for every $\delta\geq\varepsilon>0$, given $\varepsilon>0$, it suffices to find some $n_o=n_o(\varepsilon)$, such that

$$1-n\,\frac{x(\alpha+x)}{x^2+\alpha(1-\alpha)} < 0;\ \forall\ x > \delta, \delta\geq\varepsilon,\ \forall\ n > n_o \tag{5.6}$$

and

$$n-1 - n(1-\alpha)\,\frac{\alpha-y}{y^2+\alpha(1-\alpha)} > 0;\ \forall\ y > \delta,\ \delta \geq \varepsilon,\ \forall\ n > n_o \tag{5.7}$$

But both the functions $x(\alpha+x)[x^2+\alpha(1-\alpha)]^{-1}$ and $(y-\alpha)[y^2+\alpha(1-\alpha)]^{-1}$ are monotonically increasing for $x > 0$ and $y > 0$, respectively. Thus, (5.6) and (5.7) reduce to the following conditions, respectively.

$$1-n\frac{\varepsilon(\alpha+\varepsilon)}{\varepsilon^2+\alpha(1-\alpha)} < 0;\ \forall\ n > n_o \rightarrow n > \left\lceil\frac{\varepsilon^2+\alpha(1-\alpha)}{\varepsilon(\alpha+\varepsilon)}\right\rceil \stackrel{\Delta}{=} n_{o1}(\varepsilon) \tag{5.8}$$

and

$$n-1-n(1-\alpha)\,\frac{\alpha-\varepsilon}{\varepsilon^2+\alpha(1-\alpha)} > 0;\ \forall\ n > n_o \rightarrow n > \left\lceil\frac{\varepsilon^2+\alpha(1-\alpha)}{\varepsilon(1-\alpha+\varepsilon)}\right\rceil \stackrel{\Delta}{=} n_{o2}(\varepsilon) \tag{5.9}$$

For $\alpha< 1/2$, it suffices to select $n_o = \max\{n_{o1}(\varepsilon),n_{o2}(\varepsilon)\}$, to have both conditions satisfied. This completes the proof of (5.3), and the a.s. convergence with order $1/n$ holds.

[b] Subject to the conditions in the Theorem, and due to the satisfaction of (5.3), Sack's (1957) conditions hold; thus the Gaussianity of the drift process. The mean zero result holds due to the convergence in (5.2), while the conclusion on the variance is deduced from (5.1), and clearly determines the rate of convergence to the value α. □

While Theorem 1 establishes the almost sure convergence of the subsequence $\{\hat{p}_{kN+\frac{m}{2}k(k-1)}\}_{k\geq 1}$, it provides no guarantees for the convergence of the total sequence $\{\hat{p}_n\}$, whose values may generally oscillate for n indices not belonging to the set $\{kN+\frac{m}{2}k(k-1)\}_{k\geq 1}$. Indeed, while the a.s. convergence of the above subsequence imposes no conditions on the sequence $\{\hat{\lambda}_n\}$, the convergence of the total sequence $\{\hat{p}_n\}$ does. The latter are expressed in Lemma 2 below.

<u>Lemma 2</u>

For the convergence of the total sequence $\{\hat{p}_n\}$ to the value α, it is necessary that the sequence $\{\hat{\lambda}_n\}$ converge to the value λ_o determined by, $1-F(g^{-1}(\lambda_o))=\alpha$, where F(x) denotes the cumulative distribution of the i.i.d. data, at the point x. If, in addition, the two-dimensional sequence $\{\hat{\lambda}_{n+1}, \hat{p}_n\}$ is such that,

$$\inf_{|\hat{p}_n-\alpha|>\delta} (\hat{p}_n-\alpha)[\hat{p}_n-1+F(g^{-1}(\hat{\lambda}_{n+1}))] > 0\,;\ \forall\delta>0, \forall n>n_o \tag{5.10}$$

for some finite n_o, then $\{\hat{p}_n\}$ converges to α almost surely, and the sequence $\{\hat{p}_{n+1}-\hat{p}_n\}$ is asymptotically zero mean Gaussian.

<u>Proof</u>

[a] For n not belonging to the set $\{kN+\frac{m}{2}k(k-1)\}$, expression (3.7) gives:

$$E\{\hat{p}_{n+1}-\hat{p}_n \mid \hat{p}_n, \hat{\lambda}_{n+1}\} = \frac{1}{n+1}[1-F(g^{-1}(\hat{\lambda}_{n+1}))-\hat{p}_n] \tag{5.11}$$

For convergence of $\{\hat{p}_n\}$ to α it is necessary for the conditional expected drift in (5.11) to converge to zero faster than 1/n, at $\hat{p}_n=\alpha$. This clearly requires that $\{\hat{\lambda}_n\}$ converge to λ_o.

[b] If $\{\hat{\lambda}_n\}$ converge to λ_o and (5.10) is satisfied, then the conditions by Sack apply, as in the proof of Theorem 1, and, thus, almost sure convergence and asymptotic, zero-mean, Gaussianity of the increments. □

At this point, let us consider the sequence $\{\hat{\lambda}_n\}$. Since the $\hat{\lambda}_n$ estimates remain unchanged within windows of the form N+mk, it suffices to study the subsequence $\left\{\hat{\lambda}_{kN+\frac{m}{2}k(k-1)}\right\}_k$ which we will denote $\{\hat{\mu}_k\}$, for simplicity in notation. For similar simplicity, we will denote:

$$\hat{q}_k \stackrel{\Delta}{=} \hat{p}_{kN+\frac{m}{2}k(k-1)}$$

$$H(\hat{p}_n) \stackrel{\Delta}{=} \frac{\hat{p}_n(1-\hat{p}_n)(\hat{p}_n-\alpha)}{(\hat{p}_n-\alpha)^2+\alpha(1-\alpha)} \; ; \text{ for } n=kN+\frac{m}{2}k(k-1) \tag{5.12}$$

where, due to (3.6), we have:

$$H(\hat{q}_k) = -[\dot{h}(\hat{\mu}_k)](\hat{\mu}_{k+1}-\hat{\mu}_k) \tag{5.13}$$

Given cumulative distribution F(x) of the data, let us now define the sequence $\{Z_k\}$ of random variables, as follows:

$$Z_k \stackrel{\Delta}{=} \sum_{l=1}^{k-1} \frac{l}{k} \cdot \frac{2N+m(l-1)}{2N+m(k-1)} \left[F(g^{-1}(\hat{\mu}_{l+1}))-F(g^{-1}(\hat{\mu}_l))-H(\hat{q}_l)\right] \tag{5.14}$$

We now express another theorem.

Theorem 2

Let f(x) and F(x) denote the density function and the cumulative distribution of the i.i.d. data, at the point x, respectively. Let the sequence $\{f(g^{-1}(\hat{\mu}_k))/\dot{g}(g^{-1}(\hat{\mu}_k))+\dot{h}(\hat{\mu}_k)\}_k$ be absolutely bounded. Then:

(i) The sequence $\{Z_k\}$ is asymptotically Gaussian, with mean inversely proportional to k[2N+m(k-1)].

(ii) The sequence $\{\hat{\mu}_k\}$ converges to the λ_o value, such that $1-F(g^{-1}(\lambda_o))=\alpha$, if and only if the variances of the sequence $\{Z_k\}$ converge to zero. In the latter case, $\{\hat{\mu}_k\}$ converges to λ_o almost surely.

Proof

[a] Since, via Theorem 1, $\{\hat{q}_k\}$ converges almost surely to α, and due to (3.6), given $\varepsilon>0$, there exists k_o:

$$|\hat{q}_l-\alpha| < \varepsilon \; ; \forall\, l > k_o$$

$$\hat{\mu}_{l+1} - \hat{\mu}_l \approx -\frac{1}{\dot{h}(\hat{\mu}_l)}(\hat{q}_l-\alpha) \; ; \forall\, l > k_o \tag{5.15}$$

$$|F(g^{-1}(\hat{\mu}_{l+1}))-F(g^{-1}(\hat{\mu}_l))| < c_1 \,|\, \hat{\mu}_{l+1} - \hat{\mu}_l \,| \; ; \forall\, l > k_o$$

$$Z_k = \sum_{l=k_o+1}^{k-1} \frac{l}{k}\, \frac{2N+m(l-1)}{2N+m(k-1)}\, \rho(\hat{\mu}_l)[\hat{\mu}_{l+1} - \hat{\mu}_l\,] +$$

$$+\frac{1}{k[2N+m(k-1)]} \sum_{l=1}^{k_o} l\,[2N+m(l-1)][F(g^{-1}(\hat{\mu}_{l+1})) - F(g^{-1}(\hat{\mu}_l)) - H(\hat{q}_l)] \tag{5.16}$$

$$; \forall\, k \geq k_o + 2$$

where $\rho(\hat{\mu}_l)$ is absolutely bounded for all $l>k_o$, and c_1 is a finite positive constant.

Since also all the terms in the second sum of (5.16) are bounded, we easily conclude that, there exists positive finite constants c_o and d, such that:

$$| Z_k - c_o \sum_{l=k_o+1}^{k-1} \frac{l}{k} \cdot \frac{2N+m(l-1)}{2N+m(k-1)} (\hat{\mu}_{l+1} - \hat{\mu}_l) | \le \frac{d}{k[2N+m(k-1)]} \; ; \forall k \ge k_o+2 \qquad (5.17)$$

Due to Theorem 1, the sequence $\{\hat{q}_l - \alpha\}$ is asymptotically zero-mean Gaussian and converges to zero almost surely. Due to (5.15), the same holds for the sequence $\{\hat{\mu}_{l+1} - \hat{\mu}_l\}$. This, in conjunction with (5.17), proves part (i) of the theorem.

[b] From the description of the $\{\hat{q}_k\}$ sequence, in conjunction with the i.i.d. data, we first conclude, via the central limit theorem, that, given $\{\hat{\mu}_k\}$ sequence, $\hat{q}_k$ converges with rate inversely proportional to $\{\frac{k}{2} [2N+mk(k-1)]\}^{1/2}$, to:

$$\hat{q}_k \to (fa(\hat{\mu}_k) \overset{\Delta}{=} 1 - F(g^{-1}(\hat{\mu}_k)) + Z_k \qquad (5.18)$$

for Z_k as in (5.14). Due to part (i) of the theorem, for the sequence $\{\hat{\mu}_k\}$ induced by the algorithm, the sequence $\{Z_k\}$ is asymptotically Gaussian, with mean going to zero. On the other hand, via Theorem 1, the sequence $\{\hat{q}_k\}$ converges to α almost surely. Thus, in conjunction with (5.18), the sequence $\{\hat{\mu}_k\}$ will converge to λ_o: $\alpha=1-F(g^{-1}(\lambda_o))$, if and only if the variances of the sequence $\{Z_k\}$ converge to zero. Then, due to the almost sure convergence of $\{\hat{q}_k\}$, the convergence of $\{\hat{\mu}_k\}$ is almost sure as well. □

<u>Theorem 3</u>

Subject to the conditions in Theorem 2:

(i) If m=0, then the sequence $\{\hat{\mu}_k\}$ is asymptotically Gaussian, with mean converging to λ_o: $\alpha=1-F(g^{-1}(\lambda_o))$ and variance of order $O(1/N)$.

(ii) For $m\ge1$, the sequence $\{\hat{\mu}_k\}$ converges to λ_o: $\alpha=1-F(g^{-1}(\lambda_o))$, almost surely, with order $O(c_o^2/3\,mk)$.

<u>Proof</u>

Due to Theorem 3, it suffices to study the variances of the sequence $\{Z_k\}$ in (5.14) and (5.17). For m=0, we will find that this variance is of order $O(1/N)$, which proves part (i), while, for $m\neq0$, we will find that the variance is of order $O(1/mk)$, going to zero, which, in conjunction with Theorem 3, proves part (ii) of the theorem.

Due to expression (5.17), it suffices to determine the order of the variances of the sequence $\{\hat{\mu}_{l+1} - \hat{\mu}_l\}$, which, due to (5.15), is the same as the order of the variances of the sequence $\{\hat{q}_l - \alpha\}$. Now, from Theorem 1, we have that the variances of the increments $\{\hat{q}_{k+1} - \hat{q}_k\}$ are of order:

$$O\left[\frac{N+m(k-1)}{k^2[2N+m(k-1)]^2}\right] \Rightarrow \begin{cases} O(\frac{1}{4Nk^2})\; ; m=0,\ k\gg1 \\ O(\frac{1}{mk^3})\; ; m\neq0,\ k\gg1 \end{cases} \qquad (5.19)$$

Therefore, the order of the variances for the sequence $\{\hat{q}_l - \alpha\}$, and thus for the sequence $\{\hat{\mu}_{l+1} - \hat{\mu}_l\}$, is as follows:

$$\{\hat{\mu}_{l+1} - \hat{\mu}_l\} : \begin{cases} O\left(\frac{1}{4Nl}\right) ; \ m=0,\ l \gg 1 \\ O\left(\frac{1}{ml^2}\right) ; \ m \neq 0,\ l \gg 1 \end{cases} \tag{5.20}$$

Substituting the orders in (5.20) in the expressions for the variance of Z_k, for Z_k as in (5.17), we find:

$$\mathrm{Var}\{Z_k\} = \begin{cases} O\,(c_o^2/8N) ; \ m=0,\ k \gg 1 \\ O\,(c_o^2/3mk) ; \ m \neq 0,\ k \gg 1 \end{cases} \tag{5.21}$$

This completes the proof of the theorem. □

We now present a corollary, which merges the results in Lemma 2 and Theorem 3, to express convergence properties for the sequence $\{\hat{p}_n\}$.

<u>Corollary</u>

Consider the conditions in Theorem 2. Then:

(i) If m=0, the total sequence $\{\hat{p}_n\}$ does not converge. Only the subsequence $\{\hat{q}_k\}$ converges then to α almost surely.

(ii) If $m \geq 1$, the total sequence $\{\hat{p}_n\}$ converges to the value α almost surely, with order $O\left(\frac{c_o^2}{3mk}\right)$.

<u>Proof</u>

(a) Since the sequence $\{\hat{\mu}_k\}$ does not converge to λ_o, when m=0, the necessary condition in Lemma 2 is not satisfied, so $\{\hat{p}_n\}$ does not converge.

(b) Since $\{\hat{\mu}_k\}$ converges almost surely to λ_o, when $m \neq 0$, due to Theorem 3, the necessary conditon in Lemma 2 is satisfied. Condition (5.10) in the Lemma is then also satisfied due to the almost sure converge of $\{\hat{\mu}_k\}$, in conjunction with the correct direction of the $\hat{\mu}_k$ adaptations induced by (3.6). Thus, the total sequence $\{\hat{p}_n\}$ converges then almost surely to the value α. The order is due to (5.21). □

<u>Remarks:</u> As concluded from Theorem 3 and the Corollary, if the density, f(x), that generates the data, and the function, $[h(\hat{\lambda}_n)]$, in (3.6) satisfy the conditions in Theorem 2, then, the algorithm converges almost surely, if the data windows per $\hat{\lambda}_n$ value increase monotonically. The $\{\hat{\lambda}_n\}$ sequence converges then to the value λ_o, such that, $\alpha = 1 - F(g^{-1}(\lambda_o))$; that is to the correct value, and the order of convergence is proportional to the product of the data window size increment times the number of the $\hat{\lambda}_n$-value iterations. The above order is then also inverseley proportional to the square of a constant c_o,

whose value depends on the density f(x), and the functions g(x) and $[\dot{h}(\hat{\lambda}_n)]$. If something is known about the density f(x), then $[\dot{h}(\hat{\lambda}_n)]$ can be appropriately selected, to control c_o within small values, and thus to increase the rate of convergence significantly: then, the effect of the correction term in (3.7) has significant impact on the algorithmic performance. Elaborating upon this issue, let us refer to Section 3: let then $\dot{g}(x)$ be as in (4.3), let some convex and compact class, C, of density functions be known, let $f^*(x)$ be the least favorable density in the class (as found in Section 3), and let $[\dot{h}(\lambda)]^{-1}$ be selected as in (4.6). Then, given true data density f(x), and as can be deduced from the proof of Theorem 2, the constant c_o is controlled by the following sequence:

$$r(\hat{\lambda}_n) \overset{\Delta}{=} \frac{| f(g^{-1}(\hat{\lambda}_n)) - f^*(g^{-1}(\hat{\lambda}_n)) |}{c\, q\, (g^{-1}(\hat{\lambda}_n))} \tag{5.22}$$

where,

$$\dot{g}(x) = c\, q(x) \tag{5.23}$$

In fact, if the sequence $\{r(\hat{\lambda}_n)\}$ is bounded, c_o is the latter bound. Due to (5.22), we observe that the closer the two densities, f(x) and $f^*(x)$, are in the variational distance sense, the smaller is the constant c_o, where, for unknown f(x) in class C, the latter closeness implies relatively small variational class C. Such is the case, for example, when ε is very small in the outlier class C in (4.10). We note that if f(x) is somehow known, then $f^*(x)$ should be selected equal to f(x), in which case the correction term in (3.7) eliminates completely the $Z_k - d\,k^{-1}[2N+m(k-1)]^{-1}$ part of the bias in (5.18), increasing the order of almost sure converge of the algorithm to $O(\frac{1}{mk^2})$. In the case that $f^*(x)=c\,q(x)$, as in (4.12), for the existence of a finite c_o, it suffices that f(x)/q(x) be bounded everywhere, as is the case where q (x) and f(x) are both Gaussian with finite means and f(x) has smaller variance than q(x) does. In the latter case, as the constant c in (5.22) increases, the constant c_o decreases, and so does in the case when the variance of f(x) is larger than that of q(x), instead, but the ratio f(x)/q(x) is bounded for x values around the value $g^{-1}(\lambda_o)$; for λ_o: $\alpha=1-F(g^{-1}(\lambda_o))$. The constant c in (5.23) affects the convergence rate of the algorithm, but represents a network design characteristic, whose value is selected to control operational network performance issues rather than the convergence rate of the learning algorithm. For example, when g(x) is such that, g(x) = min (x,c/2) ; for $x\geq 0$ and g(-x) = - g(x), the constant c is present for the control of data outliers, (robustness), and it induces an operational network performance tradeoff at the nominal data density versus outliers, (see Halford et al (1991) and Batalama et al). If, on the other hand, the constant c in (5.22) is not necessarily the same as the constant c in (5.23), then, the former may present convergence tradeoffs: as the value of the constant then increases, the asymptotic convergence rate increases as well, but the constants d and k_o in (5.17) may also increase, slowing down the initial incremental portion of the overall convergence performance.

In summary, the algorithm converges almost surely to the unique optimum of λ, under general conditions of the data densities, and under such conditions, its convergence rate is controllable. □

6. Numerical Results

In the implementation of the algorithm, we considered $\alpha=0.3$, initial data window size N=5, and window size increment m=1. All results presented herein are averages over 50 independent Monte Carlo runs. We initially selected the per neural element operation such that $g(x)=c[\Phi(x)-1/2]$, thus $\dot{g}(x)=c\phi(x)$, where $\phi(x)$ is the density function of the zero-mean and unit-variance Gaussian random variable. For the classes of stationary and memoryless processes that generate the data, we considered: (a) The class C_1 of all possible densities, in which case we selected $[\dot{h}(\lambda)]^{-1}=-c$ with c=2, (as in (4.12)), and (b) The outlier model class C_2 with the resulting $[\dot{h}(\lambda)]^{-1}$ function in (4.10), with $f_o(x)$ the Gaussian density $\sigma^{-1}\,\phi(x/\sigma)$ and for various values of the parameters σ and ε. As to the density f(x) that generates the actual data, we selected the nominal density $f(x)=f_o(x)=\sigma^{-1}\phi(x/\sigma)$ and the mixed density $f(x)=f_o(x)=(1-\zeta)\sigma^{-1}\phi(x/\sigma)+\zeta\delta(z-x)$, where $\delta(\cdot)$ is the delta function and z is a constant.

In Figures 2, 3, 4, we exhibit the convergence of the sequence $\{\hat{\lambda}_n\}$, for the class C_1, and true data density $f(x)=\sigma^{-1}\phi(x\,|\,\sigma)$, with various values of the variance σ^2. We observe, that as theoretically expected, the convergence rate improves when the true data density is the same as the one used in g(x), (Gaussian with $\sigma^2=1$). In Figures 5 and 6, we exhibit results for the class C_1, for input data with outliers: that is, $f(x)=(1-\zeta)\sigma^{-1}\phi(x\,|\,\sigma)+\zeta\,\delta(z-x)$, where we selected $\zeta=0.01$, z = -40,40 and $\sigma^2=0.5,2$. In the latter figures, portraying the data outliers as undesirable fixtures, we study the behavior of the algorithm regarding convergence to the optimal λ value induced by the nominal density $\sigma^{-1}\phi(x/\sigma)$. Of course, due to the presence of outliers (for any contamination level ζ), the algorithm does not converge to this desirable λ value.

In Figures 7, 8, we exhibit the convergence of the sequence $\{\hat{\lambda}_n\}$, for the class C_2 with $\varepsilon=0.1$ and $\sigma^2=0.5$, 2. In all cases, we selected N=5, m=1, and the design of the algorithm was performed at the least favorable density. The true data density is the nominal density $\sigma^{-1}\phi(x\,|\,\sigma)$. From the above figures, we observe the justification of our theoretical results and arguments. In particular, a priori restriction of the class of data densities and use of this a priori knowledge in the design of the learning algorithm, leads to faster convergence only if the variational distance between the density that generates the data and the least favorable density used in the design of the algorithm is smaller than the same distance induced when the least favorable density is substituted by the density $\phi(x)$ which results in the constant value 2 of the function $-[\dot{h}(\lambda)]$. The cases in Figures 7 and 8 induce variational distances larger than those induced when $-[\dot{h}(\lambda)]=2$, especially when $\sigma^2=2$ (Figure 8); thus, the worse convergence behavior, as compared to that exhibited in Figures 2 and 4. In the present design, the behavior of the algorithm in the presence of outliers is similar to that in class C_1, (no figures on the latter case are included here).

Next, we selected a truncated odd function as the per neuron operation, defined as follows: $g_1(x)=\min(x,d);x\geq 0$, and $g_1(-x)=-g_1(x)$, for some finite positive constant d. We note that this truncated operation is asymptotically robust, while this is not the case for the previously considered g(x) (see Kazakos et al (1989)). The value of d in $g_1(\cdot)$ is then selected for outlier control : the smaller the d, the higher the resistance to outliers, but the worse the performance in their absence. For this neural operation, we make the following choices: (i) $-[\dot{h}(\lambda)]^{-1}=2d$, and data either generated by the density $\sigma^{-1}\phi(x/\sigma)$, or by the density $(1-\zeta)\sigma^{-1}\,\phi(x/\sigma)+\zeta\,\delta(z-x)$, for selected values of the parameters σ, ζ, and z.

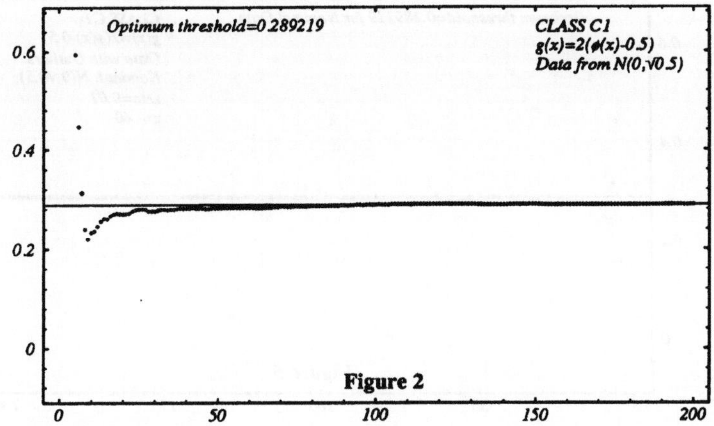

Figure 2

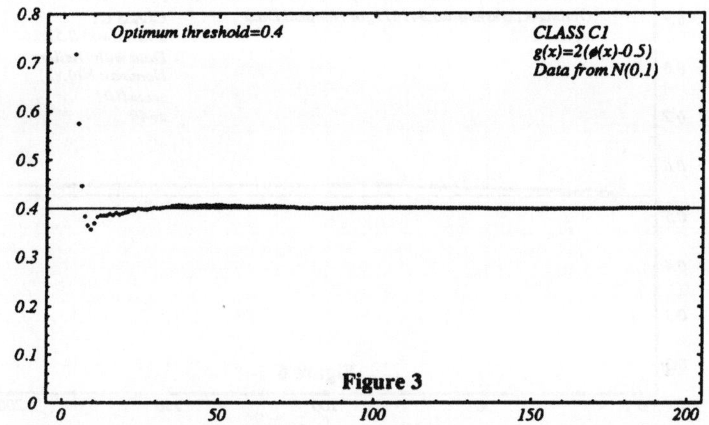

Figure 3

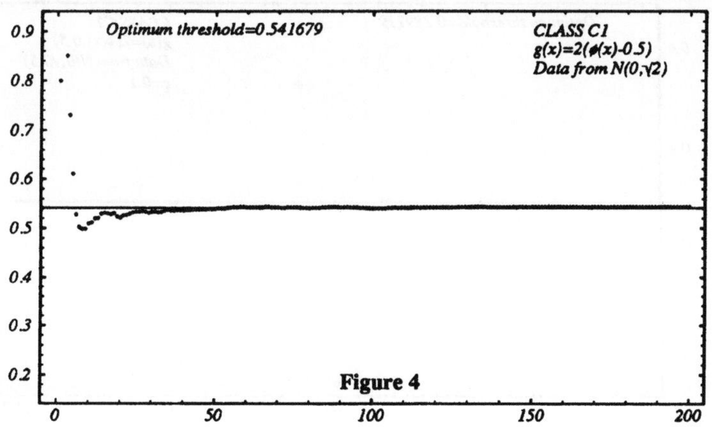

Figure 4

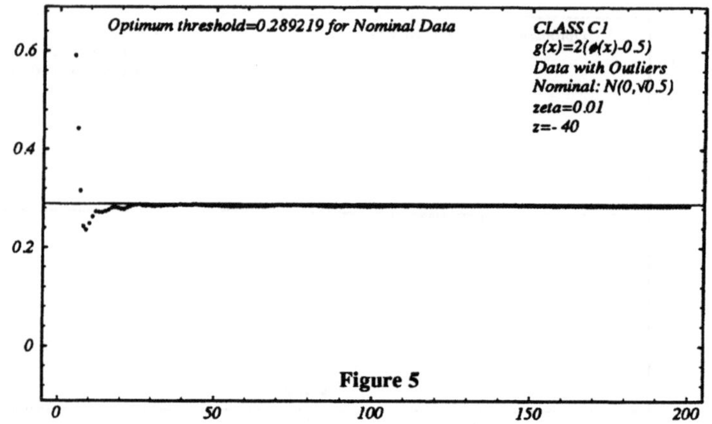

Figure 5

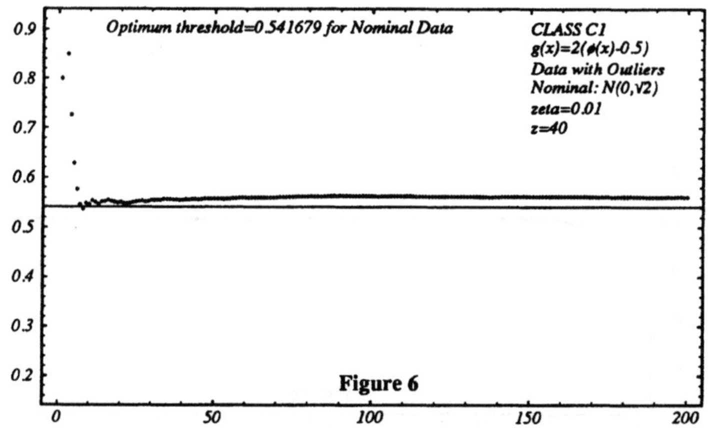

Figure 6

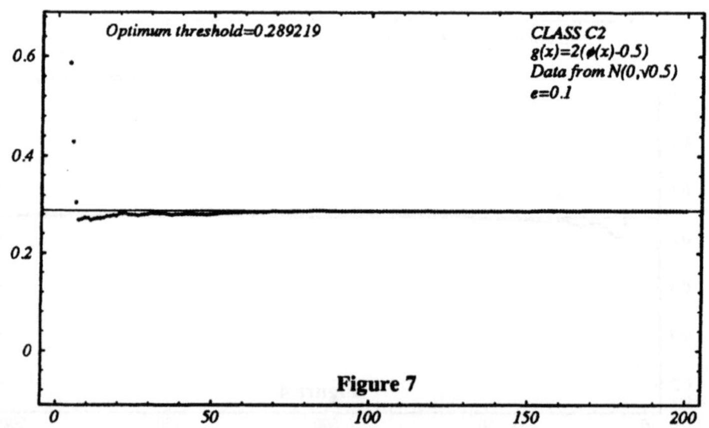

Figure 7

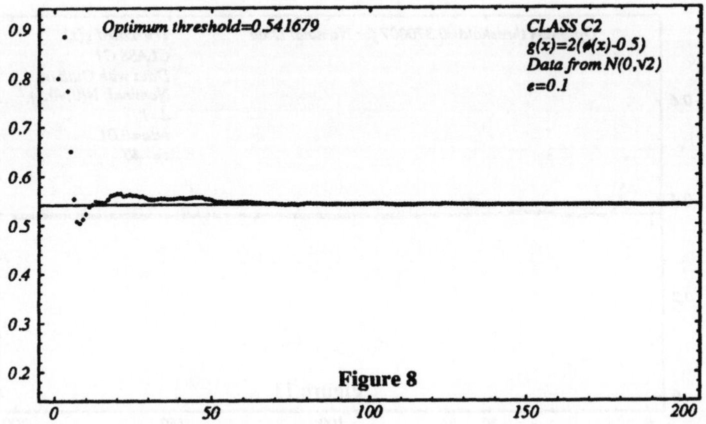

Figure 8

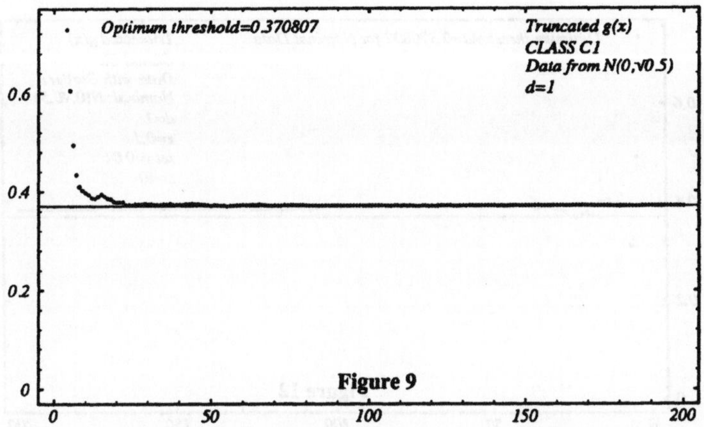

Figure 9

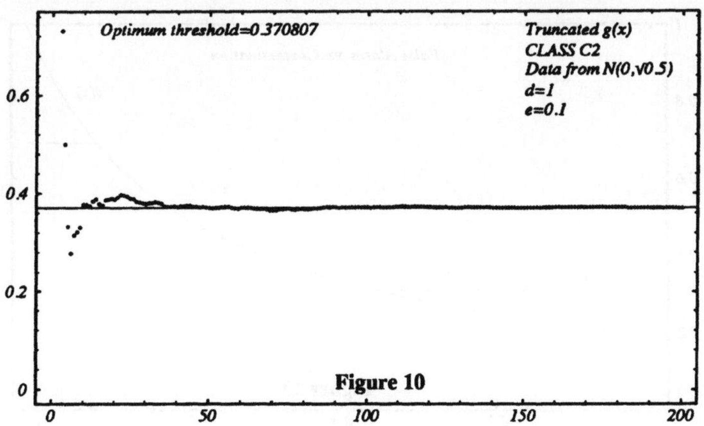

Figure 10

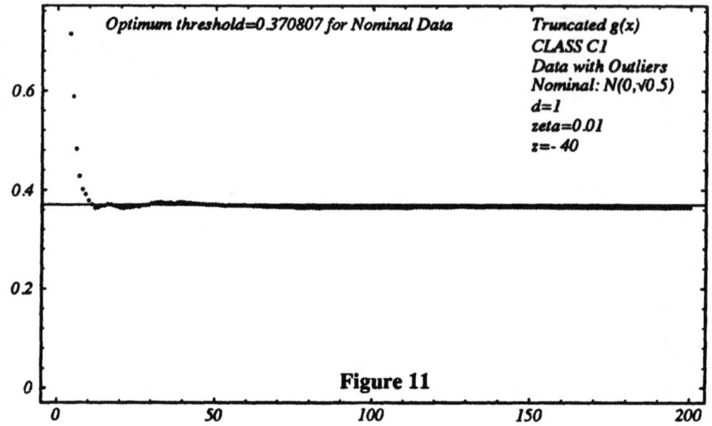

Figure 11

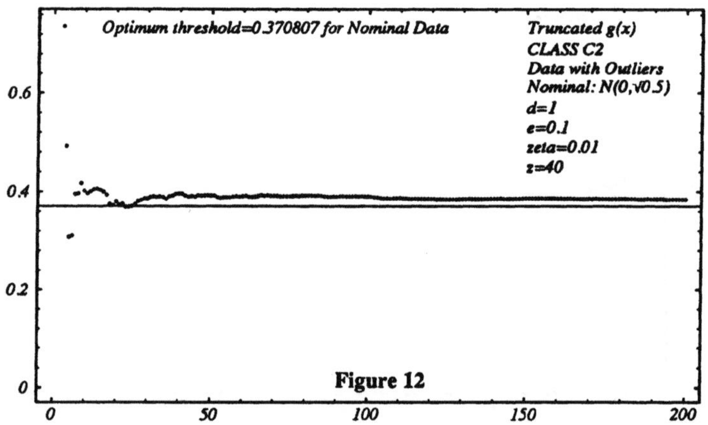

Figure 12

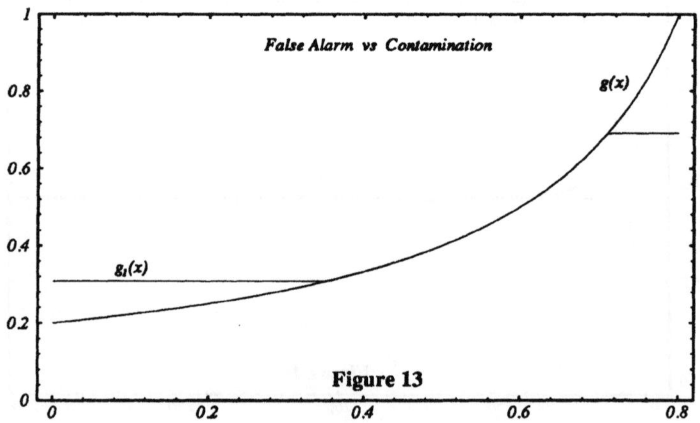

Figure 13

(ii) $-[\dot{h}(\lambda)]^{-1} = 2d\,[f^*(g^{-1}(\lambda))]^{-1}$, and data densities as in (i), where $f^*(x)$ is the least favorable density in the outlier class $F_\varepsilon = \{f: f(x) - (1-\varepsilon)\sigma^{-1}\,\phi(\frac{x}{\sigma}) \geq 0\}$. This least favorable density is given by (4.10) for $q(x) = (2d)^{-1}$; $-d < x < d$, and is specifically determined as follows, for $\frac{\sqrt{\pi}}{\sqrt{2}(1-\varepsilon)} < \frac{d}{\sigma} < \Phi^{-1}\left[\frac{2-\varepsilon}{2(1-\varepsilon)}\right]$:

$$f^*(x) = \begin{cases} (1-\varepsilon)\,\sigma^{-1}\,\phi(\frac{x}{\sigma})\,; & |x| \leq x_o \\ \frac{\mu}{2d} & ;\ x_o < |x| < d \\ 0 & ;\ \text{otherwise} \end{cases} \tag{6.1}$$

where,

$$\begin{aligned} x_o:\ \Phi\left[\frac{x_o}{\sigma}\right] + (d-x_o)\,\sigma^{-1}\,\phi\left[\frac{x_o}{\sigma}\right] &= \frac{2-\varepsilon}{2(1-\varepsilon)} \\ \frac{\mu}{2d} &= (1-\varepsilon)\,\sigma^{-1}\,\phi\left[\frac{x_o}{\sigma}\right] \end{aligned} \tag{6.2}$$

For comparison with the results in Figures 2, 5, 7, we selected here N=5, m=1, d=1, $\sigma^2 = 0.5$, z = 40 or -40, $\zeta = 0.01$, and $\varepsilon = 0.1$. In Figures 9 and 10, the data are generated by the density $\sigma^{-1}\,\phi(x/\sigma)$ with σ^2=0.5, while $-[\dot{h}(\lambda)]^{-1}$ is equal to 2d versus $2d[f^*(g^{-1}(\lambda))]^{-1}$, respectively. In Figures 11 and 12, the data are generated by the density $(1-\zeta)\,\sigma^{-1}\phi(x/\sigma) + \zeta\,\delta(z-x)$ with σ^2=0.5, $\zeta = 0.01$, and z = -40 and 40 respectively. Again, $-[\dot{h}(\lambda)]^{-1}$ is set equal to 2d versus $2d[f^*(g^{-1}(\lambda))]^{-1}$, while convergence to the optimal λ value at the nominal density $\sigma^{-1}\phi(x/\sigma)$ is studied.

In order to compare the behavior of the $g_1(\cdot)$ and $g(\cdot)$ designs in the presence of outliers, we fix a desired false alarm α and assume training data from $f(x)=(1-\zeta)f_o(x)+\zeta\delta(z-x)$. We vary the contamination level ζ and examine the deviation of the false alarm induced by those two designs from the desired value α, when they operate in the presence of nominal only data (i.e. data from $f_o(x)$). It is well known that their behavior is identical as far as $g_1(x)$ operates in its linearly increasing region (i.e. the corresponding threshold is between -d and d). The difference is in the resistance of $g_1(x)$ to high values of contamination ζ, at the expense of worse performance in nearly nominal data. This situation is depicted in Figure 13, where we used $f_o(x)$=$\phi(x)$, z=-40, α=0.2, d=0.5 and ζ varies from 0 up to 0.8.

REFERENCES

Abdelhamid, F. (1973), "Transformation of Observations in Stochastic Approximation", *Ann. of Statistics,* 1. 1158-1174, Nov.

Ackley, D. H., Hinton, G. E., and Sejnowski, T. J., (1985), " A Learning Algorithm for Boltzman Machines", *Cognitive Science,* Vol. 9, 147-169.

Akaike, H. (1974), "A New Look at the Statistical Model Identification," *IEEE Trans. Automat. Contr.*, AC-19, 716-723.

Amari, S., Kurata, K. and Nagaoka, H. (1992), "Information Geometry of Boltzman Machines", *IEEE Trans. Neural Networks,* Vol. 3, 260-271.

Bansal, R. K. and Papantoni-Kazakos, P. (1986), "An Algorithm for Detecting a Change in Stochastic Processes, "*IEEE Trans. Inform. Theory,* IT-32, 227-235.

Bansal, R. K., and Papantoni-Kazakos, P. (1989), "Outlier Resistant Algorithms for Detecting a Change in Stochastic Process", *IEEE Trans. Inform. Th.,* Vol. 35, 521-535.

Barron, A. R., va Straten, F. W., and Barron , R. L. (1977), "Adaptive Learning Network Approach to Weather Forecasting: A Summary", Proc. IEEE Int. Conf. Cybernetics and Society, 724-727.

Batalama, S., Koyiantis, A., Papantoni-Kazakos, P., and Kazakos, D., "Feed Forward Neural Structures in Binary Hypothesis Testing", *IEEE Trans. Comm.*, submitted.

Beran, R. (1976), "Adaptive Estimates for Autoregressive Process", *Annals Instit. Statist. Math.*, 28, 77-89.

Beran, R. (1977), "Robust Location Estimates," *Annals of Statistics,* 5, No. 3., 431-444.

Beran, R., (1978), "An Efficient and Robust Adaptive Estimator of Location", *Annals of Statistics,* 6, No. 2, 292-313.

Bickel, P. J. (1975), "One-Step Huber Estimates in the Linear Model", *J. Amer. Statist. Assoc.,* 70, 429-434.

Blahut, R. E. (1987), "Hypothesis Testing and Information Theory", *IEEE Trans. on Info. Theory,* Vol. IT-20, pp. 405-417.

Blum, J. R., (1954), "Multidimensional Stochastic Approximation Procedure", *Ann. Math. Stat.,* 25, pp. 737-744.

Chernoff, H. (1952), "A Measure of Asymptotic Efficienty for Tests of a Hypothesis Based on the Sum of Observations", *Ann. Math. Stat.,* 23, pp. 493-507.

Elman, J. And Zipser, D. (1988), "Learning the Hidden Structure of Speech", *Journal of the Accoustical Society of America,* 83, pp. 1615-1626.

Fabian V. (1968), "On Asymptotic Normality in Stochastic Approximation", *The Annals of Mathematical Statistics,* 39, pp. 1327-1332.

Fisher, R. A. (1922), "The Goodness of Fit of Regression Formulae and the Distribution of Regression Coefficients", *J. Roy Stat. Soc.,* 85 597i

Fisher, R. A. (1934), "Probability Likelihood and Quantity of Information in the Logic Uncertain Inference", *Proc. Roy. Soc.* A, 146 1f.

Gerencser, L. (1986), "Parameter Tracking of Time-Varying Continuous-Time Linear Stochastic Systems", in *Modelling, Identification and Robust Control,* eds., C. I. Byrnes and A. Lindquist, Amsterdam: North-Holland, pp. 581-594.

Gorman. P. and Sejnowski, T. (1988), "Analysis of Hidden Units in a Layered Network Trained to Classify Sonar Targets", *Neural Networks,* 1. 75-90.

Halford, K., Kazakos, D., Pados, D., and Papantoni-Kazakos, P., (1991), "Neural Network Structures with Feedback, in Binary Hypothesis Testing", IEEE Conf. on Systems,

Man, and Cybernetics.

Hopfield, J. (1982), "Neural Networks and Physical Systems with Emergent Collective Computational Abilities", Proc. Nat. Ac. of Science, 79, 2554-2558.

Huber, P. J. (1973), "Robust Regression: Asymptotics, Conjectures, and Monte Carlo", *Annals of Statist.* 1, No. 5, 799-821.

Huber, P. J. (1981), *Robust Statistics,* Wiley.

Kashyap, R. L. And Blaydon, C. C. (1966), "Recovery of Functions from Noisy Measurements Taken at Randomly Selected Points and its Application to Pattern Classification", *Proceedings of the IEEE,* 54, 1127-1129.

Kashyap, R. L., Blaydon, C. C. and Fu, K.S. (1970), "Stochastic Approximation", in *Adaptive Learning and Pattern Recognition Systems,* eds. J. M. Mendel and K. S. Ku, New York: Academic Press, pp. 329-355.

Kazakos, D. (1990), "Asymptotically Tight Error Bounds for Multihypothesis Testing Using Large Deviation Theory", *11th Prague Mathematics Conference,* Prague Czechoslovakis, Aug.

Kazakos, D. (1991), "Asymptotic Error Probability Expressions for Multihypothesis Testing Using Multisensor Data", *IEEE Trans. Systems, Man, and Cybernetics,* Special Issue, Sept.

Kazakos, D. (1992), "New Error Bounds and Optimum Quanization for Multisensor Distributed Signal Detection", *IEEE Trans. Commun.,* to appear.

Kazakos, D. and Papantoni-Kazakos, P., (1989), *Detection and Estimation,* Computer Science Press, New York.

Kazakos, D. and Papantoni-Kazakos, P. (1991), "Neural Networks for Binary Hypothesis Testing and Performance Studies", IEEE Conference on Neural Networks for Ocean Engineering, Aug., Washington, D.C.

Kiefer, J. and Wolfowitz, J. (1952), "Stochastic Estimation of the Maximum of a Regression Function", *Ann. Math. Statist.* 23, 462-466.

Kushner, H. (1987), "Asymptotic Global Behavior for Stochastic Approximations and Diffusions with Slowly Decreasing Noise Effects: Global Minimization via Monte Carlo", *SIAM Journal on Applied Mathematics,* 47, 169-185.

Kushner, H. and Clark, D. (1978), *Stochastic Approximation Methods for Constrained and Unconstrained Systems,* Berlin: Springer-Verlag.

Ljung, L. (1977), "Analysis of Recursive Stochastic Algorithms", *IEEE Transactions on Automatic Control,* 22, 551-575.

Ljung, L. and Soderstrom T. (1983), *Theory and Practice of Recursive Identification,* Cambridge, MA: MIT Press.

Minsky, M. and Papert, S. (1969), *Perceptrons* Cambridge, MA: MIT Press.

Papantoni-Kazakos, P. and Kazakos, D., (1991), "Fundamental Neural Structures, Operations, and Asymptotic Performance Criteria in Decentralized Hypothesis Testing", Proceedings, *IFAC,* and submitted for journal publication.

Rissanen, J. (1978), "Modelling by Shortest Data Description", *Automatica,* 14, 465-471.

Robbins, H. E. and Monro (1951), "A Stochastic Approximation Method", *Ann. Math. Stat.,* 22, 400-407.

Rosenblatt, F. (1957), "The Perceptron: A Perceiving and Recognizing Automaton", Report 85-60-1, Cornell Aeronautical Laboratory, Buffalo, New York.

Rumelhart, D. E., Hinton, G. E. and Williams, R. J. (1986), "Learning Internal Representations by Error Propagation", in *Parallel Distributed Processing: Explorations in the Microstructures of Cognition* (Vnnol. 1), eds. D. E. Rumelhart and J. L. McClelland, Cambridge, MA: MIT Press, pp. 318-362.

Wald, A. (1943), "Tests of Statistical Hypotheses Concerning Several Parameters when the Number of Observations is Large", *Transactions of the American Mathematical Society,* 54, 426-482.

Werbos, P. (1974), "Beyond Regression: New Tools for Prediction and Analysis in the Behavioral Sciences", Ph.D. Dissertation, Harvard Univ., Dept. of Applied Mathematics.

White, H. (1989), "Some Asymptotic Results for Learning in Single Hidden-Layer Feed-forward Network Models", *Journal of the American Statistical Association,* Dec. 84, 1003-1013.

Widrow, B. and Hoff, M. E. (1960), "Adaptive Switching Circuits", 1960 IRE WESCON Convention Record, 96-104.

Widrow, B. (1962), "Generalizaion and Information Storage in Networks of Adaline Neurons", *Self-Organizing Systems,* M. D. Yovits, G. T. Jacobi, and G. D. Goldstein (ed's), Spartan Books, Washington, D.C., 435-461.

Widrow, B. Winter, R. G. and Baxter, R. A. (1987), "Learning Phenomena in Layered Neural Networks", Proc. IEEE 1st Inst. Conf. Neural Networks II, 441-430.

Young, T. Y. and Westerberg, R. A. (1972), "Stochastic Approximation with Nonstationary Regression Function", *IEEE Transactions on Information Theory,* 18, 518-519.

LIKELIHOOD ESTIMATION OF DIRECTIONAL INTERACTION

M.TANEMURA
The Institute of Statistical Mathematics
4-6-7 Minami-Azabu, Minato-ku, Tokyo 106, JAPAN

1. Introduction

The spatial pattern of objects in the space has attracted statisticians and researchers in application fields, especially when the objects are 'points'. Then, the statistical methods of analyzing spatial point patterns have been developed both in the directions of likelihood-based and of nonparametric point of view.

There appear some new interesting features in the case when each object possesses some characteristics in addition to its locational information. Let us now suppose the object has a direction or an orientation as its nature. The direction may be that of the wind at each of spatially distributed observatories of weather service, or may be the orientation of the bodies or the movements of individuals of animals in a certain habitat, or it may be the orientation of magnetic property of rocks collected from various sites in an area. The direction may also be attached by a certain abstract transformation of a quantity of other nature of object. Then, the spatial pattern of directions thus obtained gives a new class of spatial process.

Our main concern here is the spatial pattern which is exhibited by the directional data of each point. The purpose of this paper is the characterization and the model fitting to such data.

The spatial patterns of such objects might be simply incorporated into the framework of 'marked point processes'. Although this point process involves many latent applicabilities, statistical techniques are rather less compared with the usual spatial point processes, especially of likelihood-based ones. The reason for this seems mainly to be due to the difficulty of getting likelihood parametrization. See, Ogata and Tanemura (1985), as an example of the likelihood analysis of marked point processes.

In this paper, we discuss the case when the mark is given as a direction of each object. Our approach here is due to the maximum likelihood by using the Gibbs canonical distribution as a likelihood function. Similar approaches are given by many authors including our selves.

2. Directional Interaction Model and Likelihood

Suppose that, in a finite region V of space, a set of coordinates $X \equiv (x_1, x_2, \ldots, x_N)$ of N points is given, and suppose that a set of directional vectors $S \equiv (s_1, s_2, \ldots, s_N)$ is attached to each spatial point ($s_i \in \Omega, |s_i| = 1, i = 1, \cdots, N$; Ω is the angular space).

H. Bozdogan (ed.), Proceedings of the First US/Japan Conference on the Frontiers of Statistical Modeling: An Informational Approach, 293–313.

Then, we assume that an interaction potential of form $\Phi_\xi(x_i, x_j; s_i, s_j)$ works between two points i and j, and that the given set of X and S obey the so-called 'Gibbs canonical distribution' of marked point processes under the above potential (see Stoyan *et al.* (1987), or Ogata and Tanemura (1985) for the reasoning and example of Gibbs distribution for the marked point processes). Here ξ is a parameter which characterizes the form of the potential.

A likelihood procedure could be formulated for estimating the general form of interaction potential $\Phi_\xi(x_i, x_j; s_i, s_j)$ in a very formal fashion. In this paper, however, we consider the potential model of the form

$$\Phi_\xi(x_i, x_j; s_i, s_j) \equiv f_\xi(s_i, s_j)$$

which indicates that the interaction works only between a pair of directional vectors s_j and s_j and that it does not depend on their positions x_i and x_j. Then we call f_ξ a directional interaction model. Note that, in this model, the information of X is assumed to be included implicitly as the lattice model, for instance.

In this case, the Gibbs distribution of S is given by the probability density

$$p(S|f_\xi) = \frac{1}{Z(f_\xi; N)} \exp\left\{-\sum_{i=1}^{N-1}\sum_{j=i+1}^{N} f_\xi(s_i, s_j)\right\}.$$

Thus, the explicit form of log-likelihood for our directional interaction model tends to be

$$\log L(\xi; S) = -\sum_{i=1}^{N-1}\sum_{j=i+1}^{N} f_\xi(s_i, s_j) - \log Z(f_\xi; N).$$

Then we can formally estimate $\xi = \hat{\xi}$ as a value which maximizes the above function. However, the second term of the above equation is the normalizing constant and is written in a complicated and multiple integration form as

$$Z(f_\xi; N) = \frac{1}{|\Omega|^N} \int_\Omega \cdots \int_\Omega ds_1 \cdots ds_N \exp\{-\sum_{i=1}^{N-1}\sum_{j=i+1}^{N} f_\xi(s_i, s_j)\},$$

and then, because of this complexity, it is usually difficult to perform the maximization of the above log-likelihood function.

Therefore, we introduce here further simplifications of directional interaction model. Firstly, we confine the model as

$$f_\xi(s_i, s_j) \equiv f_\xi(s_i \cdot s_j),$$

i.e., f_ξ is defined as a function of only the inner product of s_i and s_j. Secondly, we assume that the interaction works only between the 'contiguous pair' s_i and s_j. In other words, we assume that the interaction of a certain directional vector s_i, say, is effective only with vectors whose locations are 'contiguous' to s_i.

Then, we can evaluate the above normalizing constant in a simple form, sometimes in analytical forms. In the following, we show several interesting models with such properties.

3. Exact Likelihoods

If all vectors are located on the points of a regular lattice, the contiguous pair mentioned in the previous section will corresponds to the nearest neighbour pair, in this case. In this section and the following we consider such regular lattice models.

3.1 ISING MODEL

The so-called 'Ising model' is well known in statistical physics. In this model, the objects are the 'spins' whose directions are limited only to 'up' or 'down' characterized by 1 or -1, respectively. Moreover, the spatial position of objects are limited to the lattice sites. It will be worth discussing this model here, since it is rarely discussed in the context of likelihood procedure. For the Ising model, then, the angular space will be $\Omega = \{\pm 1\}$, and the interaction will be

$$f_\xi \equiv -\xi s_i \cdot s_j,$$

where the interacting pair (i, j) is limited to the nearest neighbour.

We mention now two cases in which exact log-likelihoods are obtained.

3.1.1. *1-D Ising model.* For a one-dimensional array of spins, where

$$s_i = \pm 1; \quad i = 1, 2, \cdots, N,$$

the log-likelihood function can be written analytically as

$$\log L = \xi \sum_{i=1}^{N-1} s_i s_{i+1} - (N-1)\log(2\cosh\xi).$$

3.1.2. *2-D Ising model.* In the case of two-dimensional Ising model, the partition function, i.e., the normalizing constant is analytically obtained by Onsager (1944). We mention here only the square-lattice case,

$$s_{ij} = \pm 1; \quad (i, j) \in Z^2 \text{ square lattice.}$$

For this model, the log-likelihood can be written by using the Onsager solution as

$$\begin{aligned}\log L &= \frac{\xi}{2}\sum_i\sum_j s_{ij}(s_{i-1,j} + s_{i+1,j} + s_{i,j-1} + s_{i,j+1}) \\ &- \frac{N}{2\pi^2}\int_0^\pi\int_0^\pi \log\left\{(\cosh 2\xi)^2 - (\cos\omega + \cos\omega')\sinh 2\xi\right\} d\omega d\omega'.\end{aligned}$$

3.2 LINEAR CHAIN OF UNIT VECTORS

We consider then an interesting class of spin models which are different from the above Ising model: the linear chain of unit vectors, i.e., one-dimensional array of unit vectors whose interaction is written as

$$\sum_{i=1}^{N-1} \sum_{j=i+1}^{N} f_\xi(s_i \cdot s_j) = \sum_{i=1}^{N-1} f_\xi(s_i \cdot s_{i+1}),$$

where the summation is taken only for the nearest neighbour pairs.

3.2.1. *2-D unit vectors (Circular unit vectors).* When the unit vectors are two-dimensional, we can put

$$s_i = (a_i, b_i), \quad a_i = \cos\phi_i, \quad b_i = \sin\phi_i, \quad |\Omega| = 2\pi.$$

Then, the normalizing constant could be reduced to

$$\begin{aligned} Z(f_\xi; N) &= \frac{1}{(2\pi)^N} \int_0^{2\pi} d\phi_1 \prod_{i=1}^{N-1} \int_0^{2\pi} \exp\{-f_\xi(\cos\phi_{i+1})\} d\phi_{i+1} \\ &= \left[\frac{1}{2\pi} \int_0^{2\pi} \exp\{-f_\xi(\cos\phi)\} d\phi\right]^{N-1}. \end{aligned}$$

Thus, the log-likelihood function will be given as

$$\log L = -\sum_{i=1}^{N-1} f_\xi(\cos(\phi_i - \phi_{i+1})) - (N-1)\log\left[\frac{1}{2\pi}\int_0^{2\pi} \exp\{-f_\xi(\cos\phi)\}d\phi\right], \quad (3.1)$$

where the multiplicity of integration is single. Then, we can easily evaluate the above function by using numerical integration method if necessary.

3.2.2. *3-D unit vectors (Spherical unit vectors).* In case the unit vectors are three-dimensional, we can put

$$s_i = (a_i, b_i, c_i),$$

where the angular space is $\Omega = \{(\theta, \phi);\ 0 \le \theta < \pi,\ 0 \le \phi < 2\pi\}$ and $|\Omega| = 4\pi$, and

$$\begin{cases} a_i &= \sin\theta_i \cos\phi_i \\ b_i &= \sin\theta_i \sin\phi_i \\ c_i &= \cos\theta_i. \end{cases}$$

Thus, the normalizing constant tends to be

$$\begin{aligned} Z(f_\xi; N) &= \frac{1}{(4\pi)^N} \int d\Omega_1 \prod_{i=1}^{N-1} \int_0^{\pi} \exp\{-f_\xi(\cos\theta_{i+1})\} \sin\theta_{i+1} d\theta_{i+1} \int_0^{2\pi} d\phi_{i+1} \\ &= \left[\frac{1}{2}\int_0^{\pi} \exp\{-f_\xi(\cos\theta)\} \sin\theta d\theta\right]^{N-1} \\ &= \left[\frac{1}{2}\int_{-1}^{1} \exp\{-f_\xi(t)\} dt\right]^{N-1}, \end{aligned}$$

where $\int d\Omega_1$ stands for the integration about whole angular space of the first unit vector. Then, the log-likelihood function becomes, in this case,

$$\log L = -\sum_{i=1}^{N-1} f_\xi(s_i \cdot s_{i+1}) - (N-1)\log\left[\frac{1}{2}\int_{-1}^{1}\exp\{-f_\xi(t)\}dt\right], \qquad (3.2)$$

again the integration be simplificated to perform numerical calculation.

It is interesting to note that for the Poisson model, where $\xi = 0$ holds,

$$\log L = 0, \quad \mathrm{AIC} = 0$$

holds in all the above cases.

3.3 SERIAL CORRELATION OF LINEAR CHAIN OF UNIT VECTORS

For the above models, the expectation of the product of neighbouring vectors, i.e. the serial correlation coefficient,

$$E[s_i \cdot s_{i+1}]$$

is also given in an analytic form. From the definition, we obtain in an integration form as

$$E[s_i \cdot s_{i+1}] = \frac{1}{Z(f_\xi; N)}\frac{1}{|\Omega|^N}\int_\Omega \cdots \int_\Omega ds_1 \cdots ds_N \; s_i \cdot s_{i+1}\; \exp\{-\sum_{i=1}^{N-1} f_\xi(s_i \cdot s_{i+1})\}.$$

This expression is simplified for the models of linear chain of unit vectors as follows.

3.3.1. *2-D unit vectors.* In the case of the linear chain of two-dimensional unit vectors, we obtain as the serial correlation coefficient,

$$\begin{aligned} E[s_i \cdot s_{i+1}] &= \frac{\int_0^{2\pi} \cos\phi \cdot \exp\{-f_\xi(\cos\phi)\}d\phi}{\int_0^{2\pi}\exp\{-f_\xi(\cos\phi)\}d\phi} \\ &= \frac{\int_0^{2\pi} \cos\phi \cdot \exp(\xi\cos\phi)d\phi}{\int_0^{2\pi}\exp(\xi\cos\phi)d\phi}, \quad \text{for} \quad f_\xi = -\xi s_i \cdot s_{i+1}. \end{aligned}$$

3.3.2. *3-D unit vectors.* When we consider the linear chain of three-dimensional unit vectors, we attain the following as the serial correlation coefficient,

$$\begin{aligned} E[s_i \cdot s_{i+1}] &= \frac{\int_{-1}^{1} t \cdot \exp\{-f_\xi(t)\}dt}{\int_{-1}^{1}\exp\{-f_\xi(t)\}dt} \\ &= \coth\xi - \frac{1}{\xi}, \quad \text{for} \quad f_\xi = -\xi s_i \cdot s_{i+1}. \end{aligned}$$

Table 1. Summary of Model fitting for Simulated Data.
Linear chain of 2-D unit vectors. ($N = 200$; No. of samples : $n = 50$)
The last column is the theoretical estimate for $\sqrt{N}\times$(standard deviation).

ξ_0	$\hat{\xi}$	$\hat{\sigma}$	$\sqrt{N}\hat{\sigma}$	$1/\sqrt{-\frac{1}{N}\mathrm{E}[\partial^2 \log L/\partial\xi^2]}$
-2.0	-2.0122	0.1840	2.6027	2.4676
-1.5	-1.4888	0.1495	2.1136	2.0113
-1.0	-0.9925	0.1157	1.6363	1.6799
-0.5	-0.5193	0.1085	1.5340	1.4806
0.0	-0.0068	0.0897	1.2683	1.4142
0.5	0.5275	0.0975	1.3795	1.4806
1.0	1.0147	0.1170	1.6545	1.6799
1.5	1.5137	0.1421	2.0094	2.0113
2.0	2.0155	0.1592	2.2514	2.4676

4. Simulation Study

In order to see the feasibility of the procedure of estimation provided so far, we have done computer simulations for the linear chains of two dimensional and three dimensional unit vectors. For both cases, we adopted the model

$$f_\xi = -\xi s_i \cdot s_{i+1}. \tag{4.1}$$

In statistical physics, this interaction model is known as 'XY model' if s_i is fixed in the X-Y plane, and is known as 'Heisenberg model' if s_i is limited to unit sphere.

After choosing the length of the chain to be $N = 200$, we performed computer simulation of this model for nine cases of parameter values, $\xi_0 = -2.0, -1.5, -1.0, -0.5, 0.0, 0.5, 1.0, 1.5$ and 2.0. According to the above model, the negative parameter values indicate the repulsive directional interaction, while the positive values the attractive interaction. The case of $\xi_0 = 0$ represents, of course, that there is no directional interaction between contiguous pairs. In other words, it means the sequence of vectors is completely independent. For each case, we obtained respectively fifty independent samples of equilibrium configuration. Table 1 shows a summarized result of model fitting for simulated data. In this table, the first column represents the true value ξ_0, the second the mean of 50 samples of estimate $\hat{\xi}$, the third the sample standard deviation of 50 values of $\hat{\xi}$. The last column is the reciprocal of square root of Fisher information rate, and it reresents the theoretical estimate for $\sqrt{N} \times \hat{\sigma}$. The resulting values indicate good coincidence both between ξ_0 and the mean of $\hat{\xi}$ and between $\sqrt{N}\hat{\sigma}$ and its theoretical estimate. The tendency of the increase of $\sqrt{N}\hat{\sigma}$ as the increase of the absolute value of ξ can be explained as follows: For the 'XY model', the theoretical Fisher information rate can be explicitly given by using Eqs.(3.1) and (4.1) as

$$-\frac{1}{N}\mathrm{E}[\partial^2 \log L/\partial \xi^2] = \frac{\int_0^{2\pi} \cos^2\phi \cdot \exp(\xi\cos\phi)d\phi}{\int_0^{2\pi} \exp(\xi\cos\phi)d\phi} - \left\{\frac{\int_0^{2\pi} \cos\phi \cdot \exp(\xi\cos\phi)d\phi}{\int_0^{2\pi} \exp(\xi\cos\phi)d\phi}\right\}^2$$
$$\equiv \int_0^{2\pi} \cos^2\phi \, g_\xi(\phi)d\phi - \left\{\int_0^{2\pi} \cos\phi \, g_\xi(\phi)d\phi\right\}^2,$$

indicating that the right hand side of the above equation is the variance of $\cos\phi$ under the probability density $g_\xi(\phi) = \exp(\xi\cos\phi)/\int_0^{2\pi}\exp(\xi\cos\phi)d\phi$. Thus, we can intuitively see that when $|\xi|$ becomes large, the interaction between contiguous vectors tends to be big, whether it is attractive or repulsive. Then, the variance of angle between contiguous vectors will tend to be small when the interaction becomes large. Hence, we conclude the standard deviation of the estimator $\hat{\xi}$, i.e. $\hat{\sigma}_\xi$ or $\sqrt{N}\hat{\sigma}_\xi$, becomes large as the increase of $|\xi|$.

Furthermore, we can detect whether the true model is selected or not for each sample data by using the model selection procedure by the value of AIC

$$\mathrm{AIC}(k) = (-2)\text{max log likelihood} + 2k,$$

where k is the number of adjustable parameters. In our case, the value of AIC for true model is AIC(0) and that of AIC for the estimate is AIC(1), because the number of adjustable parameter is ξ only in our case. Specifically, let us note for the non-interacting model, i.e. for the Poisson model, it holds AIC $\equiv 0$. According to the minimum AIC procedure, the true model is selected if AIC(0) $-$ AIC(1) is negative. By applying this procedure to the set of samples used in Table 1, 42 to 44 samples among 50 are selected as being true respectively for all cases of parameter values. Figure 1. illustrates the summary of the result of model fitting for respective samples. The horizontal axis in Figure 1. indicates the true value ξ_0 and the vertical axis the estimated value $\hat{\xi}$. Each cross in Figure 1. represents the result of fitting for each sample data and the pattern of crosses shows how the estimates scatter.

Similar summarized result for the linear chain of 3-D unit vectors of the above model (Heisenberg model) is given in Table 2. This table shows that the coincidence between the true values and the means of estimates is excellent as well as the coincidence between the experimental and theoretical standard deviations. In this case, the same order of number of samples are selected to be true as in the above 2-D vector case. Figure 2. gives the result of model fitting, correspondingly to Figure 1.

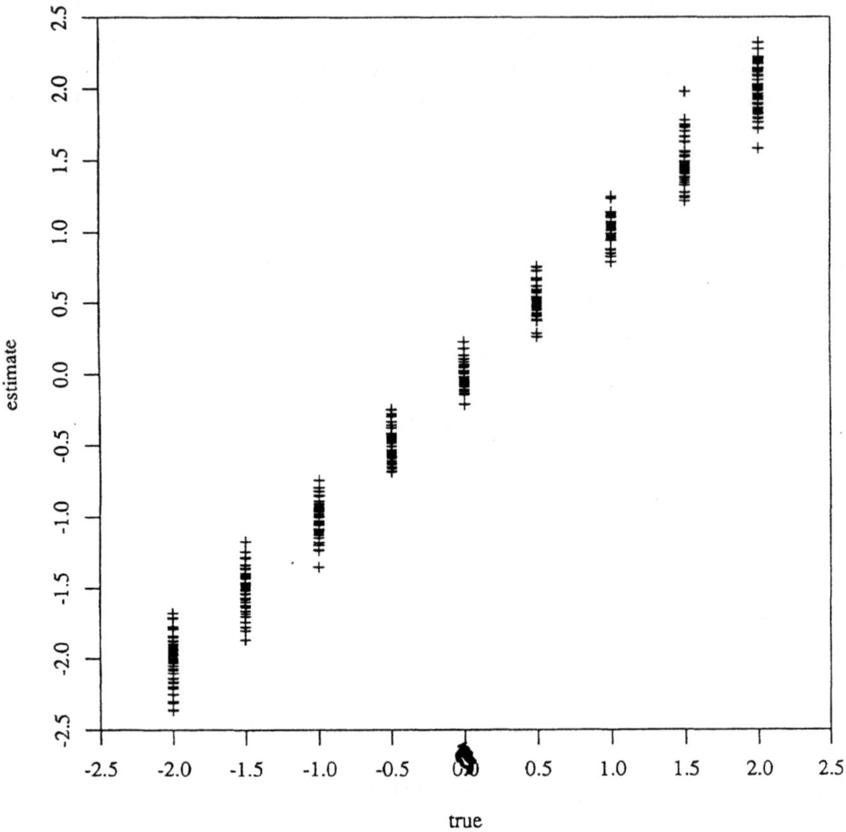

Figure 1: Model fitting for simulated data: Linear chain of 2-D unit vectors.

Table 2. Summary of Model fitting for Simulated Data. Linear chain of 3-D unit vectors ($N = 200$; $n = 50$). The last column is the theoretical estimate for $\sqrt{N}\times$(standard deviation).

ξ_0	$\hat{\xi}$	$\hat{\sigma}_\xi$	$\sqrt{N}\hat{\sigma}_\xi$	$1/\sqrt{-\frac{1}{N}\mathrm{E}[\partial^2 \log L/\partial\xi^2]}$
-2.0	-2.0003	0.1924	2.7214	2.4035
-1.5	-1.5195	0.1509	2.1346	2.1188
-1.0	-1.0062	0.1461	2.0658	1.9085
-0.5	-0.5060	0.1097	1.5507	1.7797
0.0	-0.0091	0.1150	1.6230	1.7364
0.5	0.4677	0.1231	1.7412	1.7797
1.0	0.9566	0.1337	1.8908	1.9085
1.5	1.4998	0.1318	1.8642	2.1188
2.0	1.9976	0.1646	2.3272	2.4035

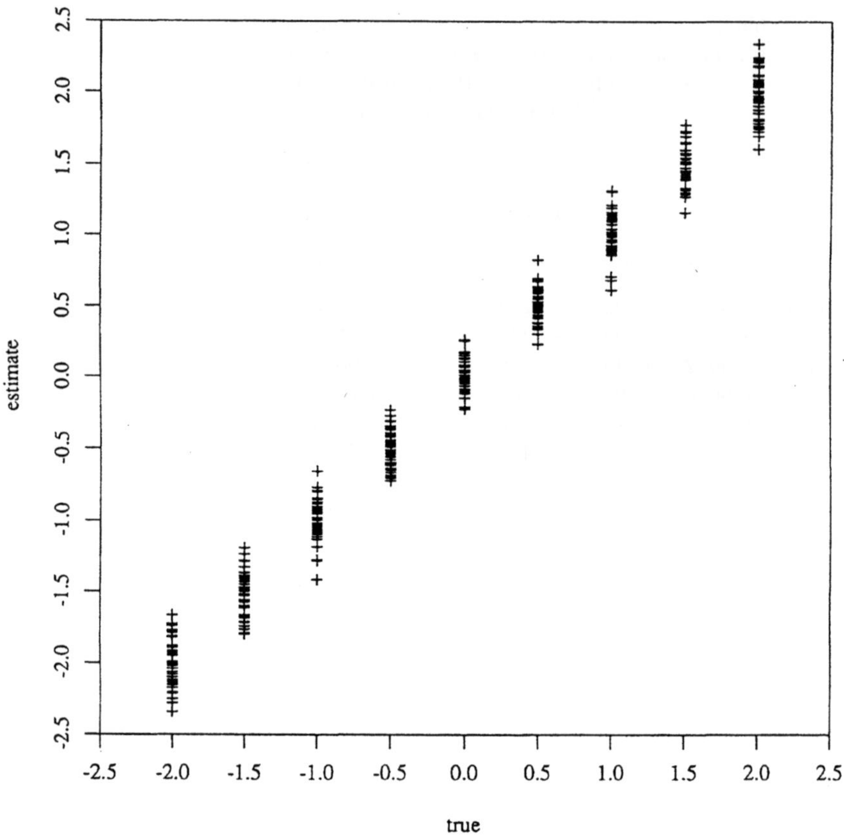

Figure 2: Model fitting for simulated data: Linear chain of 3-D unit vectors.

The behaviour of increasing variance σ_ξ as the increase of $|\xi|$, similar to the case of 2-D unit vectors, can also be explained as follows: From Eqs.(3.2) and (4.1), we obtain

$$\begin{aligned} -\frac{1}{N}\mathrm{E}[\partial^2 \log L/\partial\xi^2] &\simeq \frac{\int_0^\pi \cos^2\theta \exp(\xi\cos\theta)\sin\theta d\theta}{\int_0^\pi \exp(\xi\cos\theta)\sin\theta d\theta} - \left[\frac{\int_0^\pi \cos\theta \exp(\xi\cos\theta)\sin\theta d\theta}{\int_0^\pi \exp(\xi\cos\theta)\sin\theta d\theta}\right]^2 \\ &\equiv \int_0^\pi \cos^2\theta\, h_\xi(\theta)d\theta - \left\{\int_0^\pi \cos\theta\, h_\xi(\theta)d\theta\right\}^2 \\ &= 1 - \coth^2\xi + \frac{1}{\xi^2}, \end{aligned}$$

where the approximation in the first line is the replacement of $N-1$ by N when N is moderately large. The second expression in the above equation shows, similarly in the case of 2-D unit vectors, that it is equal to the variance of $\cos\theta$ under the probability density $h_\xi(\theta) = \sin\theta\exp(\xi\cos\theta)/\int_0^\pi \sin\theta\exp(\xi\cos\theta)d\theta = \xi\exp(\xi\cos\theta)\sin\theta/(2\sinh\xi)$. Then, the intuitive reasoning for the increase of standard deviation of the estimator $\hat{\xi}$ as the increase of $|\xi|$ will be similarly understood as in the case of 2-D unit vectors. Thus we find the feasibility of our procedure and the validity of our simulation.

5. Analysis of some real data

In order to show the applicability of our procedure, we have analyzed several real data.

5.1 MOVEMENT OF SEA STARS

Table 3(a) shows two sets of data of the directions traveled by sea stars or starfish *Astrospecten jonstoni* on successive ten days in radian (Upton and Fingleton, 1989: Table 9.47). Two individuals of sea stars listed in Table 3(a) are known to be initially located only a few centimeters apart, both travelled a considerable distance in an essentially northerly direction. It will then be interesting to see what kinds of directional interaction are found between the two individuals by applying our procedure to these data in order to characterize the movement of sea stars. Table 3(b) is the result of our analysis. This table indicates that, in both cases, the directional interactions are highly attractive, that means, sea stars had a tendency to move the similar direction during the observed ten days.

5.2 MAGNETIC REMANENCE OF ROCKS

As the second example, we consider two sets of data of magnetic remanence of rocks (Upton and Fingleton, 1989: Tables 10.1 and 10.3). Table 4(a) gives measurements of

Table 3(a). Directions traveled by two sea stars on successive days (in radian). (Upton and Fingleton, 1989).

Data No.	Sea star 1	Sea star 2
1	5.861	0.119
2	1.592	6.030
3	5.918	0.894
4	5.564	6.271
5	0.150	0.030
6	0.122	5.828
7	6.212	5.596
8	0.080	5.836
9	6.049	5.754
10	5.672	5.564

Table 3(b). Summary of Model fitting for Sea Star Data.

Data ID	Data size	$\hat{\xi}$	$\log \hat{L}$	AIC
1	10	1.624	3.97	-5.94
2	10	3.986	9.24	-16.49

Table 4(a). Magnetic remanence measurement of 17 red-bed specimens (Upton and Fingleton, 1989).

Data No.	θ	ϕ
1	1.379	5.131
2	1.030	5.009
3	0.908	4.957
4	1.117	4.922
5	0.559	4.835
6	0.698	4.154
7	1.326	5.847
8	0.942	5.550
9	0.942	5.393
10	1.152	5.376
11	0.977	5.306
12	0.960	5.219
13	1.204	5.184
14	1.152	5.184
15	0.995	5.166
16	1.047	4.869
17	0.820	4.869

three-dimensional angle (θ, ϕ) in radian for 17 Australian red-bed specimens. Table 4(b) gives similar information for 52 rock samples after they had been subjected to a temperature of 680°C. Note that these data are assumed to be the sets of random sample points on the unit sphere. Therefore, the nature of these data is different from the type of data which we considered so far in this paper. However, it will be interesting to analyze these magnetic remanence data from the view point of one dimensional array of unit vectors. As we have no information on the way of arragement of each vector, it will be quite natural to consider the randomization test. For that purpose, we have done the random rearrangement of data of Tables 4(a), (b) one hundred times each. The result is summarized in Table 4(c). Values in the columns of $\hat{\xi}$ and AIC respectively represent the means of 100 random permutations and the values in parentheses respectively represent their standard deviations. As for the red-bed data ($N = 17$), the values of $\hat{\xi}$ and AIC indicates that a strong attractive interaction may be working among unit vectors, which in other words shows almost all vectors are pointing to similar directions. The last fact will be easily seen by graphical plotting (e.g., see Figure 10.2c of Upton and Fingleton (1989)). Our result is consistent with the conclusion of Fisher *et al.* (1981) who conclude that most observations in Table 4(a) are concordant and that, however, the 6th and 7th

Table 4(b). Magnetic remanence of 52 rock specimens (Upton and Fingleton, 1989).

Data No.	θ	ϕ	Data No.	θ	ϕ
1	0.340	5.646	27	1.895	5.756
2	2.716	5.501	28	1.442	5.519
3	1.271	0.176	29	0.674	1.499
4	1.513	3.147	30	1.686	0.182
5	0.920	3.686	31	1.573	1.265
6	0.489	3.482	32	1.002	4.662
7	0.284	6.166	33	2.398	5.171
8	0.271	3.173	34	2.417	0.330
9	2.271	5.438	35	3.012	5.988
10	1.506	1.482	36	2.740	1.562
11	0.869	6.156	37	2.222	0.295
12	2.290	3.456	38	2.651	3.810
13	1.283	2.990	39	2.576	6.261
14	1.035	3.508	40	2.513	5.679
15	1.587	4.522	41	1.328	1.538
16	2.239	3.665	42	1.831	0.377
17	2.388	3.337	43	0.920	6.241
18	0.880	1.491	44	2.342	3.766
19	0.297	0.380	45	1.756	1.841
20	2.012	4.271	46	2.803	0.466
21	1.105	1.154	47	2.601	2.079
22	1.267	5.901	48	1.485	6.142
23	1.819	4.264	49	1.335	2.091
24	1.997	0.309	50	1.176	0.363
25	2.768	0.784	51	1.417	5.543
26	1.913	3.882	52	2.093	1.731

Table 4(c). Summary of model fitting for magnetic property of rocks.

Model: Linear chain of 3-D unit vectors

$$f_{\xi} = -\xi s_i \cdot s_{i+1}.$$

Values in the table are means (standard deviation) of 100 randomized permutations.

Data set	$\hat{\xi}$	AIC
Red-bed ($N = 17$)	8.659 (1.763)	-56.67 (5.97)
Remanence ($N = 52$)	0.035 (0.258)	0.88 (1.96)

observation should be treated as outliers through a method of spherical statistics.

As for the magnetic remanence data at 680°C of 52 measurements, the values of $\hat{\xi}$ and AIC in Table 4(c) indicates that there is no interaction among unit vectors since the mean of AIC is positive and then the Poisson model is selected from the discussion in the last section. This is also true for respective arrangements of randomized permutation since Poisson model was selected for 82 trials among 100 through AIC values. Our result for this data set is in accord with the analysis by Fisher and Lee (1986) who conclude that test of uniformity on this data do not reveal any departure from uniformity.

5.3 AMINO ACID SEQUENCES

The detection of similarity of amino acid sequences is known to be important in the study of protein evolution or other molecular biological studies. There are much investigations by molecular biologists in this direction. See Doolittle (1981), for instance. We give here a new method of characterizing the pattern of amino acid sequences by applying the likelihood procedure of directional interaction discussed so far.

For that purpose, we use the similarity table among elemental amino acids (see Table 5). It is known that there exist twenty elemental amino acids and that each of them uniquely corresponds to the codon in DNA or RNA (see, for example, Alberts *et al.* (1983)). Table 5 shows the degree of similarity among the amino acids derived by comparing the physical and chemical properties of respective amino acids (Miyata *et al.*, 1986). In the table, the value 1000 represents the biggest similarity, while the value 0 the most dissimilarity.

It will then be interesting to translate amino acid sequences into one-dimensional series of unit vectors whose angular information is determined by the similarity among amino acids. Accordingly, we use Table 5 to assign an angle ϕ_{ij} between two amino acids i and j by the relation

$$\phi_{ij} = \pi \left(1 - \frac{d_{ij}}{1000}\right), \tag{5.1}$$

as one possibility, where d_{ij} represents the similarity value between i and j given in Table 5. Then, for a similar pair of amino acids, a smaller angle is assigned (when the similarity value is 1000, the assigned angle is 0), while for a dissimilar pair, a bigger angle is assigned (when the similarity is 0, the angle is π).

We then consider two types of analysis, that is, the analysis within a sequence and that between a pair of sequences.

5.3.1. *Analysis of Individual Sequence.* Firstly, for a single amino acid sequence, we assign angles for a contiguous pair of amino acids by the method discussed above. Then, we can obtain a sequence of angles assigned to each pair of amino acid sites

Table 5. Degrees of similarity between amino acids (Miyata *et al.*, 1986). This table is used to characterize amino acid sequences.

		G	A	S	T	P	L	I	M	V
Gly	G	1000	740	757	514	722	0	0	45	211
Ala	A		1000	854	742	982	211	231	308	471
Ser	S			1000	745	840	131	157	237	385
Thr	T				1000	751	357	388	468	594
Pro	P					1000	228	251	325	488
Leu	L						1000	960	882	740
Ile	I							1000	917	757
Met	M								1000	822
Val	V									1000

		D	N	E	Q	F	Y	W	K	R	H	C
Gly	G	322	439	205	291	0	0	0	0	0	205	365
Ala	A	322	491	297	451	77	91	0	154	165	379	602
Ser	S	465	625	411	528	14	48	0	225	217	445	474
Thr	T	414	600	477	680	257	300	0	399	420	622	585
Pro	P	314	485	291	451	94	108	0	160	171	385	620
Leu	L	0	2	0	228	820	731	505	148	251	259	528
Ile	I	0	37	31	265	825	754	508	188	288	300	534
Met	M	0	120	105	342	765	734	459	248	345	374	582
Val	V	28	211	151	391	591	565	282	228	305	397	754
Asp	D	1000	814	742	579	0	0	0	414	331	508	5
Asn	N		1000	757	717	0	22	0	474	417	631	191
Glu	E			1000	760	0	79	0	674	585	725	68
Gln	Q				1000	197	291	22	697	677	908	291
Phe	F					1000	862	682	185	294	248	360
Tyr	Y						1000	697	308	422	351	319
Trp	W							1000	111	222	97	45
Lys	K								1000	885	774	65
Arg	R									1000	765	125
His	H										1000	268
Cys	C											1000

Table 6. Samples of amino acid sequence.

Twelve sequences are of length 90 each and taken from Miyata *et al.* (1986).

```
17.6      : PIWVVPKKQDASGKQKFRIVIDYRKLNEITVG---DRHP-IPNMDEILGKLGRC-NYFTT
            IDLAKGFHQIEMDPESVSKTAFS-------

CaMV      : PAFLVN-NEAEKRRGKKRMVVNYKAMNKATIG---DAYN-LPNKDELLTLIRGK-KIFSS
            FDCKSGFWQVLLDQESRPLTAFT-------

M-MuLV    : PLLPV-KKP-GT--NDYRPVQDLREVNKRVE----DIHPTVPNPYNLLSGLPPSHQWYTV
            LDLKDAFFCLRLHPTSQPLFAFEW-RDPEM

IAP       : PIFVI-KKKSGK----WRLLHDLRAINNQMH----L-FGPVQRGLPLLSALPQDWKLI-I
            IDIKDCFFSIPLYPRDRPRFAFTIPSLNHM

RSV       : PVFVI-RKASGS----YRLLHDLRAVNAKLV----P-FGAVQQGAPVLSALPRGWPLM-V
            LDLKDCFFSIPLAEQDREAFAFTLPSVNNQ

HTLV-I    : PVFPV-KKANGT----WRFIHDLRATNSLTI----DLSSSSPGPPDL-SSLPTTLAHLQT
            IDLRDAFFQIPLPKQFQPYFAFTVPQQCNY

HTLV-II   : PVFPV-KKPNGK----WRFIHDLRATNAITT----TLTSPSPGPPDL-TSLPTALPHLQT
            IDLTDAFFQIPLPKQYQPYFAFTIPQPCNY

BLV       : PVFPV-RKPNGA----WRFVHDLRATNALTK----PIPALSPGPPDL-TAIPTHPPHIIC
            LDLKDAFFQIPVEDRFRSYLSFTLPSPGGL

HTLV-III: PVFAI-KKKDST---KWRKLVDFRELNKRTQ----D-FWEVQLGIPHPAGLKKKKSVT-V
            LDVGDAYFSVPLDEDFRKYTAFTIPSINNE

WHV       : GVFLVDKNPNNS--SESRLVVDFSQFSRGHTRVHWPKF-AVPNLQTLANLLSTDLQWL-S
            LDVSAAFYHIPISPAAVPHLLVG-------

HBV       : GVFLVDKNPHNT--TESRLVVDFSQFSRGSTHVSWPKF-AVPNLQSLTNLLSSNLSWL-S
            LDVSAAFYHIPLHPAAMPHLLVG-------

DBHV      : KLFLVDKNSRNT--EEARLVVDFSQFSKGKNAMRFPRY-WSPNLSTLRRILPVGMPRI-S
            LDLSQAFYHLPLNPASSSRLAVS-------
```

Table 7. Model fitting to the single amino acid sequence.

Data	$\hat{\xi}$	$\log \hat{L}$
17.6	-0.552	6.42
CaMV	-0.388	3.25
M-MuLV	-0.596	7.41
IAP	-0.654	8.80
RSV	-0.454	4.43
HTLV-I	-0.446	4.27
HTLV-II	-0.467	4.66
BLV	-0.374	3.03
HTLV-III	-0.469	4.70
WHV	-0.441	4.17
HBV	-0.432	4.00
DHBV	-0.540	6.14

from left to right of the original sequence. Hence, we determine the angular value of a specific site by summing up the angle data assigned to all the sites to the left of there. Consequently, we obtain a sequence of directions from the original amino acid sequence. We can then characterize the transformed sequence by estimating the parameter of our directional interaction model.

We analyzed twelve sequences of length ninety as given in Table 6. The data were taken from Miyata *et al.* (1986). These data are the part of the amino acid sequences translated from the DNA or RNA sequences of retroviral reverse transcriptases of viruses (M-MuLV, IAP, RSV, HTLV-I, HTLV-II, BLV, HTLV-III), putative polymerases (CaMV, HBV, WHV, DHBV), or copia-like element ("17.6"). Toh *et al.* (1983) and Miyata *et al.* (1986) found by their "homology matrix" comparisons that these amino acid sequences are similar with each other. In the data in Table 6, '–' means a gap. The gap represents a site where a deletion or an insertion by a certain length of nucleotide sequence occurs. Toh *et al.* (1983) counted gaps as one substitution regardless of their length and assigned a lowest score of similarity for a gap. Accordingly, we assigned π, i.e. the lowest similarity value, as an angle between the gap and amino acids. For a pair of consecutive gaps, on the other hand, we tentatively assigned the angle 0 by assuming the consecutive gaps has a highest similarity. In Table 7, we give the result of our analysis. This result shows that, for all sequences, the directional interaction is repulsive. This indicates the dissimilar amino acids are liable to be contiguous. The result seems to be natural because, in order the specific amino acid sequence to be stable, it will be desirable that contiguous amino acids are dissimilar with each other and difficult to be substituted.

Table 8. Estimates of parameter ξ for the amino acid sequences in pair.

	CaMV	M-MuLV	IAP	RSV	HTLV-I	HTLV-II	BLV
17.6	0.820	0.447	0.111	0.106	0.251	0.205	0.132
CaMV		0.148	0.049	-0.030	0.047	-0.009	-0.035
M-MuLV			0.771	0.710	0.931	0.853	0.758
IAP				2.027	0.818	0.859	0.800
RSV					1.180	1.167	1.051
HTLV-I						4.830	2.294
HTLV-II							2.293

	HTLV-III	WHV	HBV	DHBV
17.6	0.110	0.506	0.491	0.504
CaMV	0.016	0.395	0.392	0.246
M-MuLV	0.494	0.161	0.193	0.144
IAP	0.916	0.081	0.053	-0.051
RSV	0.952	0.092	0.072	0.041
HTLV-I	0.569	0.020	0.023	0.075
HTLV-II	0.544	0.078	0.083	0.169
BLV	0.601	-0.070	-0.031	0.083
HTLV-III		-0.085	-0.072	-0.148
WHV			5.175	1.924
HBV				1.902

5.3.2. *Analysis of a Pair of Sequences.* As the next analysis of amino acid sequences, we compared two sequences in pairs for the data in Table 6. The assignment of angle is done by using Eq.(5.1) again, but in this case the angle ϕ_{ij} is given for the two amino acid elements i and j at the same corresponding site of a pair of sequences. Then we get a series of angles for the pair of sequences. Thus, by the similar construction as given above, we obtain a sequence of directions for a pair of amino acid sequences.

Table 8 shows the result of parameter estimation for 66 pairs of comparison from the data of Table 6. Table 9 presents the estimated log likelihoods corresponding to Table 8. These tables indicate, for several pairs of amino acid sequences, the estimates for ξ are large which suggests the strong similarity for these pairs. It is interesting to note that the result of our analysis is not contradicting the conclusions obtained from the method different from ours by Toh *et al.* (1983) and Miyata *et al.* (1986). For example, Miyata *et al.* (1986) concludes that the pair of HBV (Hepatitis B virus) and WHV (Woodchuck hepatitis virus) and the pair of HTLV-I (Human T-cell leukemia virus type I) and HTLV-II (Human T-cell leukemia virus type II), are in the group of the biggest similarity. In accord with this, our estimates of ξ for the

Table 9. Estimates of log likelihood for the amino acid sequences in pair.

	CaMV	M-MuLV	IAP	RSV	HTLV-I	HTLV-II	BLV
17.6	13.28	4.29	0.27	0.25	1.39	0.93	0.38
CaMV		0.48	0.05	0.02	0.05	0.00	0.03
M-MuLV			11.91	10.25	16.58	14.25	11.54
IAP				51.67	13.23	14.42	12.72
RSV					24.56	24.14	20.35
HTLV-I						101.65	59.04
HTLV-II							58.99

	HTLV-III	WHV	HBV	DHBV
17.6	0.27	5.45	5.12	5.40
CaMV	0.01	3.37	3.32	1.33
M-MuLV	5.19	0.57	0.82	0.46
IAP	16.13	0.15	0.06	0.06
RSV	17.26	0.19	0.12	0.04
HTLV-I	6.80	0.01	0.01	0.12
HTLV-II	6.23	0.14	0.15	0.63
BLV	7.52	0.11	0.02	0.15
HTLV-III		0.16	0.17	0.49
WHV			105.21	48.62
HBV				47.96

pair HBV and WHV and for the pair HTLV-I and HTLV-II in Table 8, respectively, indicate big positive values among others ($\hat{\xi} = 5.175$ and 4.830, respectively), which suggests big similarities in respective pairs. As for the pair of data "17.6" (Drosophila melanogaster copia-like element 17.6) and CaMV (Cauliflower mosaic virus), Miyata *et al.* (1986) concludes that they are classified into the same group with a moderate degree of similarity. Coincidentally, our result shows that "17.6" is more similar to CaMV than other data as can be seen from Table 8. The other features of results by Miyata *et al.* (1986) also coincide generally with our results.

Thus, we see that the likelihood method proposed in this section is applicable to the analysis of amino acid sequences.

Acknowledgement

The author is very grateful to the referees for their valuable comments.

References

Alberts, B., Bray, D., Lewis, J., Raff, M., Roberts,K., and Watson, J.D. (1983). *Molecular Biology of the Cell*, Garland Publishing Inc., New York.

Doolittle, R. F. (1981). Similar amino acid sequences: chance or common ancestry?, *Science*, Vol. 214, 149–159.

Fisher, N.I. and Lee, A.J. (1986). Correlation coefficients for random variables on a unit sphere or hypersphere, *Biometrika*, Vol. 73, 159–164.

Fisher, N.I., Lewis, T. and Willcox, M.E. (1981). Tests of discordancy for samples from Fisher's distribution on the sphere, *Appl. Statist.*, Vol. 30, 230–237.

Miyata, T., Toh, H. and Hayashida, H. (1986). Detection of genes of reverse transcriptase enzyme with the aid of computer (in Japanese), *Saiensu* (Japanese version of *Scientific American*), Vol. 16, No.2, 86–97.

Ogata, Y. and Tanemura, M. (1985). Estimation of interaction potentials of marked spatial point processes through the maximum likelihood method, *Biometrics*, Vol. 41, 421–33.

Onsager, L. (1944). Crystal statistics. I. A two-dimensional model with an order-disorder transition, *Physical Review*, Vol. 65, 117–49.

Stoyan, D., Kendall, W. S. and Mecke, J. (1987). *Stochastic Geometry and its Applications*, Akedemie-Verlag, Berlin.

Toh, H., Hayashida, H. and Miyata, T. (1983). Sequence homology between retroviral reverse transcriptase and putative polymerases of hepatitis B virus and cauliflower mosaic virus, *Nature*, Vol. 305, 827–829.

Upton, G. J. G. and Fingleton, B. (1989). *Spatial Data Analysis by Example, Vol.2, Categorical and Directional Data*, John Wiley & Sons, Chichester.

NEW APPROACH TO OPTIMIZATION OF STABLE ESTIMATION

Alexander M. SHURYGIN
Department of Mechanics and Mathematics
Moscow State University
Moscow 119899, RUSSIA

1. On Robust Estimators

Let $x_1, \ldots, x_n$ be a random sample from an unknown probability density function (p.d.f) $f(x-\mu)$, which depends on the location parameter μ. Suppose we use another p.d.f. $g(x-\mu)$ instead of the unknown p.d.f. $f(x-\mu)$. We want to obtain an estimator $\hat{\mu}$ of μ depending on g, such that its variance is smaller than the variance of the p.d.f. $f(x-\mu)$. In this case, the maximum likelihood estimator (MLE) $\mathring{\mu}$ of μ for $g(x-\mu)$ is not a good estimator. This important fact to estimate a location parameter was introduced by Tukey (1960). Tukey considered the p.d.f.

$$f(x-\mu) = (1-\varepsilon)\, \varphi\, (x-\mu) + \varepsilon h(x-\mu), \tag{1.1}$$

where $\varphi(y)$ is the p.d.f. of the standard normal distribution N(0, 1), $h(y) = h(-y)$ is another symmetric p.d.f.. Varying h, Tukey (1960) showed that the MLE $\mathring{\mu}$ of μ for $\varphi(y)$ is catastrophically unstable even for small $\varepsilon < 0$.

Huber (1964) used (1.1) to obtain optimal estimators. From

$$\Sigma\, \psi(x_i - \hat{\mu}) = 0 \tag{1.2}$$

for all $\hat{\mu}$, he found

$$\psi_H(y) = \begin{cases} y & \text{if } |y| \leq k \\ k \operatorname{sgn} y & \text{if } |y| \geq k, \end{cases} \quad k = k(\varepsilon), \tag{1.3}$$

which is minimax with the asymptotic variance of $\hat{\mu}$ given by

$$u(\psi, h) = 1E\, n\, (\hat{\mu}-\mu)^2, \tag{1.4}$$

so that

$$\psi_H(y) = \arg\min_{\psi} \max_{h(y) = h(-y)} u\,(\psi, h).$$

H. Bozdogan (ed.), Proceedings of the First US/Japan Conference on the Frontiers of Statistical Modeling: An Informational Approach, 315–340.

Here and in the sequel of the paper, 1E will be used to denote the mean of an asymptotic distribution. Huber (1964) named the estimator $\hat{\mu}$ obtained from (1.2) which resulted in (1.3), the "robust estimator". This result was the beginning of the theory of robustness. In other words, the goal here is to try to find estimators which are stable with respect to small changes in the model's p.d.f.. In what follows, the words "robust" and "robustness" will be used as the usual statistical terminology for estimators.

Hampel *et al.* (1986) considered the problem of stability of the location estimators analytically based on (1.1). They concluded that the MLE is not stable but that the sample median and Huber's (1964) estimators are.

A group of statisticians at Princeton University (see Andrews *et al.* 1974) carried out a a large scale Monte Carlo experiments to compare 68 estimators of location, when the sample had non-symmetric contamination. For a sample size n=5, 10, 20, 50, respectively, these authors generated $x_1, \ldots, x_{n-1}$ observations from the standard normal distribution N(0, 1), and studied the influence of the location of x_n on the stability of $\hat{\mu}$. They found out that neither the robust estimator given in (1.3), nor the sample median were stable. Each of the authors introduced their own estimators using $\psi(y)$ close to the sine in a neighbourhood of $y = x - \mu = 0$, for decreasing $|\psi(y)|$ and large $|y|$, and equal to zero for a very large $|y|$ ("redescending estimators"), and further depending on an unestimated parameter like k in (1.3).

The problem of optimization of estimators was practically discarded. Currently software packages have a number of robust estimators of the mean of the normal distribution implemented, but no guidance is provided to the user as to which estimator one should choose in their problem.

The reason for failure of Huber's (1964) beautiful optimization is not readily understood. The majority of statisticians suspect that it is the condition h(y) = h(-y). Jaeckel (1971, p. 1029) specified the Tukey's problem as follows: "The amount of asymmetric contamination is large enough to effect the performance of the estimator, but it is too small to be measured accurately at the given sample size." Using asymptotics $\varepsilon \sim n^{-1/2} \to 0$, Jaeckel obtained the consistency of an estimator without the condition h(y) = h(-y). Then he discarded this condition, resolved the minimax problem, and still obtained the same result as Huber! Jaeckel further showed the asymptotic equivalence of M-, L- and R-estimators. Therefore, his work enlarged the domain of significance of this problem, which turned out to be important for all estimators.

There were attempts to construct robust estimators for the p-dimensional normal distribution of $\mathbf{x} \sim N_p(\mathbf{m}, \mathbf{C})$. Meshalkin (1973) proposed exponentional weighting and proved the consistency of estimators $\hat{\mathbf{m}}$ and $\hat{\mathbf{C}}$ which can be written as solutions of the system of equations:

$$\begin{cases} \Sigma\,(\mathbf{x}_i - \hat{\mathbf{m}}) \exp(-\lambda q_i/2) = 0, \\ \Sigma\,[(\mathbf{x}_i - \hat{\mathbf{m}})(\mathbf{x}_i - \hat{\mathbf{m}})^t + (1 + \upsilon)^{-1}\hat{\mathbf{C}}] \exp(-\upsilon q_i/2) = 0, \end{cases} \tag{1.5}$$

where $q_i = (\mathbf{x}_i - \hat{\mathbf{m}})^t\hat{\mathbf{C}}^{-1}(\mathbf{x}_i - \hat{\mathbf{m}})$ and λ, $\upsilon > 0$ are appropriate costants, and "t" denotes the transposition. For the case when $p = 1$ (the univariate case), Meshalkin (1973) recommended the values $\lambda = \upsilon = 1/2$. Later Marrona (1976) considered the system:

$$\begin{cases} \Sigma\,(\mathbf{x}_i - \hat{\mathbf{m}})\, \xi\,(q_i) = 0, \\ \Sigma\,(\mathbf{x}_i - \hat{\mathbf{m}})(\mathbf{x}_i - \hat{\mathbf{m}})^t\, \xi^2\,(q_i) = 1_p\hat{\mathbf{C}}, \end{cases}$$

where

$$\xi(q) = \begin{cases} 1 & \text{if } \sqrt{q} \le k, \\ k/\sqrt{q} & \text{if } \sqrt{q} \ge k. \end{cases} \tag{1.6}$$

In (1.6), $\xi(q)$ corresponds to Huber's (1964) weighting function in (1.3). Note that 1_p is a function of p in the system of equations above. Both Meshalkin and Maronna complained of instability of the estimators for large p, the number variables. Other weighting functions in the multivariate case were not adopted.

In the problem of multiple regression of $y \varepsilon R^1$ on $x \varepsilon R^p$, we have $y = r(x, \theta) + z$, $\theta \varepsilon R^q$, and for noise z with $E(z) = 0$, $E(z^2) = \sigma^2$. We want to obtain the estimator $\hat{\theta}$ of θ such that the variance

$$1E\, n[\hat{y} - r\,(x, \theta)]^2 \tag{1.7}$$

of regression prediction $\hat{y} = r(x, \hat{\theta})$ would be minimal. If $z \sim N(0, \sigma^2)$, the MLE $\mathring{\theta}$ of θ is best, and it minimizes the sum of squares $\Sigma\,(y_i - \hat{y}_i)^2$. One can easily see the formal analogy of this with the case of a sample from the normal distribution $N(\mu, \sigma^2)$, where the MLE $\mathring{\mu}$ minimizes the sum of squares $\Sigma\,(x_i - \mathring{\mu})^2$. This analogy can be applied to robust regression very readily using a robust estimator $\hat{\mu}$ of μ as follows: if one uses some function $\psi(y)$ minimizing the sum $\Sigma\psi(x_i - \hat{\mu})$, then minimizing the sum $\Sigma\psi(y_i - \hat{y}_i)$ will yield the corresponding regression.

Thus, in the literature many robust regressions have appeared as there are number of authors in the Princeton experiment. In addition, Meshalkin proposed the exponental weighted regression corresponding to (1.5). Yurečkova (1971) suggested $\psi(y) = |y|$ corresponding to the sample median, and Forsythe *et al.* (1973) suggested

$$\psi(y) = |y|^{\xi}, \quad 1 < \xi < 2. \tag{1.8}$$

In the papers devoted to robustness, many statisticians obtained interesting results both from theoretical and practical points of view. But these statisticians were unable to resolve the general problem of optimal stable estimation due to the obstacles which are more difficult than just purely analyzing statistical models. These obstacles can be overcome. But, before we do this, it is important to briefly review as to what is lacking in the literature on this problem starting from the beginning (see Huber 1964) till recent papers.

1. Besides the mean of the normal distribution, we have many other parameters of different distributions to estimate. Analytical obstacles have prevented consideration of them in the theory of robustness. Researchers could not find the best ψ for a parameter differing from the location parameter. But again obstacles can be overcome with analytical methods.

2. It is not clear how one can get a point estimator of a parameter, when the estimator depends on unestimated parameter ε or $k(\varepsilon)$ in (1.3) and (1.6), λ and υ in (1.5), and ξ in (1.8). But the dependence of estimator on ε disappears for a realistic asymptotics $\varepsilon = \varepsilon_n \sim n^{-\zeta/2} \to 0$, $0 < \zeta < 1$.

3. The scheme of mixture (1.1) is very suitable as a test for estimator stability, and Tukey (1960) used it just as that. But it does not fit within the optimization framework with nonsymmetrical contamination. Indeed, we do not know the sign of the bias, and the worst contamination h(y) turns out to be symmetrical, because such is our ignorance. So the best $\psi(y)$ will automatically be the best one for a symmetrical $h(y) = h(-y)$. We need to change the fundamental scheme (1.1) by a scheme with non-symmetrical deviation.

4. It is difficult for a user to choose a stable estimator for the mean of the normal distribution, because many estimators are currently available, and each one of them is backed by the name of a famous statistician. The way out of this, is to find an

optimal estimator.

5. If an estimator, minimizing $\Sigma\Psi(x_i-\hat{\mu})$ for the mean μ of the normal distribution, is good in some sense, then it does not mean that minimization of $\Sigma\Psi(\hat{y}_i-y_i)$ will be also good for regression, since (1.4) and (1.7) are different functions. So the problem of stable estimation and the problem of stable regression must be considered separately.

The theory of robustness is an excellent illustration of Gödel's famous theorem if we formulate it in terms of Tukey's mixture in (1.1). However, the problem has no solution based on the mixture and as soon as the fundamental scheme is changed (so as to avoid Jaeckel's paradox).

Since 1979, this author has carried out extensive research on the theory of robustness in these directions. The results of these investigations are not well-known beyond the Russian community. In the remainder of this paper, we review some of these results with the hope that this paper can serve to provide a broader exposure for the findings in other parts of the world.

2. Estimating a Parameter

2.1 MINIMUM CONTRAST ESTIMATORS (MCE)

Let $x_1, ..., x_n \varepsilon \chi \subset R^1$ be an independent random sample from the p.d.f. $f(x) = f(x, \theta)$, $\theta \in \Theta \subset R^1$, and let $\hat{\theta}$ be an estimator of θ obtained from the equation

$$\Sigma\psi(x_i, \hat{\theta}) = 0. \tag{2.1}$$

According to Huber (1964) $\hat{\theta}$ is M-estimator , or minimum contrast estimator (MCE) according to Pfanzagl (1964). The symbol 1E will be used for a mean in the asymptotic distribution. In the asymptotic expansion of $\hat{\theta}$ by powers of $n^{-1/2}$ (see Chibisov 1973), we denote $\bar{o}(n^{-k})$ to be a member if $n^k\bar{o}(n^{-k})$ converges to a non-degenerate distribution with first two moments. Otherwise, the symbol $\bar{o}(n^{-k})$ will be used if $n^k\bar{o}(n^{-k})$ converges to zero. In these notations, $\hat{\theta} = \theta + \bar{o}(1)$ is a consistent estimator of θ, $\hat{\theta} = \theta + \bar{o}(n^{-1/2})$ is a consistent estimator of θ "of order $1/\sqrt{n}$ ".

Theorem 1. Let for all $\theta \in \Theta$

(1) $\hat{\theta} = \theta + \bar{o}(n^{-1/2})$,

(2) in a neighbourhood of $t = \theta$, $n^{-1}\Sigma\psi(x_i, t) = E\psi(x, t) + \bar{o}(n^{-1/2})$,

$E\psi(x, t)$ have two derivatives with respect to t,

$$\frac{\partial}{\partial\psi}\frac{d}{dt} E\psi(x, t) = E\frac{\partial}{\partial\psi}\frac{d}{dt}\psi(x, t),$$

(3) the MLE $\mathring{\theta} = \theta + \bar{o}(n^{-1/2})$ exist.

Then

(a) $E\psi(x, \theta) = 0$,

(b) $E\psi^2(x, \theta) < \infty$,

(c) $\frac{d}{dt}E\psi(x, t)\,|_{t=\theta} \neq 0$,

(d) the asymptotic variance of $\hat{\theta}$ is

$$1E\, n(\hat{\theta}-\theta)^2 = E\psi^2(x, \theta)\, [\frac{d}{dt}\, E\psi(x, t)\,|_{t=\theta}]^{-2},$$

(e) the variational equality

$$\frac{\partial}{\partial\psi}[\frac{d}{dt}\psi(x, t)\, f(x, \theta)\,|_{t=\theta}] = c\frac{d}{d\theta}f(x, \theta)$$

holds, c is a constant.

Supposing (a), (b),(c), many authors obtained (1) and (d) for continuous $\psi(x, t)$. The theorem allows to consider $\hat{\theta}$ with interrupted ψ like the sample median. Result (d) was established by L. Eiler only for the case $\theta = \mu$ (when $c = 1$), and it restricted objects of the robustness theory with the location parameter. New result (d) for an arbitrary parameter gives possibility to obtain the best ψ for any case (see Shurygin 1985a).

2.2 ESTIMATOR STABILITY

Estimator stability is a freely used term in statistics. But it can be concretized according the general idea of stability in mechanics. We call the trajectory of an object stable, if any small influence leads to small alteration of the trajectory (a billiard ball over a table). A trajectory is called unstable, if a small influence leads to non-small alteration (the billiard ball over a knife blade). The same idea is in the definition of differential equation stability by

Liapunov: small deviation in initial conditions oughts to lead to small deviation in the solution trajectory. Applying this idea to estimator stability, we ought to require that any small alteration of the p.d.f. leads to small alteration of the estimator variance. The variance is the function of integrals depending on the p.d.f. f(x, θ), so Eiler's variation of the variance to f ought to be finite. We have the following definition.

Definition 1. Under conditions of Theorem 1, estimator $\hat{\theta}$ of parameter θ in the p.d.f. $f = f(x, \theta)$ is called stable, if $\frac{\partial}{\partial f}$ 1E n $(\hat{\theta} - \theta)^2$ exists, and unstable in the opposite case.

The variation is a linear combination of two integrals (Shurygin 1992):

$$\int_{\chi} \psi^2(x)dx, \qquad \frac{d}{dt}\int_{\chi} \psi(x, t)dt \Big|_{t=\theta}, \tag{2.2}$$

and is finite if and only if both integrals exist. So stability turns out to be dependent on χ and independent of the p.d.f. f(x, θ). Thus an estimator is stable or unstable for all continuous distributions on χ. One can see that for infinite χ existence of the frst integral in (2.2) leads to existence in the second one for a location parameter.

Applying criteria (2.2), one can be sure that all popular estimators are unstable on infinite χ: the sample mean with $\psi(x-\mu) = x-\mu$, the sample median with $\psi(x-\mu) = \text{sgn}(x-\mu)$, the mean squared deviation with $\psi(x, \sigma) = x^2 - \sigma^2$, the quartil estimator of variance (see Example 2), Huber's robust estimator in (1.3), and Marrona's multivariate estimators in (1.6). But Meshalkin's estimator in (1.5) and "redescending" estimators in the Princeton experiment are stable.

It is not clear how did Meshalkin obtain his estimators. It seems that he has simply guessed the solution and that we cannot generalize it. But the method in the Princeton experiment (see Andrews *et al.* 1972) is clear. These authors modeled non-symmetric point contamination of a sample by x_n, then studied the variance of the estimators in the dependence of distribution of x_n based on different sample sizes. This method can be formalized.

2.3 POINT BAYESIAN CONTAMINATION

Let each of independent observations x_1, ..., x_n be distributed according to the p.d.f. $g(x,\theta)$ with probability 1-ε, or be equal to some contaminating value x* with probability ε, so that the conditional p.d.f. of x be

$$f(x, \theta \mid x^*) = (1-\varepsilon)g(x, \theta) + \varepsilon\delta(x-x^*),$$

where δ(y) is the delta-function of y. In a more wider experiment (with different sample sizes), let x* be random variable with the p.d.f. s(x*, θ), so that the conditionless p.d.f. of x be

$$f(x, x^*, \theta) = (1-\varepsilon)g(x, \theta) + \varepsilon\delta(x-x^*)\, s(x^*, \theta) \qquad (2.3)$$

(Shurygin 1982a, 1982b). The scheme principally differs from the mixture (1.1) in that there is the distribution not of x in χ but of (x, x*) in χ^2. So optimization will give different result.

The sample contaminated in the point would not be considered literally. In this case for large n, contaminating observations would stick together and the best decision is to throw out these points and to use the MLE. To avoid deleting, we could change δ(x-x*) for example by $\varphi((x-x^*)/\sigma_n)/\sigma_n$ which converges to δ(x-x*) if $\sigma_n \to 0$ when $n \to \infty$. But we shall not introduce this bulky model here, because we understand scheme (2.3) as the simple mathematical scheme of the Princeton experiment, suitable for optimization of estimators.

Consider asymptotics (Shurygin 1985a):

$$\varepsilon^2 = \varepsilon^2(n) = \gamma n^{-\zeta} \to 0, \;\; 0 < \zeta < 1, \;\; 0 < \gamma = \text{const}. \qquad (2.4)$$

We shall use the following usual notation of a mean for some p.d.f. b(x)

$$E_b u(x) = \int_\chi u(x) b(x) dx.$$

Now we can find the best $\psi = \psi(s)$ for an arbitrary parameter θ.

Theorem 2. Under conditions of Theorem 1 for g(x) instead of f(x), let for all $t \in \Theta$

(1) $\{x: g(x, t) > 0\} = \{x: s(x, t) > 0\} = \chi$, $g(x, t) \equiv s(x, t)$,

(2) $E_s\psi(x, t)$ and $E_s\psi^2(x, t)$ exist and be continuous on t,

(3) ψ be the set of functions $\Psi(x, t)$ satisfying conditions of Theorem 1.

Then

(a) $\hat{\theta} = \theta + \bar{o}\,(1)$ for $f(x, \theta)$ in (2.3)

(b) the asymptotic variance of $\hat{\theta}$ is

$$v(\psi, s) = 1E\, n^{\zeta}(\hat{\theta} - \theta)^2 = \gamma E_s\psi^2(x)\, [\frac{d}{dt} E_g\psi(x, t)\,|_{t=0}]^{-2},$$

(c) the asymptotic variance has minimum on $\psi(x)\in\psi$ when

$$\psi^*(x, \theta) = \arg\min_{\psi}\, v(\psi, s) = c[\frac{d}{d\theta}\log g(x, \theta) - \beta]\, g(x)/s(x),$$

where $c \neq 0$ is an arbitrary constant, β: $E_g\psi(x) = 0$.

2.4 OPTIMAL ESTIMATOR

Having the best $\psi(x)$ in Theorem 2.(c), we can find the worst contamination with

$$s^*(x) = \arg\max_{s\in S}\min_{\psi} v(\psi, s),$$

but it depends on S. Let H be the set of functions s(x) for which the best $\psi(x)$ exists, G be the parametric family of p.d.f. and $g(x)\in G \subset H$.

Definition 2. An optimal estimator will be called median if S=H.

Definition 3. An optimal estimator will be called stoyki (~steady) if S=G.

Theorem 3. For the median estimator $\psi(x, \theta) = c\ \mathrm{sgn}\, [\frac{d}{d\theta}\log g(x, \theta) - \beta]$, where c and β are from Theorem 2.

This theorem shows that the median estimator is unstable for infinite χ.

Example 1. (Shurygin 1989a). The exponential distribution with p.d.f.

$$g(x, \sigma) = \sigma^{-1}\exp(-x/\sigma), \quad x \geq 0.$$

The median estimator $\breve{\sigma}$ of σ satisfies the equation

$$\Sigma \operatorname{sgn}(x_i - \breve{\sigma} \log 2) = 0, \quad \text{leff } \breve{\sigma} \approx 0.48.$$

The stoyki estimator $\tilde{\sigma}$ satisfies the equation

$$\Sigma(x_i - \frac{2}{3}\tilde{\sigma}) \exp(-x_i/2\tilde{\sigma}) = 0, \quad \text{leff } \breve{\sigma} \approx 0.63.$$

Here and in the sequel leff denotes the limit efficiency for the pure sample with the p.d.f. $g(x, \theta)$.

Example 2. (Shurygin 1980a, d). The normal distribution with p.d.f.

$$g(x, \mu, \sigma) = (\sqrt{2\pi}\,\sigma)^{-1} \exp(-(x-\mu)^2/2\sigma^2), \quad -\infty < x < \infty.$$

The median estimator $\breve{\mu}$ is the sample median, leff $\breve{\mu} = 2/\pi$. The median estimator $\breve{\sigma}$ satisfies the equation

$$\Sigma \operatorname{sgn}(|x_i - \mu| - \alpha\breve{\sigma}) = 0, \quad \alpha \approx 0.676: \int_\alpha^\infty \varphi(t)dt = 1/4, \quad \text{leff } \breve{\sigma} \approx 0.37;$$

this estimator is often recommended in the theory of robustness. Stoyki estimators $\tilde{\mu}$ and $\tilde{\sigma}^2$ satisfy the following system of equations:

$$\begin{cases} \Sigma(x_i-\tilde{\mu}) \exp(-(x_i-\tilde{\mu})^2/4\tilde{\sigma}^2) = 0, & \text{leff } \tilde{\mu} \approx 0.94, \\ \Sigma[(x_i-\tilde{\mu})^2 - \frac{3}{5}\tilde{\sigma}^2] \exp(-(x_i-\tilde{\mu})^2/3\tilde{\sigma}^2) = 0, & \text{leff } \tilde{\sigma}^2 \approx 0.60. \end{cases}$$

the system coincides with Meshalkin's one (1.5) with $p=1$, $\lambda=1/2$, $\gamma=2/3$.

Obviously these values of leff for the stoyki estimators are greater than for the median ones when there is a pure sample.

Consider the reverse problem of estimator efficiency in the sample with point Bayesian contamination, depending on contaminating $s(x^*)$, for the simplest case of estimating the mean for the standard normal distribution $N(0, 1)$. For the stoyki estimator the worst $s(x) = \varphi(x/\sqrt{2})/\sqrt{2}$. It belongs to the one-parametric family of contaminating distributions $N(0, \omega^2)$. For any ω there is the best estimator $\dot{\mu}$, and the stoyki one $\tilde{\mu}$ is $\dot{\mu}$ with the maximal variance, when $\omega^2 = 2$. The ratios $v(\tilde{\mu}) / v(\dot{\mu})$, i.e. efficiency of $\tilde{\mu}$ in the contaminated sample, for the MLE $\mathring{\mu}$, the sample median $\breve{\mu}$ and the stoyki estimator $\tilde{\mu}$ are given in Table 1.

Table 1. The Efficiency of $\tilde{\mu}$ in the Contaminated Sample

$v(\tilde{\mu})/v(\dot{\mu})$ for $\omega=$	1	2	3	4	5	10	∞
$\mathring{\mu}$	1	0.29	0.10	0.04	0.02	0.003	0
$\breve{\mu}$	0.64	0.74	0.55	0.43	0.35	0.18	0
$\tilde{\mu}$	0.86	0.96	0.91	0.87	0.86	0.85	0.59

For the median estimator the worst $s(x) = |x| \exp(-x^2/2)/2$. Consider the one-parametric family of contaminating p.d.f. $s(x, \omega) = |x| \exp(-x^2/2\omega^2)/2\omega$ and let $\dot{\mu}$ denote the best estimator in such contamination. Then the ratios of variances are given in Table 2.

Table 2. The Efficiency of $\tilde{\mu}$ in the Contaminated Sample

$v(\tilde{\mu})/v(\dot{\mu})$ for $\omega=$	1	2	3	4	5	10	∞
$\mathring{\mu}$	0.78	0.86	0.02	0.006	0.002	0.0002	0
$\breve{\mu}$	1	0.88	0.63	0.49	0.39	0.20	0
$\tilde{\mu}$	0.93	0.64	0.54	0.51	0.49	0.47	0.46

One can see very easily the catastrophic instability of the MLE. The sample median is some what better, because in (2.2) the first integral averages more slowly and the second one exists. The stoyki estimator has the clear advantage. Huber (1981) said that the normal distributions with heavy tails were typical. Then values ω would be large and the stoyki estimator would be many times better in the variance than the MLE.

3. Multivariate Normal Distribution

Let $\mathbf{x} \sim N_p(\mathbf{m}, \mathbf{C})$ in $\mathbf{R}^p$ with the p.d.f. given by

$$(2\pi)^{-p/2} |\mathbf{C}|^{-1/2} \exp\{-(\mathbf{x} - \mathbf{m})^t \mathbf{C}^{-1} (\mathbf{x} - \mathbf{m})/2\}.$$

First consider the problem of stability of $\hat{\mathbf{C}}$ when p increases.

3.1 REGULARIZATION OF ESTIMATOR $\hat{\mathbf{C}}$

The MLE $\mathring{\mathbf{C}}$ of the covariance matrix $\mathbf{C}$ degenerates when p increases. When the covariance matrix $\mathbf{C}=\mathbf{I}$, the identity matrix, then

$$E|\mathring{C}| = 1 - p(p-1)/2n + 0(n^{-2}).$$

The ridge estimator

$$\hat{C}_r = \mathring{C} + \alpha I,$$

where $\alpha > 0$ is small, has no such drawback, but it depends on the units of measurements of the variables. To avoid this scale dependence, this author (see Shurygin 1983) proposed the diagonal estimator:

$$\hat{C}_d = (1-\beta)\mathring{C} + \beta\mathring{C}_{dg},$$

where $\mathring{C}_{dg}$ is the diagonal matrix of the MLE covariance matrix $\mathring{C}$. Now we need a criterion to determine the values of the constants α and β. We can require that the term $0(n^{-1})$ in the asymptotic expansion of $E|\hat{C}|$ be equal to zero (Shurygin 1985c).

Definition 4. The estimator $\check{C}$ of the covariance matrix **C** will be called *stipulated* if it is consistent and

$$E|\check{C}| = |C| + 0(n^{-2}).$$

We use this to obtain the values α and β.

Example 3. For the *stipulated ridge estimator*

$$\alpha = p(p-1)[2n \operatorname{tr} C^{-1}]^{-1}.$$

Example 4. For the *stipulated diagonal estimator*

$$\beta = p(p-1)[2n(\operatorname{tr} \check{R}_{dg}^{-1} - p)]^{-1}, \tag{3.1}$$

where $\check{R}_{dg} = \check{C}_{dg}^{-1/2}\ \check{C}\ \check{C}_{dg}^{-1/2} = R$ is the corresponding correlation matrix.

In the vicinity of a real p=12-dimensional two-sample geochemical distribution, these estimators were compared in the linear discriminant analysis problem with equal covariances when the sample sizes $n_1 = n_2 = n$ shown in Figure 1. For a sample size n=15 and 10, discrimination error using the MLE covariance matrix $\mathring{C}$ is appreciably more

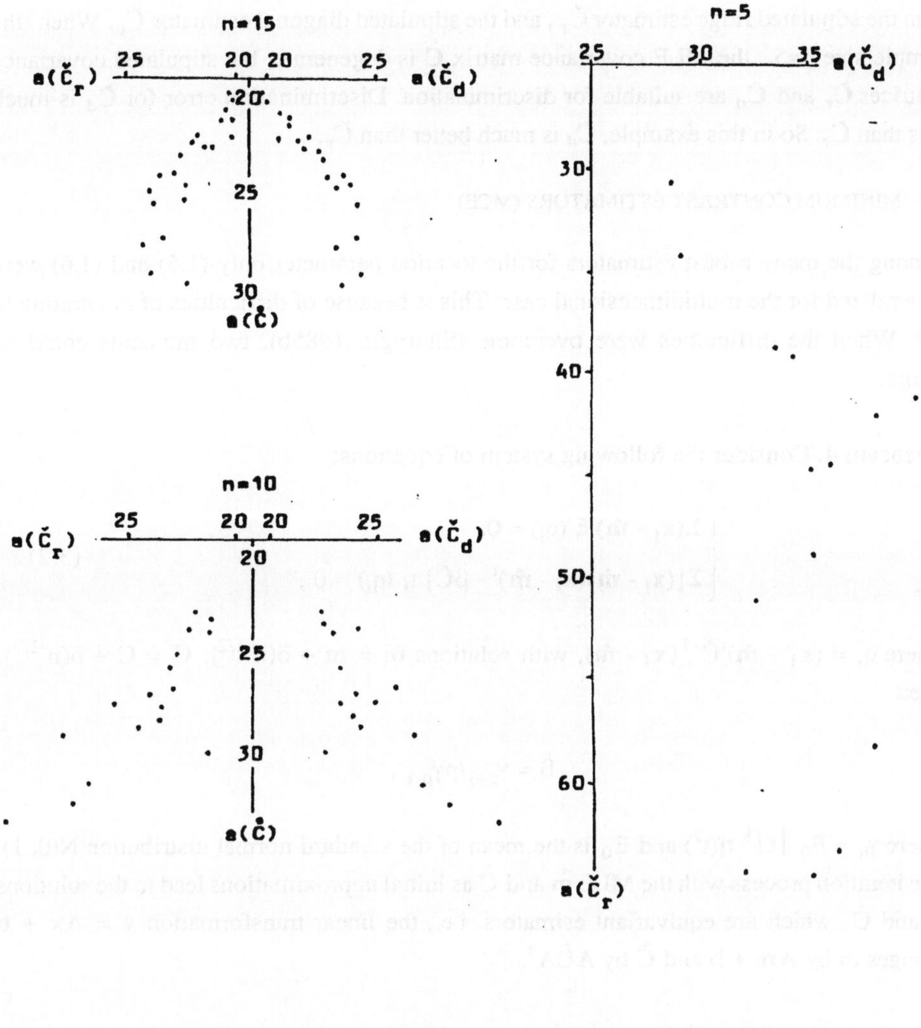

Figure 1. **Errors $a(\hat{C})$ of the linear discrimination of two p=12-dimensional populations modeled by the Monte Carlo method in the vicinity of a real geochemical problem (Shurygin 1980b) using (1) the sample sizes $n_1=n_2=n=15$, 10, and 5; (2) estimators of the covariance matrix C: $\mathring{C}$ the MLE, $\check{C}_r$ the stipulated ridge estimator, and $\check{C}_d$ the stipulated diagonal estimator.**

than the stipulated ridge estimator $\check{C}_r$, and the stipulated diagonal estimator $\check{C}_d$. When the sample size n=5, the MLE covariance matrix $\mathring{C}$ is degenerate, but stipulated covariance matrices $\check{C}_r$ and $\check{C}_d$ are suitable for discrimination. Discrimination error for $\check{C}_d$ is much less than $\check{C}_r$. So in this example, $\check{C}_d$ is much better than $\check{C}_r$.

3.2 MINIMUM CONTRAST ESTIMATORS (MCE)

Among the many robust estimators for the location parameter only (1.5) and (1.6) were generalized for the multidimensional case. This is because of difficulties of integrating in $\mathbf{R}^p$. When the difficulties were overcome (Shurygin 1985b), two moments could be found.

Theorem 4. Consider the following system of equations:

$$\begin{cases} \Sigma(\mathbf{x}_i - \hat{\mathbf{m}})\, \xi\, (q_i) = 0, \\ \Sigma[(\mathbf{x}_i - \hat{\mathbf{m}})(\mathbf{x}_i - \hat{\mathbf{m}})^t - \beta\hat{\mathbf{C}}]\, \eta\, (q_i) = 0\,, \end{cases} \tag{3.2}$$

where $q_i = (\mathbf{x}_i - \hat{\mathbf{m}})^t\hat{\mathbf{C}}^{-1}(\mathbf{x}_i - \hat{\mathbf{m}})$, with solutions $\hat{\mathbf{m}} = \mathbf{m} + \bar{o}(n^{-1/2})$, $\hat{\mathbf{C}} = \mathbf{C} + \bar{o}(n^{-1/2})$. Then

$$\beta = \gamma_{p+1}/p\gamma_{p-1}\,,$$

where $\gamma_h = E_0 \,|t|^h \eta(t^2)$ and E_0 is the mean of the standard normal distribution N(0, 1). The iteration process with the MLE $\mathring{\mathbf{m}}$ and $\mathring{\mathbf{C}}$ as initial approximations lead to the solutions $\hat{\mathbf{m}}$ and $\hat{\mathbf{C}}$ which are equivariant esimators, i.e., the linear transformation $\mathbf{y} = \mathbf{Ax} + \mathbf{b}$ changes $\hat{\mathbf{m}}$ by $\mathbf{A}\hat{\mathbf{m}} + \mathbf{b}$ and $\hat{\mathbf{C}}$ by $\mathbf{A}\hat{\mathbf{C}}\mathbf{A}^t$.

Corollary. Coefficient β does not depend on p if and only if the weighting function $\eta(q)$ is exponential (1.5).

Theorem 4 gives possibility to make the multidimensional generalization of any one-dimensional robust estimator. The corollary explains why Meshalkin (1973) easily obtained his generalization in (1.5).

Both Meshalkin (1973) and Marrona (1976) complained that estimators' errors increased with p, the number of variables. In what follows, we give some asymptotic expansions of estimators in the case when $\mathbf{y} \sim N_p(\mathbf{0}, \mathbf{I})$. The general case can be obtained by the linear transformation according to Theorem 4. Denote

$$\chi = \chi_p = \begin{cases} 1 & \text{if } p=1, \\ 2^{p/2-1}\pi^{-1/2}p\Gamma(p/2) & \text{if } p \geq 2, \end{cases}$$

where $\mathbf{r}_i = \mathbf{y}_i^t \mathbf{y}_i$ and for a matrix $\mathbf{A}$ $\mathbf{A}_{\overline{dg}} = \mathbf{A} - \mathbf{A}_{dg}$.

Theorem 5. Let all equations of system (3.2) meet the conditions of Theorem 1 and $\mathbf{y} \sim N_p(\mathbf{0}, \mathbf{I})$. Then the etimators $\hat{\mathbf{m}}$ and $\hat{\mathbf{C}}$ have asymptotic expansions:

$$\hat{\mathbf{m}} = \gamma_{p+1}^{-1} \frac{1}{n} \sum \mathbf{y}_i\, \xi(\mathbf{r}_i) + \bar{o}(n^{-1}),$$

$$\hat{\mathbf{C}} = \mathbf{I} + 2\chi \frac{n^{-1}\sum [(\mathbf{y}_i\mathbf{y}_i^t)_{dg} - \beta\mathbf{I}]\, \eta(\mathbf{r}_i)}{3\gamma_{p+1}(p+2)^{-1} - \beta\, \gamma_{p+1}} + \chi \frac{p+2}{\gamma_{p+3}} \frac{1}{n} \sum (\mathbf{y}_i\mathbf{y}_i^t)_{\overline{dg}}\, \eta(\mathbf{r}_i) + \bar{o}(n^{-1}),$$

where γ_h is given in Theorem 4.

The expressions for variances and covariances have been obtained, but they are very bulky to report here. Below we just give the expressions for the exponential weighing since these are more compact.

Corollary. Matrices $\sqrt{n}\,(\hat{\mathbf{C}} - \mathring{\mathbf{C}})$ and $\sqrt{n}\,(\mathring{\mathbf{C}} - \mathbf{I})$ are asymptotically independent when $\eta(q)$ is exponential in (1.5). In this case for $\hat{\mathbf{m}} = (\hat{m}_1, \ldots, \hat{m}_p)^t$ and $\hat{\mathbf{C}} = \{c_{\alpha\beta}\}$ if $\lambda, \upsilon > -1/2$

$$1\mathrm{E}\; n\hat{m}_\alpha^2 = (1+\lambda)^{p+2}\,(1+2\lambda)^{-p/2-1},$$

$$1\mathrm{E}\; n\hat{c}_{\alpha\beta}^2 = (1+\upsilon)^{p+4}\,(1+2\upsilon)^{-p/2-2}, \quad \alpha \neq \beta$$

$$1\mathrm{E}\; n(\hat{c}_{\alpha\alpha}-1)^2 = (2+4\upsilon+3\upsilon^2)\,(1+\upsilon)^{p+2}\,(1+2\upsilon)^{-p/2-1},$$

$$1\mathrm{E}\; n(\hat{c}_{\alpha\alpha}-1)\,(\hat{c}_{\beta\beta}-1) = \upsilon(1+\upsilon)^{p+2}\,(1+2\upsilon)^{-p/2-1}, \quad \alpha \neq \beta,$$

vector $\sqrt{n}\,\hat{\mathbf{m}}$ and matrix $\sqrt{n}\,(\mathring{\mathbf{C}} - \mathbf{I})$ are asymptotically independent.

Using the above formulas, one can be convinced that indeed the errors increase almost linearly with p.

3.3 STOYKI ESTIMATORS

To avoid the instability, let us write the stoyki estimators. They are generalized to the multivariate case automatically.

Theorem 6. Stoyki estimators $\hat{\mathbf{m}}$ and $\hat{\mathbf{C}}$ of the parameters $\mathbf{m}$ and $\mathbf{C}$ for the multivariate normal distribution $N_p(\mathbf{m}, \mathbf{C})$ satisfy the system (1.5) when $\lambda = (p+1)^{-1}$ and $\upsilon = 2(p+2)^{-1}$.

Corollary. For the stoyki estimators

$$\lim_{p\to\infty} \text{leff}\ \hat{m}_\alpha = \lim_{p\to\infty} \text{leff}\ \hat{c}_{\alpha\alpha} = \lim_{p\to\infty\ \alpha\neq\beta} \text{leff}\ \hat{c}_{\alpha\beta} = 1.$$

This corollary shows that the stoyki estimators are free of growth of errors when the number of variables p, increase.

3.4 PRINCIPAL COMPONENTS

The solution $\hat{\boldsymbol{\eta}}$ of the equation $(\hat{\mathbf{C}} - \pi\mathbf{I})\hat{\boldsymbol{\eta}} = 0$, where $\hat{\mathbf{C}}$ is the stoyki estimator of **C**, is stoyki, i.e., it minimizes $E(\hat{\boldsymbol{\eta}} - \boldsymbol{\eta})^t(\hat{\boldsymbol{\eta}} - \boldsymbol{\eta})$ in the model of the point Bayesian contamination.

3.5 STOYKI REGULARIZED ESTIMATOR OF C

In the asymptotic expansion of the stoyki estimator $\hat{\mathbf{C}}$ by powers of $n^{-1/2}$, the main term $\bar{o}(1)$, chosen according to Theorem 4, ensures the consistency of the estimator. The second term $\bar{o}(n^{-1/2})$, according to Theorem 6, secures the stability of the estimator to deviations from normality. Now, we can choose the third term $\bar{o}(n^{-1})$ to make the estimator stable to growth in p. In constructing the stipulated diagonal stoyki estimator, the coefficientβ can differ from β given in (3.1), but the Corollary of Theorem 6 shows that this difference is not too large. So, we can use the simple β in (3.1) for the regularization of the stoyki estimator $\hat{\mathbf{C}}$ without the cumbersome second moment expressions in the Corollary of Theorem 5.

When $p > n$, the MLE $\mathring{\mathbf{C}}$ is not reliable as an initial approximation because it is degenerate. In this case one can use $\overset{\circ}{C}_{dg}$ which is suitable for all $p \gtrless n$.

3.6 A CLUSTER-TEST STATISTIC

If there is a mixture of normal samples differing in locations, after some iterations the stoyki estimator $\hat{m}$, converges to the mean of the largest sample, and the stoyki estimator $\hat{C}$ approaches to the covariance matrix of this sample. This property can be used for testing the homogeneity of a sample by the value

$$\kappa = \frac{1}{np} \Sigma(x_i - \hat{m})^t \hat{C}^{-1}(x_i - \hat{m}) ,$$

where $\hat{m}$ and $\hat{C}$ are exponentially weighted estimators (see Shurygin 1979, 1980c).

Theorem 7. If $x_1, ..., x_n$ have the distribution $N_p(m, C)$ and estimators $\hat{m}$, $\hat{C}$ satisfy the system (1.5), with $\lambda, \upsilon > -1/2$, then

$$\sqrt{np}\,(\kappa-1) \xrightarrow{d} N(0, \tau^2),$$

$$\tau^2 = (2+4\upsilon+3\upsilon^2)\,(1+\upsilon)^{p+2}\,(1+2\upsilon)^{-p/2-2} - 2 + (p-1)\upsilon^{p+2}\,(1+2\upsilon)^{-p/2-2} .$$

This theorem is true when $\hat{m}$ is the stoyki estimator and $\hat{C}$ is the stipulated diagonal stoyki estimator.

The distribution of the test has strong positive asymmetry, but the Monte Carlo experiment showed that if p=1 and $0 < \upsilon \leq 0.5$, the 95% confidence interval is appropriate even when n=10 (Shurygin 1980).

The test is very sensitive to the presence of superfluous quantity of points (extreme values) on a periphery of the distribution. So in cluster analysis we classify the observations into clusters until the hypothesis of homogenity is satisfied or passes for all clusters.

4. Regression

Multiple regression of a variable $y \in \mathrm{y} \subset R^1$ on a vector $x = (x^{(1)}, ..., x^{(p)})^T \in \chi \subset R^p$ with regression function $r(x, \theta)$ depending on a vector of parameters $\boldsymbol{\theta} = (\theta_1, ..., \theta_p)^t \in \Theta \subset R^q$, is defined by the equation

$$y = r(x, \theta) + z,$$

and for z $Ez = 0$, $Ez^2 = \sigma^2$. In contrast to the ordinary regression model, where only z is random, we shall consider the case where both x and z are random. So, in this case the joint p.d.f. of x and y is

$$f(x, y, \theta) = f_1(x)f_2(y-r(x, \theta)). \tag{4.1}$$

Suppose we have a sample $(x_1, y_1), ..., (x_n, y_n)$, and the values z_i are independent. We want to find an estimator $\hat{\theta}$ of θ such that the variance

$$1E\, n[\hat{y} - r(x,\theta)]^2$$

of prediction $\hat{y} = r(x, \hat{\theta})$ of a new observation y by a known value of x would be minimal (Shurygin 1990a, 1990b).

4.1 MINIMUM CONTRAST REGRESSION

Consider the minimum contrast regression estimator $\hat{\theta}$ obtained from the system

$$\Sigma\, \psi_j(x_i, y_i, \hat{\theta}) = 0, \quad j=1, ..., q, \tag{4.2}$$

or if we introduce the vector-function $\psi = (\psi_1, ..., \psi_q)^t$, obtained from the equation

$$\Sigma\, \psi_j(x_i\, y_i, \hat{\theta}) = 0\,. \tag{4.2'}$$

There is a function $\psi(x, y, \theta)$ such that $\psi_j = \frac{d}{d\theta_j}\psi$ and the solution of the system (4.2) minimizes the sum $\Sigma\, \psi(x_i, y_i, \hat{\theta})$. This fact will be denoted in such a way that: $\Sigma\, \psi(x_i, y_i, \hat{\theta}) \Rightarrow \min$. For derivatives, we shall use ordinary notations in vector analysis. If f is a function depending on components of a vector $v = (v_1, ...)^t$, then $\frac{df}{dv} = (\frac{df}{dv_1}, ...)^t$ is a vector. If f is a vector-function $f = (f_1, ...)^t$, then $\frac{df}{dv} = \{df_i/dv_j\}$ is a matrix.

The prediction variance for the minimum contrast regression is given in the following theorem.

Theorem 8. Let for almost all $x \in \chi$ and all $\theta \in \Theta$

(1) $\hat{\theta} = \theta + \bar{o}(n^{-1/2})$ and be the solution of (4.2),

(2) functions ψ_j in (4.2) meet the conditions of Theorem 1,

(3) regression function r(x, t) have two derivatives in the neighbourhood of t=θ,

(4) matrices $A = E \frac{d}{d\theta} r(x, \theta)(\frac{d}{d\theta} r(x, \theta))^T$ and $B^{-1} = \frac{d}{dt} E\, \psi(x, y, t)\,|_{t=\theta}$ be positively determined.

Then

(a) value $\sqrt{n}\,[\hat{y}-r(x, \theta)]$ is asymptotically normal

(b) with the asymptotic variance

$$\mathrm{lE}\, n[\hat{y}-r(x, \theta)]^2 = 1\mathrm{E}\, n(\hat{\theta}-\theta)^t A(\hat{\theta}-\theta) = E\, \psi^t B^t AB\psi,$$

(c) which has minimum when

$$\psi(x, y, \theta) = C \frac{d}{d\theta} \log f_2(y-r(x, \theta)),$$

where C is a non-degenerate matrix,

(d) the variational equality

$$\frac{\partial}{\partial \psi} \{[\frac{d}{d\theta} \psi(x, y, \theta)]\, f(x, y, \theta)\} = D \frac{d}{d\theta} f(x, y, \theta),$$

is true up to a non-degenerate matrix D.

This theorem is a regression analog of Theorem 1.

4.2 STABILITY OF REGRESSION

Definition 1 of estimator stability fits for the regression problem after changing x and χ by $u = (x, y)^t$ and $U = \chi \times Y$. Existence of the integrals

$$\int_u \psi^2(u)du, \qquad\qquad (4.3)$$

$$\frac{d}{dt} \int_u \psi(u, t)du\,|_{t=0}$$

is the condition of stability. The classical regression does not depend on the distribution of x (see Theorem 8.c). The confluent analysis considers the normal distribution of x. In our construction, if χ and (or) Y are infinite, the integrals diverge for all previously suggested regressions.

4.3 POINT BAYESIAN CONTAMINATION

Let $u = (x, y)^t$ have the p.d.f. be given by

$$f(u, u^*, \theta) = (1-\varepsilon)\, g(u, \theta) + \varepsilon\delta(u-u^*)\, s(u^*, \theta), \tag{4.4}$$

which is analogous to (2.3), and asymptotics (2.4) hold. Now we can find the best $\psi = \psi(s)$.

Theorem 9. Let

(1) $\hat{\theta} = \theta + \bar{o}(n^{-1/2})$ for g(x, y) and be the solution of system (4.2),

(2) p.d. g(x, y), s(x, y) and functions $\psi_j(x, y)$ meet the conditions of Theorem 2.

Then

(a) $\hat{\theta} = \theta + \bar{o}(1)$ for p.d. (4.4),

(b) there exist

$$w(\psi, s) = 1E\, n^{\xi}[\hat{y} - r(x, \theta)]^2 = \gamma\, E_s\, \psi^t B^t AB\psi,$$

where matrices A and B are as in theorem 8.4 after changing E by E_g,

c) $w(\psi, s)$ has a minimum in ψ when

$$\psi^*(x, y) = \arg\min_{\psi} w(\psi, s) = C\,[\frac{d}{d\theta} \log g(x, y, \theta) - \beta]\, g(x, y)/s(x, y),$$

where C is an arbitrary non-degenerate matrix and β: $E_g\psi = 0$.

4.4 OPTIMAL REGRESSION

Having the best ψ in Theorem 9.c we can find the worst s:

$$s^*(x, y) = \arg\max_{s\in S} \min_{\psi} w(\psi, s).$$

Let H be the set of functions s(u), for which the best $\psi(u)$ exists, G be the parametric family of p.d. and $g(u) \in G \subset H$.

Definition 4. An optimal regression will be called median, if S=H.

Definition 5. An optimal regression will be called stoyki, if S=G; when G is not defined "naturally," it will be specially determined.

These definitions are regression equivalents of Definitions 2 and 3.

Theorem 10. For the median regression

$$\psi(x, y) = D\left(\frac{d}{d\theta}\log g - \beta\right)\left[\left(\frac{d}{d\theta}\log g - \beta\right)^t F^t B^t ABF\left(\frac{d}{d\theta}\log g - \beta\right)\right]^{-1/2},$$

where D, F are arbitrary non-degenerate matrices and β: $E_g\psi = 0$.

This theorem shows that the median regression is unstable on infinite *U*.

Example 5. The multiple linear normal regression is the case when

$$x \sim N_p(m, C),\ r(x, \theta) = \theta_0 + \sum_i^p \theta_j x^{(j)},\ z = y - r(x, \theta) \sim N(0, \sigma^2).$$ **We shall denote**

$\Delta_i = y_i - r(x_i, \hat{\theta})$. **For the median regression**

$$\Sigma\,|\,\Delta_i\,|\ [1 + (x_i - m)^t C^{-1}(x_i - m)]^{-1/2} \Rightarrow \min.$$

Setting the quadratic form equal to zero, we obtain the regresssion by Yurečkova (1971). Defferentiating the sum by components $\hat{\theta}_j$ of $\hat{\theta} = (\theta_0, ..., \theta_p)^t$, we get the equations for summands

$$\psi_j(x, y) = \theta_j(\text{sgn}\,\Delta)\ [1 + (x_i - m)^t C^{-1}(x_i - m)]^{-1/2}, \quad j=0, 1, ..., p,$$

for which integral (4.3) does not exist. For the stoyki regression

$$- \Sigma \exp (-\eta_p(\mathbf{x}_i - \mathbf{m})^t \mathbf{C}^{-1}(\mathbf{x}_i - \mathbf{m})/2 + \Delta_i^2/4\sigma^2) \Rightarrow \min,$$

where $\eta_p = \sqrt{(p+5)/(p+1)} - 1)/2$. If the quadratic form would be equal to zero, the sum looks like Meshalkin's regression with $\lambda=0.5$ in the first equation of (1.5). The limit efficiency of the regression in a pure sample is 0.76 when p=1 and increases when p grows. The stoyki regression is stable on infinite χ and Y.

In contrast to the classical and robust regressions, the median and stoyki ones weigh not only deviations of y_i from $r(x_i, \theta)$ but deviations of x_i from **m**.

4.5 A COMPARISON OF DIFFERENT REGRESSIONS

A comparison of different regressions was made by Petrovich and Shleg (1987). They imitated contaminated samples and compared the classic regression, six robust methods and the stoyki regression in the normal linear case. For the stoyki regression they used the stoyki estimators $\hat{\mathbf{m}}$, $\hat{\mathbf{C}}$ and the stoyki estimator $\hat{\sigma}^2$ with $\mu=0$ and the multiplier $n(n-p-1)^{-1}$, which is usual for correction in the regression problems. The stoyki regression had minimal error of the prediction. In Figure 2, there are different lines of the linear regression for the two-dimensional population of medical data with two artificial contaminated points. One can see that the stoyki regression is the only one that is practically stable and that robust regressions are even more unstable than the classical one. So based on the theoretical reasons above and the conclusions on instability of all regressions but the stoyki one, we have a practical confirmation here.

4.6 EXTRAPOLATION OF LOCAL-LINEAR TREND

Extrapolation of local-linear trend, where new observations have to be entered into the solution with more weights than earlier ones, arises in many applications. The problem can not be formulated in the classical regression model of equivalent observations, but can be resolved by the Bayesian contamination (Shurygin 1989b).

Let $x \in R^1$ have the uniform distribution on (0,1), $r(x, \theta) = \theta_0 + \theta_1 x$, $z \sim N(0, \sigma^2)$ and the p.d.f. of contamiation s(x, y) differ from g(x, y) by the distribution of $z \sim N(0, \sigma^2\omega^2(x))$. We need to find the maximin estimator of Ey at x=1. So we have to change the minimizing integral $w(\psi, s)$ in Theorem 9.b by

$$w^*(\psi, s) = 1E\, n^{\xi}[\hat{y} - r(x, \theta) \mid x=1]^2 .$$

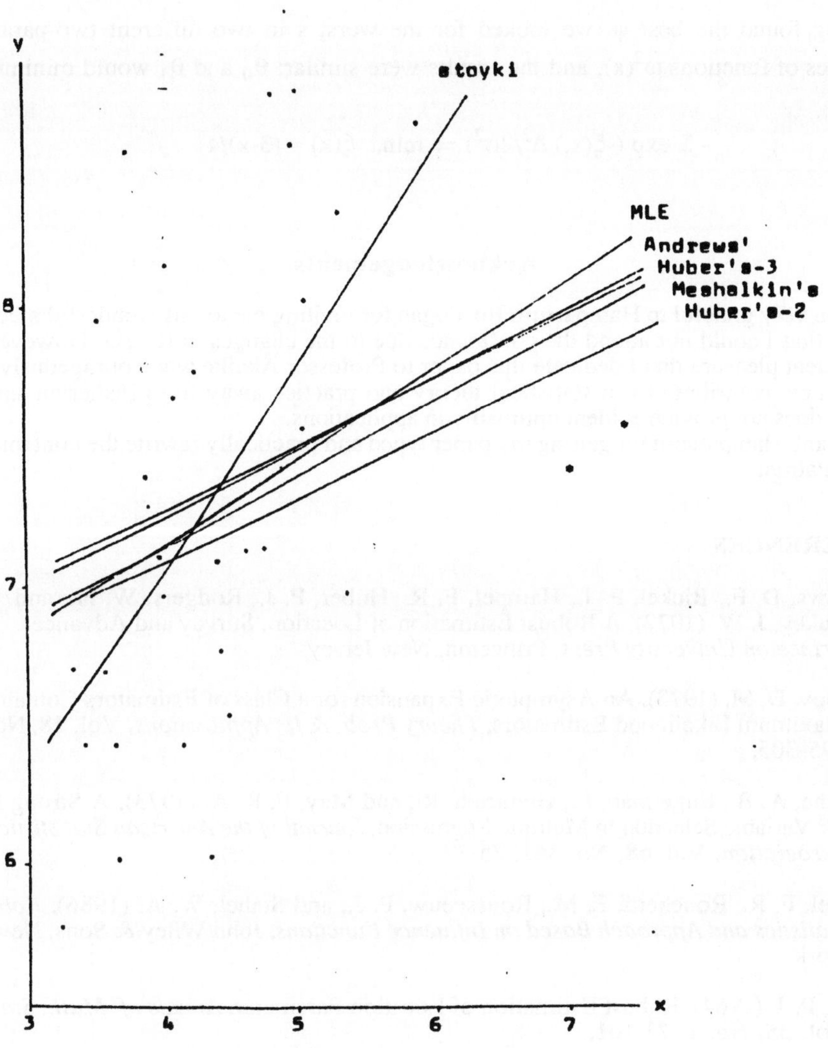

Figure 2. Comparison of different regressions of y on x for a data set with two artificially contaminating points (asterisks) (Petrovich and Shleg 1987): the stoyki regression, the MLE, and four robust methods.

Having found the best ψ, we looked for the worst s in two different two-parametric families of functions $\omega^2(x)$, and the results were similar: θ_0 and θ_1 would minimize the sum

$$- \Sigma \exp (-\xi(x_i)\ \Delta_i^2/2\sigma^2) \Rightarrow \min, \quad \xi(x) \approx (3-x)/4 .$$

Acknowledgements

I am very grateful to Hamparsum Bozdogan for inviting me to this wonderful meeting. I regret that I could not attend the conference due to the changes in Russia. However, it is with great pleasure that I dedicate this paper to Professor Akaike who courageously made many new contributions to statistical theory and practice away from Fisherian statistics which does not provide evident optimality in applications.

I thank Hamparsum for getting my paper typed and practically rewrite the content of the presentation.

REFERENCES

Andrews, D. F., Bickel, P. J., Hampel, F. R., Huber, P. J., Rodgers, W. H., and Tukey, J. W. (1972). A Robust Estimation of Location: Survey and Advances. *Princeton University Press*, Princeton, New Jersey.

Chibisov, D. M. (1973). An Asymptotic Expansion for a Class of Estimators Containing Maximum Likelihood Estimators, *Theory Prob. & Its Applications*, Vol. **18**, No. 2, 295-305.

Forsythe, A. B., Engelman, L., Gennrich, R., and May, F. R. A. (1973). A Strong Rule for Variable Selection in Multiple Regression, *Journal of the American Statististical Association,* Vol. **68**, No. 341, 75-77.

Hampel, F. R., Ronchetti, E. M., Rousseeuw, P. J., and Stahel, W. A. (1986). *Robust Statistics and Approach Based on Influence Functions*, John Wiley & Sons, New York.

Huber, P. J. (1964). Robust Estimation of Location Parameter, *Annals of Math. Statist.*, Vol. **35**, No. 1, 73-101.

Huber, P. J. (1973). Robust Regression: Asymptotics, Conjectures and Monte Carlo. *Annals of Statist.*, Vol. **1**, No. 5, 799-821.

Huber, P. J. (1981). *Robust Statistics*, John Wiley & Sons, New York.

Jaeckel, L. A. (1971). Robust Estimates of Location: Symmetry and Asymmetric Contamination, *Annals of Math. Statist.*, Vol. **42**, No. 3, 1020-1034.

Marrona, R. A. (1976). Robust M-Estimators of Multivariate Location and Scatter, *Annals of Math. Statist.*, Vol. **4**, No. 1, 51-67.

Meshalkin, L. D. (1973). Some Mathematical Methods for the Study of Non-Communicalbe Diseases, in the *Proc. 6-th Intern. Meeting on Uses of Epidemiol. in Planning Health Services*, Primosten, Yugoslavia, 1971, Vol. **1**, 250-256.

Pfanzagl, J. (1969). On Measurability and Consistency of Minimum Contrast Estimates, *Metrica*, Vol. **14**, 248-278.

Petrovich, M. L., and Shleg, G. K. (1987). Study of Robust Regression Methods by Statistical Experiments, Factory laboratory, No. 3, 41-48 (in Russian).

Shurygin, A. M. (1979). Estimators of the Normal Distribution Parameters in Contaminated Samples and Statistical Cluster-Test, *Theory Prob. & Its Appl.*, Vol. **24**, No. 1, 233-234.

Shurygin, A. M. (1980a). Estimators of the Normal Distribution Parameters with Exponential Weighting of Observations: Asymptotic Theory, in *Algorithms and Software for Applied Statistical Analysis*, Nauka (Ed.), Moscow, 241-259 (in Russian).

Shurygin, A. M. (1980b). Comparison of the Efficiency of Different Algorithms in the Discrimination of Real Objects, *Item*, 354-360.

Shurygin, A. M. (1980c). The Statistical Cluster-Test, *Item*, 360-366.

Shurygin, A. M. (1982a). Stable Estimators of Distribution Parameters. Factory Laboratory, No. 2, 88-93 (in Russian).

Shurygin, A. M. (1982b). Robustness and Optimization of Asymptotic Estimator Stability, in the 15-th European Meeting of Statisticians, Palermo, Italy.

Shurygin, A. M. (1983). The Linear Combination of the Simplest Discriminator and Fisher's One, in *Applied Statistics*, Nauka (Ed.), Moscow, 144-158 (in Russian).

Shurygin, A. M. (1985a). Robustness and Stability of Statistical Estimators, in *Statistics, Probability and Economics*, Nauka (Ed.), Moscow, 90-98 (in Russian).

Shurygin, A. M. (1985b). Stable Estimators for Parameters of the Multivariate Normal Distribution, *Item*, 314-317.

Shurygin, A. M. (1985c). Ways for Perfecting of Discriminant Analysis in the Normal Case, *Item*, 379-382.

Shurygin, A. M. (1986). The Estimators that are Robust in Series of Samples, in the *1-st Bernoulli Congress*, Vol. **1**, Tashkent-Moscow, 22.

Shurygin, A. M. (1989a). Stoyki Estimators for the Parameter of the Exponential Distribution, in the *4-th Conference on Multivar. Statist. Analysis its Appl. in Economics*, Tartu, Estonia, 23 (in Russian).

Shurygin, A. M. (1989b). Extrapolation of the Local Linear Trend, *Item*, 61-62.

Shurygin, A. M. (1990a). Parametric Minimax Optimization of Regression in Series of Samples, in *Multvariae Statistical Analysis and Real Process Modeling*, Nauka (Ed.), Moscow, 107-117 (in Russian).

Shurygin, A. M. (1990b). Non-Parametric Minimax Optimization of Regression in Series of Samples, *Item*, 163-166.

Shurygin, A. M. (1992). Stable Estimating of a Parameter, *in the 3-th Intern. Workshop on Model Oriented Data Analysis*, St. Petersburg, 31.

Tukey, J. W. (1960). A Survey of Sampling from Contaminated Distibution, in *Contribution to Probability and Statistics*, I. Olkin (Ed.), Stanford Univ. Press, 448-486.

Yurečkova, J. (1971). Nonparametric Estimates of Regression Coefficients, *Annals of Math. Statist.*, Vol. **42**, No. 4, 1328-1338.

INDEX

CONTENTS OF VOLUME 1

CONTENTS OF VOLUME 2

MIX
Papier aus verantwortungsvollen Quellen
Paper from responsible sources
FSC® C105338

If you have any concerns about our products,
you can contact us on
ProductSafety@springernature.com

In case Publisher is established outside the EU,
the EU authorized representative is:
Springer Nature Customer Service Center GmbH
Europaplatz 3, 69115 Heidelberg, Germany

Printed by Libri Plureos GmbH
in Hamburg, Germany